Two Rivers Entangled

TWO RIVERS ENTANGLED

*An Ecological History of
the Tigris and Euphrates in
the Twentieth Century*

Dale J. Stahl

STANFORD UNIVERSITY PRESS
Stanford, California

Stanford University Press
Stanford, California

© 2026 by Dale James Stahl. All rights reserved.

No part of this book may be reproduced or transmitted in any form or by any means, electronic or mechanical, including photocopying and recording, or in any information storage or retrieval system, without the prior written permission of Stanford University Press.

Library of Congress Cataloging-in-Publication Data
Names: Stahl, Dale author
Title: Two rivers entangled : an ecological history of the Tigris and Euphrates in the twentieth century / Dale J. Stahl.
Description: Stanford, California : Stanford University Press, [2026] | Includes bibliographical references and index
Identifiers: LCCN 2025026460 (print) | LCCN 2025026461 (ebook) | ISBN 9781503640177 cloth | ISBN 9781503644731 paperback | ISBN 9781503644748 ebook
Subjects: LCSH: Human ecology—Middle East—History—20th century | Watershed ecology—Middle East—History—20th century | River engineering—Middle East—History—20th century | Euphrates River—History—20th century | Tigris River—History—20th century | Middle East—Politics and government—20th century
Classification: LCC DS79.89.E863 S73 2026 (print) | LCC DS79.89.E863 (ebook) | DDC 304.20956/0904—dc23/eng/20250827
LC record available at https://lccn.loc.gov/2025026460
LC ebook record available at https://lccn.loc.gov/2025026461

Cover design: Lindy Kasler
Cover: View in Eğil of the Tigris River, Shutterstock
Typeset by: Newgen in 9.75/14.5

The authorized representative in the EU for product safety and compliance is: Mare Nostrum Group B.V. | Mauritskade 21D | 1091 GC Amsterdam | The Netherlands | Email address: gpsr@mare-nostrum.co.uk | KVK chamber of commerce number: 96249943

Contents

	Acknowledgments	vii
	Map of the Tigris-Euphrates River Basin	xi
Introduction	**Rivers and the New Materialism**	1
One	**Water**	23
Two	**Salt**	52
Three	**Rock**	95
Four	**Reservoir**	134
Conclusion	**Regeneration and a New Century**	182
	Notes	197
	Bibliography	243
	Index	267

Acknowledgments

This book took shape over many years and across many landscapes—all marked by history, upheaval, and the generosity of the people who made this work possible. First and foremost, I am deeply grateful to the archivists and librarians in Damascus, Ankara, and Istanbul, whose dedication to scholarship endured despite the immense challenges of civil strife and political repression. Their work safeguarded knowledge in moments when knowledge itself was under threat, and without them, these pages would not exist.

The research for this book required travel to several countries and many months of language learning and archive rummaging. I enjoyed and endured these periods because of good friends and brilliant interlocutors, who not only helped me master verb conjugations but also kept me sane along the way. Aimee Genell never balked at a many-mile detour to a dam or canyon, proving herself the kind of companion every historian of rivers needs. My thanks also to Oscar Aguirre-Mandujano, Jared Manasek, Carey Kasten, Jordan Matthews, Shane and Julia Stratton, and Dan Auger. And, speaking of languages, Burcu Çakıruylası, Sibel Erol, Taoufik Ben-Amor, Michael Cooperson, Ramzi Salti, and Waddah al-Khatib—this would have been impossible without your teaching and, I am certain, your patience.

I began developing the intellectual framework for this book as a Humanities Research Fellow at New York University Abu Dhabi. I owe a debt of gratitude to the other fellows—Marilyn Booth, Andrew Bush, and Matthew MacLean—and to the leadership and staff, Reindert Falkenburg and

Alexandra Sandu. My time in Abu Dhabi was also enriched by conversations with Nora Barakat, Mark Swislocki, Justin Stearns, and Pascal Menoret. Thank you for listening, especially when I went down a rabbit hole, and for offering good counsel.

This book owes its existence to the generous support of institutions that believed in this research before it had fully taken shape. Grants and fellowships from the Whiting Foundation, the Institute of Turkish Studies, the Academic Research Institute in Iraq, and the David L. Boren Fellowship sustained my research and writing, while conferences and workshops organized by Onur İnal, Giacomo Parrinello, G. Mathias Kondolf, and Serkan Karas introduced me to new ideas and invaluable intellectual communities.

Mentors, readers, and fellow travelers at Columbia University contributed to this work's initial development. I was lucky to learn from Rashid Khalidi, Timothy Mitchell, Richard Bulliet, Susan Pedersen, Christine Philliou, and Elizabeth Blackmar. And I was lucky to learn with Seth Anziska, Rosie Bsheer, Arunabh Ghosh, Toby Harper, Shehab Ismail, Abhishek Kaicker, Mari Webel, and Adrien Zakar.

I am equally indebted to my colleagues at the University of Colorado Denver, who offered mentorship and encouragement while helping to protect my time. My thanks to the Department of History, the College of Liberal Arts and Sciences, and to the Office of Research Services for financial support of my work. I am especially grateful to the Office of Undergraduate Research and Creative Activities and to the three undergraduate research assistants who contributed so much to this work: Şükrü Karaoğlu, Raphael Angoulvant, and Khalid Mhareb. Thank you, also, to the staff at the Center for Faculty Development, who supported my application to the Faculty Success Program (FSP) at the National Center for Faculty Development and Diversity. My motivation and verve for this writing project survived the pandemic in large part due to my FSP coach, Michelle Teti.

The book itself would not have come about without the feedback of several people who helped me hone a proposal, revise toward publication, and work through the complexities of publishing. At the University of Colorado Denver, I am grateful to Christopher Agee, Cameron Blevins, Esther Cohen, Ryan Crewe, Rachel Gross, Amy Hasinoff, Peter Kopp, Pamela Laird, Marjorie Levine-Clark, and John Tinnell. Thank you, also, to Christopher Low and Paul Sutter for their timely advice.

Acknowledgments

At Stanford University Press, my deepest appreciation to Kate Wahl, who saw the promise of this work and shepherded the manuscript. Also, thank you to the press's staff and to the two anonymous reviewers for their time and constructive commentary.

Above all, I am grateful to family and long-time friends, who believed in me throughout this undertaking. A special thank you to Molly Shea, who endured years of half-baked ideas, incomplete arguments, and writing headaches with empathy and good cheer, never once—well, maybe once—suggesting that I should have picked an easier topic. My parents, James and Dawn, and my brother Todd never seemed to doubt this would happen, even when I did. Their quiet certainty carried me through more than a few moments of exhaustion and self-doubt. My partner, Ty Bradford, probably wished at times I had chosen a profession that didn't involve disappearing across the world for months at a time, but he met each departure and return with unwavering patience, humor, and, crucially, dinner—proving once again that love and sustenance are sometimes the same thing.

Finally, those who've worked around cats know that their presence in any intellectual endeavor is both inscrutable and inevitable. To Suki, the gray and white one who kept me company, and to Max, the orange one who specialized in pushing things off my desk—thanks, buddies. You were, in your own mysterious way, part of this process too.

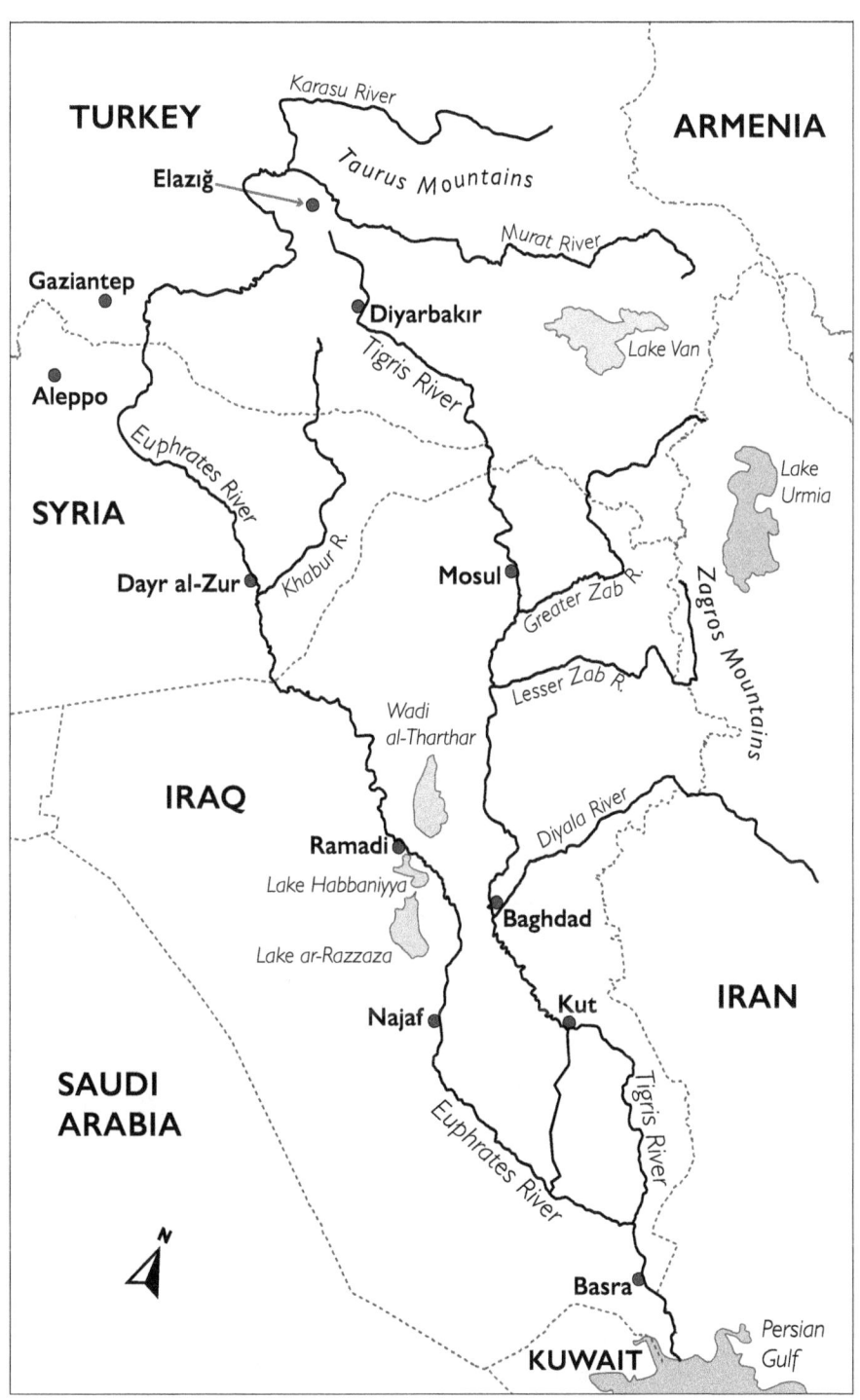
FIGURE 1. Tigris-Euphrates River Basin.

Introduction

RIVERS AND THE NEW MATERIALISM

This is a book about two rivers, and the ways that they have shaped modern states, cultures, and politics. The story, then, begins with the rivers themselves. The Tigris and Euphrates rise in the Taurus Mountains of eastern Anatolia. Their headwaters are near one another. The Euphrates begins as two large tributaries, the Kara Su and the Murat Su, which each flow westward from their origins in eastern Anatolia to meet at a confluence near the contemporary city of Elazığ in Turkey. Where their waters mingle, a single river begins: the main stem now known as the Euphrates. This confluence is about 70 kilometers from Lake Hazar, the headwaters of the Euphrates's neighboring stream, the Tigris. Each river flows a substantial distance before they meet again in southern Iraq. The Euphrates is the longest river in Asia west of the Indus (2,800 km) and nearly 1,000 km longer than the Tigris.

In the mountains of modern-day Turkey, the rivers have cut deep gorges into the terrain, much like those of the American Southwest—and, as in the Southwest, people built settlements into the canyon walls. One such site, Hasankeyf, obtained its present-day name after the Arab conquests but may have been inhabited as far back as the Akkadian Empire (third millennium BCE). Today, though, it is no longer inhabited. The ancient town currently sits beneath the Tigris River waters held back by the Ilısu Dam.[1] The

1

submerged town is a testament to the great changes along the rivers during the last century. From the gorges—many now transformed into reservoirs—of the Taurus Mountains, the rivers eventually flow into the wide plains of the Syrian steppe. These plains are home to other cities of great antiquity: Edessa, the Seleucid settlement that became Urfa the pilgrimage city; black-walled Diyarbakir, once Roman Amida; and Mardin, a city important to early Christians. It is here that one truly gains a sense of "Mesopotamia," the land between the rivers. In Arabic, the region is known as al-Jazira, literally "the island," an area defined by the water around it.[2]

The Tigris and Euphrates Rivers enter their common delta within the borders of the modern state of Iraq, just above the city of Baghdad. In the context of Mesopotamia, Baghdad is a young city, founded in 762 CE by the second Abbasid caliph al-Mansur. Here the rivers begin to meander—bowing, braiding, and curving as they make their way toward the Persian Gulf. In this area, known as al-Sawad in Arabic, the rivers have rarely remained in a single, defined channel. Seasonal floods sent them spilling across the plain, carving new paths to the sea. The land itself is their legacy, a rich alluvium deposited over millennia. In southern Iraq, well before reaching the sea and up until the late-twentieth century, the rivers regularly formed a vast area of marshes. The marshes were arguably the most important wetland environment in southwest Asia and an important bird migration habitat between northern Asia and eastern Africa. The Maʿdān, or Marsh Arabs, lived in this area for centuries before war, civil conflict, and changes to the rivers disrupted the marshes and their ways of life.[3] Finally, unlike most other rivers at their delta, the Tigris and Euphrates merge into a single channel, the Shatt al-Arab, and pass by the city of Basra before emptying their combined waters into the Persian Gulf.

Due to the proliferation of water control structures in the twentieth century, the rivers at present are roughly the same size: the mean monthly discharge (1985–2007) of the Euphrates River at Haditha Dam was 553 cubic meters per second (cumecs), while the Tigris River at Sarai Baghdad from 2000–2012 averaged 520 cumecs. Cubic meters of water per second may be hard to visualize; the two rivers today are comparable to the Colorado River or to the Oder River in its lower reaches. In the past, the rivers' unencumbered flow was much greater, with the Euphrates nearly double the size and the Tigris two and half times larger. Together, the Tigris and Euphrates Rivers carry on average nearly as much water as the Nile River.[4]

Descriptions of a river's path, size and flow can conceal wide variability from season to season. In flood, the Tigris can grow enormously, increasing tenfold in size. In summer, the Euphrates can become a relative trickle, about half the size of the Potomac River. Conditions upstream of the rivers' delta largely shape this variability. The Euphrates River receives no significant tributaries after its confluence with the Khabur River near Dayr al-Zur in modern-day Syria. The Tigris River meanwhile collects water during most of its course, including from several tributaries flowing out of the Zagros Mountains in northeastern Iraq and western Iran. As a result, snow melt and rain in the mountains of the upper parts of the basin could cause massive flooding nearly 1,000 kilometers away. In the best of times, this variability was not too wide or too overwhelming. The rivers provided sufficient water for agriculture through the winter months, did not destroy cropland in the spring, and retained sufficient flow through the summer. In the worst of times, the opposite of these three scenarios could happen: the rivers could remain low into the fall and winter, swell rapidly and flood on a biblical scale in the spring, and become a pair of meager runnels in the summer.[5]

Along with seasonal variation, substantial ecological diversity characterizes the rivers' natural regime. In the lower basin, marshlands, deserts, mountains, the sea, and irrigated landscapes all exist relatively near one another. Historians of the ancient Near East suspect this diversity gave rise to agriculture and underpinned the development of some of the first cities in human history.[6] Human communities formed in areas where inhabitants could take advantage of multiple ecological zones, for instance on the borders of the marshlands where fish and reeds could be obtained while still accessing irrigated agriculture and the steppe's domesticated animals. The combination of dry land, irrigation, and the great biodiversity of wetlands made possible large cities, and the lower basin has been home to such places for thousands of years. In the upper basin, where forms of rain-fed agriculture are possible, cities have also formed though generally of a smaller size. Upper basin inhabitants often lived in smaller communities based around agriculture or practiced forms of seminomadism, cultivating land near the rivers when possible and migrating across the steppe with their animals when not.

Out of these cities the first states emerged. Some extended their rule into the upper basin and beyond, creating well-organized empires: Akkad and Babylon, the Sassanian and the Abbasid. Such cities, states, and empires modified the rivers' environment by building massive works. These societies piled earth high to redirect floods and dug into the alluvium to create canals and bring water to fields. Such was the scale of these waterworks—the elaborate Nahrawan canal was 400 feet wide and 200 miles long—that British observers in the 1920s could trace their paths across Iraq by airplane, more than a millennium after they had first been constructed.[7] These huge constructions demonstrate how human societies have adapted to and altered the riverine ecologies of the rivers' basin over many centuries.

However, even these formidable works cannot compete with the scale of change brought about in the twentieth and twenty-first centuries. Ecologies that have supported human and nonhuman lives across the basin for millennia have been transformed almost beyond recognition over the past hundred years. Wide desert plains have been made into irrigated farmland. Marshes and wetlands have been drained and dried. Canyons and depressions have been turned into vast lakes. There is, then, something of a "before" and an "after" to the story of the Tigris and Euphrates Rivers, first a tale of a river basin shaped by pre-industrial empires, cultures and technologies, and then a twentieth-century story in which the two rivers became subject to industrial change.

Still, the industrial world wasn't a wholesale departure. Many of the ecological, economic, and cultural factors that distinguished the river basin throughout history continue to exist, from the date palm to Chaldean Christianity. The rivers still support cities, states, and empires. One can still find generals and their armies traversing the rivers like Alexander of Macedon, who traveled along and across the two rivers to defeat the Achaemenid ruler Darius III near today's Erbil in 331 BCE. And, critically, the rivers still change on their own. While the water engineering built after World War II diminished the rivers' greatest floods, significant areas along the rivers' course still suffer periodic inundation.[8]

One of the challenges of historical writing is making a fair account of such continuities while attending to big changes.[9] The industrial transformation of the rivers is only one example of this conundrum. At the same time as the rivers' industrial remodeling, Southwest Asia participated in the global shift in the twentieth century from imperial governance to that of the nation-state. A modern conceit suggests that these processes operated separately—that modern, industrial, technological societies exist above and apart from ecological change despite the long historical record alluded to above during which ecology, state formation, economic vitality, and cultural dissemination influenced one another. Indeed, the academic study of history and ecology for some time supported and reinforced this conceit; history focused on the human and ecology focused on the nonhuman.[10] The purpose of this book is to weld these factors back together and to consider the physical and sociopolitical transformation of the river basin in the twentieth century as a linked process, one rife with continuity and with rupture. To show why that approach is useful and important in today's world, let us start with a brief journey into a more recent past.

On August 7, 2014, a paramilitary organization captured a great industrial monument on the Tigris River: the Mosul Dam. The capture of this massive piece of hydroelectric infrastructure presented several thorny issues, among them that the Mosul Dam remains unfinished, and will likely never be finished.[11] In 1981, the Iraqi government under President Saddam Hussein contracted several international companies to begin building the installation. However, prior to the start of construction, no less than nine engineering companies from Britain, the United States, the Soviet Union, Finland, and Yugoslavia had investigated the Mosul dam site. They had all reached a similar conclusion—any dam built in the area would confront a complex geology characterized by water soluble rock and landslide.[12] The dam was built anyway. In 1986, it became operational and was known for a time as the Saddam Dam and as the third-largest dam in the Middle East. However, the 370-foot-tall, 2.1-mile-wide structure requires regular infusions of concrete to prevent its collapse, a disaster that would send a sixty-five-foot wave of water crashing into Mosul, a city of more than 1.5 million people. In 2014, concerns grew that the paramilitary organization now in control of the Mosul Dam might use the installation as a source of revenue, extorting surrounding communities for electricity and water, or worse, fail to complete the ever-ongoing construction required to keep the dam standing. In the end, the organization known as the Islamic State (IS) only held the Mosul Dam for ten days before US airstrikes helped a coalition of Kurdish and Iraqi military forces to retake the dam.

The Mosul Dam is thus not only a monument to industrial technology, but also a testament to a multi-layered illusion of modern order and control, an order that requires continuous construction and reconstruction. This applies both to the construction of dams and of states. These processes are material—truckloads of concrete—and cultural—the dam as a representation and service of an Iraqi (or other) state. Moreover, these processes occur in relation to and sometimes because of an independently acting nature. The more than 11 billion cubic meters of Tigris River water held back by the Mosul Dam is the solvent undermining the structure—a natural, continuous process of dissolving rock that draws trucks, concrete, concerns, and, in this case, military intervention to preserve a city and statehood.

The Mosul Dam was not the only huge dam the Islamic State captured in 2014: the group also took control of al-Tabqa Dam in Syria, which blocks the flow of the Euphrates River. The dam remained in the hands of IS fighters for nearly three years. During that time the group retained necessary technical workers and operated the dam much as the Syrian government had—selling power and distributing irrigation water to downstream users. Unlike the Mosul Dam, which could have collapsed due to natural processes, al-Tabqa Dam nearly collapsed due to military actions. On March 26, 2017, a US Special Operations unit dropped some of America's largest bombs on the dam, causing a fire and destroying equipment. The dam stopped working—the Euphrates River nearly overtopped it. Authorities in Turkey reduced the flow of the river into Syria to buy time for engineers to open its floodgates manually. Some of these local engineers were killed in airstrikes as they tried to approach the dam site. Investigations later revealed that one of the bombs, a 2000-pound bunker buster, had punched through five floors of the dam's control tower. By some miracle the American bomb did not explode. Had it detonated in that location, experts say the dam might have failed. A former director at the dam noted, "The number of casualties [from a collapse] would have exceeded the number of Syrians who have died throughout the war."[13] Put another way, with that one bomb the United States military could have destroyed more Syrian lives than the Assad regime or the Islamic State had over the preceding years of civil war.

More than the Mosul Dam, the perilous struggle for control of al-Tabqa Dam underscores the importance of these installations for state formation, or, in this case, state prevention. Al-Tabqa Dam's significance in this regard was known to its builders. At the dam's inauguration in 1976, the Assad regime renamed the town at the dam's base *ath-Thawra*—the Arabic word for "revolution"—to commemorate the 1963 coup that had brought the Baʻth Party (and eventually the Assad regime) to power. Revolution was precisely what IS had sought in taking the dam, namely the establishment of a new Sunni Muslim-majority state in the territory of Iraq and Syria. Such a state would redraw at least one of the borders imposed by European powers on Arab-majority territories after World War I.[14]

Moreover, the deadly battles over dams on the Tigris and Euphrates Rivers demonstrate how the Islamic State's revolutionary ideological aims required more than just terrorist violence and florid propaganda. Political existence (not just legitimacy) equally demanded extraction, engineering, and the provisioning of basic needs.[15] In addition to operating dams and other infrastructure the Islamic State established "Islamic Service Committees" to oversee state-owned utilities. At its height in late 2014, IS controlled an area about the size of Bulgaria with twice the population, 11 million people. The group's control of oil fields and various forms of infrastructure provided a treasury of $2 billion in assets.[16] The *New York Times* in 2014 referred to the Islamic State as "a rogue state along two rivers" and drew up a map of the group's domains.[17] Due to its control over territory, resources, and infrastructure, the Islamic State was not just seeing like a state—it was seen as a state, despite the lack of official diplomatic recognition.[18]

As a result, the Islamic State's material origins are as important to consider as its sociopolitical ones. Yet, we are used to describing state construction in sociopolitical terms. What became IS started as a clandestine organization in Jordan with ties to al-Qaida. The US-led invasion of Iraq in 2003 and the chaos that followed transformed that group into the Islamic State of Iraq, which participated in the post-invasion insurgency. The Syrian Civil War, which began in 2011 and involved multiple foreign powers, transformed the group again, from an insurgent organization into a state-builder. Had the United States not invaded Iraq, had that foreign intervention unfolded differently, and had the Syrian Arab Spring protests not devolved into an internationalized civil conflict, it is difficult to see how a tiny clandestine group based in Amman could have come to rule large parts of Syria and Iraq for several years. As Darryl Li has noted, IS emerged "in the wake of not one but two adjacent and prolonged processes of partial state collapse, in regions deemed peripheral from both Damascus and Baghdad."[19]

That sort of sociopolitical story is a familiar one, relying on a set of causal connections repeated in many historical narratives—there are many stories of human groups inventing or creating states in the wake of a previous state's collapse. Yet, if we turn to the story of the Islamic State's material origins, there is another set of connections at work. In this story interventions in the ecological processes of the basin provided important mechanisms for transforming a small paramilitary group into a state-builder.[20] In other words, the environmental presence of stateness was as important to the Islamic State's transformation as an originating idea or ongoing imagination.[21] For the Islamic State, capturing the dams was not necessarily vital in a purely military sense—the airbase near al-Tabqa Dam held greater strategic importance, for instance. However, the dam's massive remaking of the landscape and the useful resources it could produce allowed the Islamic State to become—both literally and symbolically—what it aspired to be, "the (rogue) state along the two rivers." The interventions of water engineering, then, were just as critical to the production of this state as the political and military interventions that initially produced it as a terrorist organization. The Islamic State's claiming of the dams was the modern equivalent of a Christian army capturing a mosque in Córdoba and turning it into a cathedral or a Muslim army capturing a cathedral in Constantinople and turning it into a mosque. The Islamic State changed the flags on al-Tabqa dam and repurposed a commemoration of a Baʿthist revolution as a monument to the state-building evolution of a paramilitary organization.

The dynamic (and violent) connections between rivers, dams, and would-be states in this story serve to illuminate the central problem of this book: how to understand the connected sociopolitical and environmental processes shaping states, cultures, politics, and ecologies along the Tigris and Euphrates Rivers. This book seeks to illuminate these connections by asking a set of important questions about the twentieth-century transformation of the rivers' basin. Why and how did altering riverine ecologies become a crucial component of modern state formation? How did different aspects of the rivers' natural regime—the way they worked as rivers—shape twentieth-century processes of state-making in the basin? How did different groups of people involved in transforming the rivers—engineers, politicians, poets and writers, diplomats, and scientists—make sense of or explain those transformations and their dramatic effects? Finally, how can we draw together the histories of multiple interventions—ecological and political—to comprehend more fully what the rivers have become in the twenty-first century?

These questions will help further our understanding of how water infrastructure along the two rivers has shaped the histories of the principal basin states: Iraq, Syria, and Turkey. Historians and other scholars have long recognized the importance of the rivers to the development of premodern societies in Mesopotamia, but the same cannot be said for later periods. With the notable exception of Faisal Husain's recent investigation of the Tigris and Euphrates Rivers in the context of the Ottoman Empire, the two rivers remain relatively understudied in post-1500 histories in comparison to other river basins, such as the Nile or the Mississippi.[22] Much of what has been written about the rivers' twentieth-century existence approaches the subject from the perspective of international water law, transboundary water management, and water security.[23] Little of this work has considered the historical relationship between the rivers and state formation across the basin. The more historically minded texts in Arabic and Turkish are largely concerned with the construction of the various works and international water politics, not the co-construction of the states themselves.[24]

There are reasons for this dearth of studies. Logistically, studying the history of the basin implicates several languages, and war and civil strife have made archival material inaccessible.[25] Conceptually, and more broadly, national historiography and colonial exploitation—both of which loom large over Middle Eastern environmental history—have impeded these kinds of studies. Nation-states at pains to industrialize and modernize rely on narratives of progressive, industrial societies that transcend and conquer the natural world. To acknowledge the environment as a significant factor in such histories would in some cases undermine the very conceit of modernity.[26] In addition, Middle Eastern societies have long been subjected to orientalist accusations of environmental degradation as well as determinist analyses of societal change. Geographer Diana K. Davis has demonstrated how several factors influenced colonial officials: imperial aspirations, supposedly "scientific" evaluations by specialists unfamiliar with arid lands, and a willingness to read historical sources such as travelogues literally. These factors, she argues, placed local inhabitants in a double bind, simultaneously tying local inhabitants to the land—marking them as primitive and in need of colonial intervention—and discrediting the ways they managed the environment to justify the imposition of colonial policies.[27]

The purpose of this book is to restore the rivers to modern historical view, and consequence, by attending to ecological factors in the histories of Iraq, Syria, and Turkey.[28] This book also juxtaposes historical processes across the basin, particularly in the post–World War II period, to suggest common—and problematic—visions for relations between human societies and nonhuman nature; in particular, the use of Mesopotamia's vaunted past as the "cradle of civilization." This new reading of the basin's twentieth-century history does not aspire to revise the extant nation-state historiography about Iraq, Syria, and Turkey. Indeed, it would be difficult to revise a historiography that mostly ignores this book's primary subject. Instead, *Two Rivers Entangled* shows how an ecological approach to familiar topics such as state formation and economic development can reveal different causal relationships. Demonstrating such relationships—for example, why salt is as vital as oil to economic development in Iraq—is important not only because environmental change should play a larger role in analyses of political and social change, but also because attention to the environment in history affords an opportunity to think about history itself in more spatially and temporally complex ways.

In what has become known as the "new materialism," scholars in multiple disciplines have advanced innovative ways of thinking about the relations between human and nonhuman, between thought and matter.[29] New materialism comprises various projects in multiple fields but has two general aims—to articulate a new ontology where matter is not merely the inert subject of human action, and to describe the workings of material processes based on that ontology using a wide range of sources and techniques.[30] This book is primarily concerned with the latter aim, but a short discussion of the former demonstrates how a new historical materialism has influenced the depictions of agency and the causal arguments at work in the chapters that follow.

At the heart of new materialist ontology is the assertion that material things are not just objects but contain a dynamism of their own, a "vitality" as Jane Bennett describes it. Rather than a historical materialism emanating from Marxist theory with its emphases on production, exchange, and structure—the materialism that underpinned other writings on rivers, such as those by Karl Wittfogel or Donald Worster—a vital materialism focuses on the powers inherent to things themselves.[31] As a result the new materialists set forth a diverse concept of agency, one that draws from aspects of Bruno Latour's influential actor-network theory. As in Latour's theory, agency becomes relational such that "the efficacy or effectivity to which [agency] has traditionally referred becomes distributed across an ontologically heterogeneous field, rather than being a capacity localized in a human body or in a collective produced (only) by human efforts."[32] In other words, the familiar agents of history—human beings or human groups such as social classes or governments—are not the only drivers of change, and the ways of existing most associated with humans are not the only ones that shape our common world.

In the new materialism, what once was assigned only to those familiar agents becomes distributed to other living and non-living entities such that the characteristics of a purely human agency—intention, consciousness, responsibility—are seen as happening in concert with matter, not above or outside of it. Bennett and other scholars refer to this heterogeneous field as an assemblage, a model of "confederate agency" where human and nonhuman are always interacting. This model works in contrast to forms of agency where, according to Latour, "objects do nothing, at least nothing comparable or even *connectable* to human social action".[33] The projects of new materialism emerged as a response to the excesses of the cultural turn, which in a focus on text and discourse drove a "discontent with ... the pitfalls of relativism and the prevailing preference of culture in favour of other analytical categories."[34] The cultural turn was, of course, a turn away from an older materialism. What makes the new materialism "new" is that it broadly embraces the techniques and emphases of the cultural turn—questions about the history of the body and the construction of identity, for instance—while seeking to read those histories with "the goal ... to think the social and the natural together."[35]

New materialism, then, is concerned not only with elucidating a form of agency attributable to the nonhuman, but also with how that form of agency relates to human action. Scholars have adopted different approaches to accomplish that goal, and environmental history has contributed to the development of these approaches. In 1995, Richard White described the Columbia River as an "organic machine." White sought to interpret how human labor and nonhuman energies came together at the Columbia, producing over time a river that was not purely natural and not purely human. White's formulation, "to find the natural in the dam and the unnatural in the salmon," has influenced scholars seeking to weave environmental histories with histories of technology in telling the story of a river.[36] Enviro-technical analyses combine the insights of the two fields, most notably in Sara Pritchard's work on the sociopolitical and industrial remaking of the Rhone River. Pritchard adds theoretical heft to White's shorter work, arguing that "both nature and technology are at once material and discursive."[37] Both exist as much in the material realm as in the cultural, which means "sometimes holding apart these concepts, other times merging them." Drawing from several fields including landscape studies and techno-politics, Pritchard asserts that technology must be seen in "its environmental dimensions," while nature "can simultaneously be technological."[38] More than merely an amalgamation of the organic and the mechanical, Pritchard's story of the Rhone River complicates the very categories used to depict technical transformation.

Just as Pritchard seeks to reframe both nature and technology, theorists in new materialism aim to dissolve the conceptual boundaries between material objects and humans, placing them within a common field of activity. New materialism as a whole has not settled on a single method for doing so, giving rise to a rich debate about how to conceptualize an agency that includes the nonhuman. Latour, for instance, advocates for a "flat" social world, where human and nonhuman entities operate on the same plane. He introduces the term "actant" instead of actor or agent, thereby expanding the realm of entities capable of effecting change. This approach challenges traditional hierarchies by suggesting that objects, ideas, and beings all possess the potential to influence events in the world.[39] Diana Coole, in a similar vein, refers to "distributed agentic capacities," where the contingency and indeterminacy typically associated with human agency are "diffused across many different types of material entity."[40] Her approach suggests that agency is not the exclusive domain of humans but rather a quality that emerges from human-nonhuman interactions. Karen Barad takes this emergent agency another step, drawing inspiration from quantum physics. She proposes an "agential realism" where the act of making knowledge—traditionally seen as a purely human endeavor—is reimagined as a material engagement with the world. In Barad's view, agency is inseparable from the material processes that constitute reality, highlighting the interconnectedness of human and nonhuman forces.[41] Timothy Mitchell adds another layer by reminding us that "human agency, like capital, is a technical body, is something made." He argues that human agency has its own history as a method for analyzing change in the world, suggesting the concept of "hybrid agencies," which, to this author, has always suggested the need to consider and analyze hybrid histories of agency.[42] Mitchell's concept emphasizes the historicity of agency itself, such that the ways historians and other scholars think about and narrate agency are products of historical time and expectation.

The common thread among these theorists is the recognition that the traditional concept of agency—as limited to humans or even to human minds and as unreflexively a product of modernity—needs to be reconsidered. Easier said than done, the saying goes, for as Jane Bennett has noted it "seems necessary and impossible to rewrite the default grammar of agency."[43] The proliferation of neologisms as new materialist scholars seek that revised grammar is evidence enough. One promising method of revision is to transform the "issue of power and agency" from a predetermined answer into an open question.[44] And, by tracing the emergence of agency as a question, Coole argues we may come to see agency as "more partial, contextual, and provisional" than previous theories have maintained.[45]

If agency becomes "partial, contextual, and provisional," then narratives of causality, indeed much of what we commonly understand as "history," face a similar reckoning. A new historical materialism in which agency is diffuse, distributed, and hybrid not only alters the types of agents under consideration but also renders the fabric of history more complex. Rather than a "world that somehow seems the outcome of human rationality and programming," or, in the case of rivers, one subject to a "stagnating" hydraulic determinism, the pages that follow relate a story of two rivers emphasizing the "agentic capacity" of the nonhuman as it interacts with human intention and meanings.[46] The movement of salt molecules becomes as consequential as the movement of armies. The rivers' paths and flows manifest as much in geological time as in human history. And the ideas, plans, narratives, and visions about the rivers—all the words in the archives—happen not as purely cultural representations separate from material change but as "material-discursive practices" involved in "(re)configurings of the world."[47]

Such a model of "agentic capacity" and of spatial and temporal complexity complicates and challenges dominant methods of historical narrative, namely linearity and causation. Challenging those dominant methods is not limited to new materialism. Sara Pursley, for example, in her book *Familiar Futures* takes aim at simple, stable temporalities in Iraqi history through an examination of development discourses. In her book, development practices reframed, repurposed, and repeated imaginings of past, present, and future to construct "familiar" forms of modernity and progress. The layering of past, present, and future undermines linear chronology in favor of a modernity characterized instead by "timelessness."[48]

Pursley's work focuses on discourses, particularly on imaginings of the present and future, while new materialism attempts to combine cultural modes with material ones in a critique of linear narratives of modern progress and civilization. In *The Mushroom at the End of the World,* Anna Tsing addresses the ruination of ecosystems and human societies by industrial capitalism. Tsing notes how progressive models of historical change focusing on modernization, scientific advancement, and economic growth turned "both humans and other beings into resources" and left aside critical inquiry into ongoing ecological and economic devastation. The problem, Tsing argues, is that by not facing the truth of industrial capitalism, we have neither developed adequate accounts of life amidst environmental degradation nor narratives that would correct many years of progressive history-telling. The way out of the "modern human conceit" as the sole masters and arbiters of life on earth is to interpret and portray the gathering of different time-making and world-making projects, both human and nonhuman.[49] Like Bennett, Tsing uses the concept of assemblage, though she distinguishes her usage by adding a musical modifier—"polyphony," she writes, "is music in which autonomous melodies intertwine." Thus, Tsing's idea of assemblage is one where multiple, autonomous rhythms come together—human and nonhuman species, non-living ways of being, multiple scales and temporalities—to compose "indeterminate and multidirectional" histories less fixated on linear progress.[50]

Tsing's notion of intertwining rhythms led me to the title and framework of *Two Rivers Entangled.*[51] Entanglement in this book operates as a material concept and references a longer genealogy of thinking about materiality and the human. As a concept, entanglement avoids the more common mechanistic terminology so often found in historical and social scientific accounts. Terms such as "making," "building," and "inventing" have tended to this day to foreground human action, even as scientists and other scholars have sought to reclaim them for other species and things.[52] Entangled is also in more common usage than assemblage which has disadvantages as jargon and its relation to another mechanistic verb that recalls hours putting together "ready-to-assemble" furniture in cramped graduate housing. Further, in expressing both an action (to entangle) while more often registering in English as a condition (being entangled), and in both cases expressing a mostly undesired, unintended state, entanglement invokes new materialism's "partial, contextual, and provisional" forms of agency.

Entanglement also references an intellectual history of considering the relationship between human societies and the material world. My first encounter with the term as an undergraduate history major came in reading Nicholas Thomas's *Entangled Objects*, a book that considers the social lives of objects traded between Pacific Islanders and Europeans.[53] In Thomas's account, things are situated in shifting, contingent, and sometimes indeterminate sociocultural and historical contexts, such that "objects are not what they are made to be but what they have become."[54] That process of becoming is a tangled one. Thomas relates several stories of objects passing between continents and through cultures, along the way complicating simple binaries such as colonizer-colonized and production-consumption. Humans and nonhumans are caught up with one another and in larger historical structures—a state of "promiscuity," as Thomas describes it—requiring a methodology that does not presume responsibility and culpability and that works to trace the tangle itself.[55]

In emphasizing process over a stable state, becoming over being, entanglement in this book serves as a way of conceptualizing how material, society, and culture shape one another across time and space. In this new historical materialism, the Tigris and Euphrates Rivers cannot function as mere background or a setting waiting within society to be imagined or socially constructed. The goal is not the opposite, to recapitulate a deterministic vision where control over water leads inexorably to despotism, but to trace the tangle and to wonder at the richness of interconnected histories.

Methodologically, *Two Rivers Entangled* does not unspool a linear, mono-causal story of progress (or decline) but works instead to show how multiple histories have been implicated in the immense reshaping of a pair of rivers. I have sought to demonstrate the entanglement of the river in a few different ways. First, *Two Rivers Entangled* is principally about the unwanted or unintended. In contrast to William Cronon's story of Chicago, where the problem is getting what is wanted—timber, cattle, grain—into a great and growing city, these are stories of getting rid of what is not wanted: too much water, too little water, water going to the wrong place, water carrying undesirable things such as salt, and vast hopes dashed by equally vast volumes of water.[56] These are stories less about what was constructed by human ingenuity and more about what was deconstructed, often through an amalgam of different agencies and capacities.

Second, the histories that follow work across temporal, spatial, and disciplinary scales to entangle stories that have been, mostly arbitrarily, disentangled. In some cases, this has meant demonstrating environmental connections to more familiar social and political histories. In other cases, entanglement has meant reconceiving what goes together. Historians have been doing something of this nature when considering, for instance, transnational flows or by tracing time on a vast scale with "deep" or "big" histories.[57] The challenge in this work has been to conceptualize across multiple boundaries at once, and so one of the chapters in this book (Chapter 3) offers a socio-geological history of the poetry of engineering. While that précis might seem more like the premise of an absurdist play than an academic monograph, the point is to hold in one's mind the influences of multiple time scales (geological and sociocultural) and multiple modes of meaning making (poetic and scientific)—the very stuff that has been disaggregated in pursuit of the homogenized nation or the controlled environment.

Finally, the histories in this book operate according to varying modes of causation. A typical historical narrative with a linear chronology will tend to make an argument about who caused what, assuming in the process that a historically constructed human agency exerted control over events. In a narrative where human elements are not the central focus, where agency is distributed and hybrid, and where material and social factors are entangled, that type of causation is difficult to justify. An example from this research is the notion of a completed engineering project. As noted above, Iraq's government supposedly completed the Mosul Dam in 1984, but the facility requires significant ongoing construction, more than the usual maintenance required at such installations. The dam is often referred to as "unstable," a term that refers as much to the stories told about the dam as to the ground on which it is built. One could argue that the Mosul Dam has been under construction since 1981 and will be under construction for the foreseeable future. The myth of an all-controlling human agency asserts a completion date of 1984 (the Orwellian reference is appropriate), while a narrative that considers the materiality of rock and water fundamentally questions both the myth and the assertion.

Committing to a pure human agency and its causal connections makes for an odd story about the Mosul Dam. Engineers and politicians caused the dam to be completed by contracting the work, filling the reservoir, spinning up the turbines, and communicating that it was complete. Karst, a form of limestone, in interaction with water caused the dam to leak and thus to be incomplete. Which one is it? Must we by virtue of an obsession with human agency and linear progress always understand the Mosul Dam as complete-incomplete? A correlative causation offers an alternative approach. By showing where various threads of history—engineering plans and concrete, rock formations and water pressure—rub against one another, the focus shifts from pure responsibility to measuring and demonstrating interdependence. There is no question of completion or control in the endless (un)becoming of a dam entangled in a material and cultural world.

Finally, an unentangled human agency can make stories, particularly morality tales, satisfying because they place some on the "right side" of history, but such stories rarely consider the agentic capacity of water or rock. It seems to me, though, that to do some measure of historical justice to a place with the longest recorded human history—a place with some of the earliest forms of writing that made possible history as we know it, from which the very notion of civilization emerged, and that has suffered horribly in recent decades under war, sanctions, civil strife, and now the ravages of climate disaster—the least we can do is abandon a progressive, linear notion of history. The vision of history presented here is entangled and deeply ecological. Stories of confederate agencies must then satisfy another way, perhaps in the intellectual challenge of an ongoing puzzle. How else to describe the historical significance of a never-ending dam?

This book is structured around four key ecological elements—the flow of water, the intrusion of salt, the movement of rocks, and the rise of reservoirs. Each is essential not only to understanding the twentieth-century physical transformation of the Tigris and Euphrates Rivers but also to comprehending the profound social and political changes that occurred alongside. The four chapters of the book explore these elements, examining how these features of the rivers' environment interacted with society and politics. The chapters cover different—sometimes overlapping—periods but generally follow a chronological progression from the early twentieth century and a geographical path from the lower basin to the upper, mirroring the pattern of the most significant changes in the rivers' natural flow. Rather than offering a comprehensive century-long history of the entire basin, however, this book is meant as an investigation into the intricate entanglement of human and nonhuman forces within the river basin's ecologies.

There are three major themes that emerge from this analysis of entanglement which are developed across the book's chapters. The first theme concerns the power of mythologies and invocations of civilization to shape economic development. From the beginning to the end of the twentieth century, a vision of civilization based on a glorified Mesopotamian past informed the engineering of a new river basin environment. This concept of civilization contained at its heart an imagining of how people ought to act and behave in relation to the natural world. Further, ideas for a "regeneration" of Mesopotamian civilization—what I call a form of "civilizational dreaming"—were strikingly similar whether one was standing along the banks of the Euphrates in 1909 or 1999. Historian Kate Brown has noted a comparable similarity in her analyses of the Soviet Union and United States. She shows how the spatial arrangements born from industrial endeavors resemble one another, no matter that they were undertaken in one place by a socialist nation-state and in another place by a capitalist one.[58] This suggests a common logic, a shared perception, or a habitual relation underlying these cases. For Brown, the US and USSR were both centralizing, industrializing states keen to absorb hinterlands, perhaps for different ends but by similar means. In the case of the Tigris and Euphrates Rivers the same observation could be made—were the visions of a Muslim empire for the two rivers that different from a Christian one? What about a multi-party democracy versus an authoritarian dictatorship? And, indeed, the US and USSR both played their own roles in reshaping the river basin.

The second overarching argument considers the role of engineers and engineering knowledge. In recent years engineers as a social group and as historical actors have received more attention from historians and other scholars. These studies have situated engineers ideologically in relation to histories of technology, colonialism, or the specific politics of a given state.[59] In contrast,

Two Rivers Entangled approaches engineers alongside other kinds of knowledge producers who, using the conventions and rules of their own disciplines, present a paper imitation of environmental change. Sometimes engineering processes have been referred to as calculation and abstraction; this book portrays engineering as a linguistic act. Engineering knowledge presented in technical documents is thus considered akin to a literary genre, and this research employs the same discourse analysis techniques used on art forms such as poetry, storytelling, and film. Using such "techno-poetics" to analyze engineering knowledge allows for forms of contextualization and interpretation that technical language often appears to resist. Moreover, approaching engineering in this manner places engineers and their environmental visions on the same historical continuum as other forms of knowledge.

The third major finding of this study is the importance of local contexts in the consideration of large-scale historical processes. What happened to the Tigris and Euphrates Rivers in the twentieth century happened to other rivers in other parts of the world—the Mekong, the Colorado, and the Columbia are among the more prominent examples. Scholars have taken up these similarities to offer synthetic histories on grand scales. For example, Geographer Christopher Sneddon calls the proliferation of dams in the twentieth century a "concrete revolution" and traces the networks of power and politics centered in the United States that facilitated dam building across the world.[60] Other scholars, such as Joseph Morgan Hodge, have investigated the rising authority of technical experts in colonial settings, using imperial history to show the broad impact of scientific and technical approaches on agrarian policies, water and land use, and disease mitigation.[61] Still other scholars have examined the global spread of certain discourses, from liberalism to Green Revolution developmentalism.[62] There is no question that the global and transnational trends illuminated in the aforementioned books affected and in some cases propelled the ecological, social, and political changes wrought along the Tigris and Euphrates Rivers.

However, two features of this book counter the tendency to rely on global trends as comprehensive explanations and instead enrich our understanding of how large-scale historical shifts played out. The first is an attention to specific ecological processes, to how these specific rivers act within their own dynamic ecologies. No other place on earth is exactly like the Tigris-Euphrates River Basin in its natural, geographical, and historical configuration.

The second feature is language. Poetry has provided a rich and useful method for illustrating and comprehending ecological change and its associated meanings. Along with its evocation of the natural world and importance in Arabic and Turkish-speaking cultures, poetry emphasizes the specificity of experience and the granular quality of observation in small moments in time. Poetry offers support and context for several of this book's arguments, and as a result the reader will encounter poetic language in every chapter.

On the subject of language, the research presented here draws on a wide variety of sources in four—Arabic, English, French, and Turkish—gathered from archives in multiple countries. Civil strife in Iraq during the research period precluded a visit and only preliminary studies were possible in Syria before the outbreak of civil war. However, significant insights were gleaned from archives and libraries in Britain, France, India, Turkey, and the United States. These sources include government reports, embassy dispatches, private papers, planning documents, and maps. As with any historical study, archival sources shaped the stories and actors portrayed. To expand the study's scope and provide additional context, published sources in Arabic and Turkish, such as journalism, biographies, memoirs, poetry, literature, and film, are woven throughout the analysis.

The Ottoman and Republican archives in Istanbul and Ankara, which have been invaluable for other studies of the rivers, proved less useful for this study due to the sensitive nature of dam building during the twentieth century. Instead, more relevant materials were found in libraries and agency depositories, such as the State Hydraulic Works (*Devlet Su İşleri*) library in Ankara and the Beyazıt State Library in Istanbul, where local newspapers, literary journals, and scientific and social scientific studies provided important insights into local views of dam construction. Indeed, the juxtaposition of such sources with engineering studies supports one of the methodological interventions of this book; this discussion may be found in Chapter 3.

The first of the book's four chapters begins on the banks of the Euphrates River in the early twentieth century. A bifurcation in the Euphrates and the frequent flooding of the rivers in the lower basin prompted the Ottoman government to initiate plans for new methods of water engineering. After World War I, the British obtained over the lower basin a "Mandate"—the League of Nations' assent given to victorious states to govern a former Ottoman (or German) possession. The area was known to its inhabitants as Iraq, but the country's new rulers at first called the lower basin by another name—Mesopotamia. British officials inherited the Ottoman plans for the rivers but argued about their implementation, with a group of engineers suggesting that "regeneration" should be pursued instead to preserve the fabled rivers of civilization against their uncivilized users. These arguments eventually gave way before the power of the rivers' floods, and the analysis turns to how these widespread events affected the governance of the new state. Chapter 1 argues that the annual floods played a critical role, mostly overlooked in the literature, in shaping British views on how to manage politics and the environment in their new possession. Particularly devastating floods in the mid-1920s led to a series of legal rather than physical responses, primarily land tenure and settlement laws, and the deployment of irrigation technology.

Chapter 2 turns to salt and remains focused on the lower basin. This chapter considers the relationship between Iraq's two most significant underground deposits—oil being the other, and by far the more familiar—in the country's post-World War II economic development plans. Both are brought to the surface at this time, and salt was the more consequential of the two in some important ways. By the 1950s, the longer-term effects of earlier decisions on irrigation and settlement had become clearer. In this chapter, salinity appears not only as an environmental factor to be dealt with using drainage and other techniques but also as a driver of social, political, and economic change.

Chapter 3 travels to the upper basin (Turkey) and into the 1960s and 1970s to consider the confluence of geology, environmental engineering, and the arts in the construction of a great dam on the Euphrates River. A group of Western companies worked with the Turkish government to build the Keban Project in eastern Anatolia. This chapter, above all, complicates chronological and spatial relationships by demonstrating the entanglement of rocks, culture, and science. Alternative ways to conceive of the history of dam building emerge in this chapter, as the narrative compares Turkish cultural production about the rivers to engineering plans conducted by local and international consortia. By analyzing geological, scientific, and cultural forms together, linear concepts of change and progress are undermined to be replaced by an understanding less certain and more contingent.

By Chapter 4, large dams have transformed the Euphrates River into a long, sinuous series of reservoirs and the Tigris River is well on its way to the same fate. Here, at the close of the twentieth century, I seek in the profusion of reservoirs connections between the past, present, and future of Mesopotamian civilization. In this chapter, the analysis attends to how watery and non-watery reservoirs played a role in a supposed "civilizing" of nature itself. What ideas, practices, and dreams for the future underpinned the notion that turning a river into a reservoir would bring about civilization? The biographies and writings of Turkish prime minister and president Süleyman Demirel (1924–2015) and a large collection of Arabic and French-language documents on Syria help to answer this question. In this chapter, I compare the Keban Dam in Turkey to Syria's al-Tabqa Dam, showing how often dams and weapons of war were conceived together, as a kind of civilization-building package. The reservoirs behind the dams also became sites of resistance to such civilizational dreams, and this chapter closes with an examination of two Syrian dissenters and the stories they told along a dammed Euphrates.

The conclusion of *Two Rivers Entangled* introduces the Southeast Anatolia Project, a huge multi-sector development program based on hydraulic engineering underway in Turkey. Here, the idea of "regeneration" first advocated in Iraq in the 1920s found new life in Turkey in the late 1980s and 1990s, again as a way to evoke a civilizational concept that could overcome, or at the least sideline, other visions for collective life in the rivers' basin. Both the book and century end with a question of what this repurposed idea of regeneration means for understanding the violent and often-troubled twentieth century, and the perils and opportunities for "regenerating nature" in this century.

One

WATER

The Tigris and Euphrates Rivers created the land of what is today central and southern Iraq. Over many centuries, the rivers dragged silt and mud—fifty to one hundred million tons per year—down from the Taurus and Zagros Mountains.[1] This alluvium accumulated, producing a wide plain along the rivers' path to the sea. Wind sculpted some of this alluvium into dunes and basins. Water carried and pushed the silt, in some areas depositing natural levees along the riverbanks and in others spreading the silt into shallow lakes and marshes. Within this plain the rivers changed course numerous times in a process known as avulsion, leaving their original banks entirely or creating new channels. The plain was always moving and changing at various scales; it was never a static thing as a map or photograph might depict.

As historian Faisal Husain notes, one way to comprehend how the rivers have changed paths is to look at the ancient remains of cities. Sumerian towns, once built along the banks of the Euphrates River, now "stand forlorn in the middle of the desert, tens of miles away from the modern river channel."[2] These forlorn towns show how avulsion was more than just a natural process—it could strip settlements of their water source, constituting a significant disaster for inhabitants. People living in the rivers' upper reaches, where they are hemmed in by mountains and bluffs, need not have worried so much about the rivers adopting a new path to the sea. However, in their lower reaches, where the rivers' own sediment has helped to shape the land itself, the rivers are more apt to break their banks.

The landscape of the lower basin, then, has undergone a shaping process for hundreds of kilometers and for hundreds of centuries by the rivers' action, assisted at times by climate and changes in sea level. One of the landscape's most noteworthy characteristics is its flatness. The rivers' alluvial plain extends from Ramadi on the Euphrates and Samarra on the Tigris south to Basra and the Persian Gulf. In this area, the Euphrates River descends at a leisurely average of 10 centimeters per kilometer. Baghdad is over 500 kilometers from the Gulf but only thirty-four meters above sea level.[3]

The flatness of the terrain in the lower basin contributed to the immense scale of the floods on the two rivers. With little topography to hem in or direct the rising waters, floods could spread in all directions, covering hundreds of square miles. No wonder, then, that the lower basin is the site of the archetypal flood in the religious literature of southwest Asia's monotheistic traditions. Some scholars believe the "great affliction" Noah and his family survived refers to the cataclysmic potential of the two rivers rising together in the lower basin.[4] When twentieth-century engineers contemplated controlling the rivers with modern works, they were warned, "Every engineer beginning work in the delta for a canal or railway must always keep before him the memory of Noah's flood."[5] It is not, then, an overstatement to call the Euphrates and Tigris floods "biblical."

Size is not the only notable feature of the two rivers' floods. The Tigris and Euphrates Rivers can be very different from season to season. While a normal high-water period generally occurs from March to early May, severe floods occurred (and still occur) as early as November and as late as early June, sometimes lasting only days but at other times a couple of months or more. Such changeability has complicated the human use of the rivers' waters. The wide seasonal window for flooding has meant that catastrophes have often struck at the worst possible time for the cultivator, either too early, causing the destruction of the winter harvest, or too late, disrupting the spring sowing season. To cope with the variability of the two rivers, rulers and landholders used corvée labor (*'awna*)—a form of intermittent, unpaid and forced work—to construct dikes, called bunds, along the river's course to protect settlements and crops (see Figure 2). This form of labor continued through much of the first half of the twentieth century.[6]

FIGURE 2. View of a bund (flood embankment) protecting a military camp along the Tigris River, 1943. Pedestrians promenade on top of the bund while boats dock at the riverbank. Original in watercolor by Edward Bawden. Image courtesy Imperial War Museums.

In the spring of 1919, as peace delegates in Paris struggled to reach terms after the Great War, one of these severe floods tested the bunds constructed along the rivers' banks. The flood swept through the Ottoman provinces of Mosul, Baghdad, and Basra, which were then under British military occupation. The British referred to the area as Mesopotamia, a designation that substituted Iraq's Ottoman and Islamic history with British visions of resurrecting the region's ancient status as a center of human civilization.[7] As peace negotiations progressed in 1919, the civilization that remained in Mesopotamia after years of famine and military conflict was profoundly menaced by the Tigris and Euphrates. Vast quantities of water flowed from the Taurus Mountains and arrived in the lowlands of what would later become the state of Iraq. The rivers surged over their banks, ripping through protective works and wrecking irrigation canals. The Tigris and Euphrates inundated thousands of square miles and destroyed over 100 square miles of prime cropland. A breach in the Tigris bunds north of Baghdad brought floodwaters within four miles of the city.

Southern Iraq was not so lucky and "the countryside and the river combined to form a huge lake."[8]

We will return to the history of this flood, but first let us examine how the peace negotiations happening at the same time have shaped our understanding of Iraq's history, including that of its two great rivers. Many histories of the Middle East in the twentieth century focus on Versailles as the location where new nation-states—Iraq was only one of several—were made and where, in many tellings, much of their subsequent history was determined.[9] In the case of Iraq, such narratives emphasize the state's juridical position over its geographic or cultural existence, discounting what was happening along the two rivers where other processes of formation and destruction had been and were already (and in fact were *simultaneously*) at work. One of the stories left out is how huge, nature-made lakes shaped not only the physical landscape but also the political one, and how, over time, these nature-made lakes were transformed into human-made reservoirs.

Floods like those in 1919 swept through the region several times in the 1920s as Britain sought to consolidate its newly won political power in Iraq. Another flood in 1923 devastated large parts of the country. This was followed by the great flood of 1926 when the Tigris nearly wiped out the city of Baghdad. In 1929, the Euphrates wreaked havoc on the countryside—the river overflowed its banks to such an extent that it started flowing into the Tigris near Baghdad, rather than in its usual spot some 250 miles to the south.

The floods of the 1920s threatened the functioning and legitimacy of the British administration in Iraq, yet they receive little attention in the several texts devoted to the history of Britain's tenure in Iraq. These works focus on other topics: the administration of the League of Nations "mandate," the international authorization that handed Britain control over Iraq after the war; the causes of the 1920 uprising against British rule; and the formation of an Iraqi national identity.[10] Of the major historical studies of Iraq, only Hanna Batatu's *Old Social Classes* suggests that these recurrent natural disasters had any significance in modern Iraq's social and political organization.[11] In most histories of twentieth-century Iraq, then, the country's most salient geographical feature—its two rivers—appears to be of little importance in shaping the nation-state that materialized in this part of the Middle East.

Human action instead dominates in stories of Iraq's "invention," a metaphor for state formation that occludes as much as it illuminates. By these readings human actors—Iraq's kings, politicians, British high commissioners, union members, activists, and military men—constituted, contested and molded the government and society that, with the destruction of the Ottoman Empire, came to orbit around a center at Baghdad. There are several problems with this way of reading Iraq's history, one being the limited agents allowed to participate. Moreover, the idea of Iraq's invention appears often as a shorthand for everything that is wrong with the country: the arbitrariness of its borders, the sectarian and ethnic conflicts animating its politics, and the country's sometimes-violent relations with its neighbors.[12]

Yet recent research on the creation of the Iraqi nation-state has described the same historical processes that helped produce other states elsewhere. For example, Priya Satia, in her investigations of development in Iraq, describes British ideas about Mesopotamian nature, which produced an "environmental imaginary" that "inspired an understanding of colonialism as a vehicle for technocratic developmentalism."[13] Satia's connection of the environment, colonialism, and technocratic development would sound familiar to a scholar of Egypt or India. Like several other scholars of Iraq, Toby Dodge, in his 2003 book *Inventing Iraq*, notes the importance of land policy and land tenure in the founding of the state. Dodge sees a clear connection between British conceptions of "European land-tenure regimes" and the modern processes of abstraction and simplification, which, when imposed on Iraqi society, dramatically altered the social meaning of land.[14] These processes would not sound out of place in other territories administered by European colonial powers, much less in Europe itself. Yet, while other nation-states "emerged" or were "made" or "built," Iraq is uniquely and distinctively approached as a case of invention, as historian Sara Pursley has detailed.[15]

Why would this be the case? What does the invention trope offer? For one, conceiving of Iraq's history in this way marks the country as a place for *intervention*. A state that the great powers, and Britain in particular, improperly invented might, with lessons properly learned, be reinvented by the Americans.[16] The term invention also implicates questions of legitimacy. If Iraq were invented—unlike other states with supposedly sturdier historical roots—then its legitimacy could always be questioned, no matter that Iraq holds all the regular trappings of statehood, such as international recognition, juridical sovereignty, and a central government.

Ultimately, out of the notion of invention a distinction crystallizes between methods of state creation understood as legitimate and those understood as illegitimate. Some forms of social construction—a war waged by erstwhile colonial subjects, perhaps—appear as legitimate modes of state building, while others—a territorial designation negotiated from afar—appear as questionable and contingent. For this study, the problems inherent to the invention trope point to the perilously narrow and restrictive ways that social construction may be construed. Iraq, like other states, was not *only* socially constructed. Several other agents were involved, including nonhuman actors often left out of history. What might become of our notions of "state power," of imperialism and technocratic development, if agents other than high commissioners and kings predominated in Iraqi history? What might emerge if we asked not about the social meaning of land and imported notions of land tenure but about the way land, water, and people interacted? Iraq may then appear less as a place perpetually "invented" and "reinvented" by the powerful, and more like any other—shaped by historical forces and natural features beyond human control, which various historical actors have, at times, sought to appropriate and redirect.

By examining the history of Iraq's two major rivers from such a standpoint, this chapter tells a new story of the founding of the Iraqi state by considering how human and natural systems interacted. While recognizing the state as a construction, the Mandate of Iraq was not entirely of British invention, nor even of human invention. In the story that follows, the connections between human and nonhuman at the heart of the state building project appear more as a reorganization than an invention, and less a clean break with the Ottoman past than a continuation of earlier influences.

The two rivers were, and one could argue still are, the lifeblood of Iraqi society, providing the vital ingredient for human settlement and the country's pre-oil economy. Consequently, governance in Iraq has always had to deal with the rivers, something that Iraqi statesmen and cultivators, and Iraqi, Indian, and British scientists and engineers understood and debated. At the same time, the rivers were resistant features of the natural landscape, upending human designs at every turn. A sizable flood could paralyze the country for months at a time, cutting rail and telegraph lines, rendering roads impassable, and surrounding cities with colossal moats. For the inhabitants of the alluvial plain, floods wrecked dwellings and farms, filled canals, and spread disease—the incidence of malaria, "the greatest single influence ... on Iraq's health and [population] numbers," rose precipitously after every flood.[17] A duality thus emerged in the history of Iraq as a British mandate; the rivers figured significantly in British aims to produce a pliant Iraqi state, even as they also posed the greatest nonhuman threat to that project.

To deal with the entanglement of the rivers with governance, the colonial state in Iraq faced a paradox of benefit and danger. The government sought to harness the rivers' potential while containing their threats, working to shape environmental change even as the state was compelled to adapt to it. Managing such a paradox required remaking both human and natural systems in Iraq, forging interdependencies that made the state a product of hybrid agencies. The analysis below follows the formation of a colonial state through the interaction of these two systems—one a river, the other a government. The resulting history locates the creation of the Iraqi state not in the mind of its British overlords, but in the relations between thought and matter, an imbrication of human intention and nonhuman natural force.

The hydraulic paradox of Iraq was nothing new. Management of the variable Tigris and Euphrates Rivers had long been a dilemma for rulers seeking to make a productive realm out of the lands of the two rivers.[18] Nineteenth-century Ottoman governors in Baghdad had struggled to shape the rivers in ways that would facilitate political and economic goals. Administrators such as Mehmed Reşid Paşa (1852–1857), Mehmed Namık Paşa (1861–1868) and Midhat Paşa (1869–1871) all worked to control the rivers and improve navigation and irrigation.[19] Not until the early twentieth century, however, did Ottoman authorities make a concerted effort to regulate the rivers with the new, modern engineering methods demonstrated in India and Egypt.[20] The man responsible for some of the most prominent engineering works in Egypt, Sir William Willcocks, became the key figure in Ottoman efforts to remake the Tigris and Euphrates.[21]

Willcocks unfolded his schemes for the rivers within a larger strategic context that also involved the construction of a railroad connecting the Persian Gulf with the imperial center at Istanbul. This railway was known as the Berlin-Baghdad because it would connect northward into eastern and central Europe. The scope of the project prompted British concerns about German commercial penetration into the Persian Gulf region.[22] When in 1907 the

Ottoman government accepted the designs of a French engineer for a Euphrates barrage, to be known as the Hindiyya, London took note. The British government began to see water infrastructure in Mesopotamia as a potential means to fortify British influence and counter expanding Ottoman-German amity.[23] British officials hurried to capture the contract for the Hindiyya Barrage, as the project would be "of great utility as a foundation for British enterprise in Mesopotamia."[24]

With the construction of the Hindiyya Barrage, the strategic game over the Baghdad railway could include a new and potentially lucrative prize—a greatly expanded Mesopotamian agriculture. For the Ottomans, Willcocks's plans would project power into the empire's Arab provinces and increase revenues to pay for the railway and other development projects. After the Young Turk revolution in the summer of 1908, the new Ottoman government secured Willcocks's cooperation in undertaking an engineering survey of the two rivers. Ottoman Grand Vizier Kamil Paşa invited the British irrigation engineer to Istanbul in the early autumn of that year, and Willcocks accepted a five-year contract to generate a detailed plan to restore the ancient irrigation works of Mesopotamia.[25]

Willcocks spent two and a half years in the territory that would later become Iraq, resigning halfway through his contract. Conditions in Mesopotamia were difficult. The Ottoman governor in Baghdad, Nazim Paşa, used funds that Willcocks intended for his projects to support other parts of the cash-strapped Ottoman administration. Access and safety were also ongoing problems. Willcocks's survey involved a sizable staff, which moved ponderously across a country that was under only tenuous government control. Willcocks's work moved just as slowly and, by 1911, the engineer gave up and submitted a program of works that included detailed plans and drawings only for projects on the Euphrates.[26]

Willcocks's program sought to address the essential tension for cultivation and settlement inherent in the rivers' natural regime. Too little river water in the spring meant that by the hot summer months the rivers were too low for effective irrigation, with the lucrative rice plantings in southern Iraq especially affected. Meanwhile, the fertility of the Mesopotamian soil was partly contingent on a situation of too much water—the floods spread nutrient-rich silt far from the rivers' banks. Yet, the floods posed significant danger to Iraq's cities and towns. Due to the floods' frequency, many settlements along the path of the rivers needed protection. Baghdad, positioned along the Tigris near its confluence with another river, the Diyala, has been threatened with inundation throughout its history. Since the city's founding, Baghdad has been equipped with redundant dikes and floodwalls, such that an opening or breach in one would not overwhelm the entire city. One tactic to protect the city during a flood carried its own dangers. Intentional breaches to riverbank fortifications might be made upstream of Baghdad to reduce the height of the flood at the city and lessen the danger to the population. This tactic sacrificed productive lands to the city's north and in large floods could turn the city into an island surrounded by flood water. Willcocks's program, then, aimed to utilize the normal high-water period to the benefit of agriculture while minimizing the danger from severe floods.

To accomplish this goal, Willcocks's survey called for two flood escapes, one on each river, that would protect both cities and agricultural lands by offloading excess water into nearby depressions. These depressions would hold that flood water for irrigation use during low water periods. Willcocks identified a depression south of the city of Ramadi, known as al-Habbaniyya, that could serve as a reservoir for the Euphrates. On the Tigris River he intended to channel the river's floodwaters into a depression known as Wadi ath-Tharthar, some thirty miles southwest of Samarra. Willcocks again added his support for a project that had been on the agenda of nearly every Ottoman governor in Baghdad for sixty years—the ultimate construction of a barrage at al-Hindiyya to rectify a split in the Euphrates that had desolated prime agricultural areas.[27]

Willcocks's plans were ambitious and controversial, but the Committee of Union and Progress government in Istanbul moved forward with his plans on the Euphrates River.[28] In December 1913, the Ottoman government hired the British firm of Sir J. Jackson, Ltd., of London, to construct the Hindiyya Barrage and the Habbaniyya flood escape channel. The firm completed the barrage before World War I, but the flood escape channel into the Habbaniyya depression was left unfinished. The barrage restored the Euphrates River to its earlier course and controlled important canals but did little to protect the country from flooding, a fact that would almost cause disaster in 1919.[29]

When the British captured Baghdad in 1917, the engineers and technicians traveling with the army took control of more than just a new barrage, impressive piece of water infrastructure though it was. They knew of Willcocks's plans and of his vision, expressed in articles like "The Garden of Eden and Its Restoration," published in a prominent journal and engendering much discussion.[30] Willcocks envisioned a realm where railroad and irrigation canal met to make a productive territory of geostrategic importance. Moreover, British commanders could rely on more than just visions and imaginings—they had fought three hard years of war against the Ottomans in southern Iraq. They knew of the rivers' floods and the importance of bending the waterways to human design. During the war, Ottoman armies had cut breaches in river dikes as they retreated up the Tigris, inundating the areas behind their retreating forces. Meanwhile, British gunboats had plied the rivers, protecting vital supply shipments.[31] In addition, many British officers had extensive experience in India. For these officers, agricultural development in Mesopotamia meant food for the empire, especially the British protectorates along the Persian Gulf and Gulf of Oman (the Trucial States), and a solution to what many British officials perceived as India's population problems.[32]

But before a full-fledged administration styled after India's could be imposed in Iraq, two important developments altered British administrators' visions and aspirations. One of these developments we know a good deal about—the imposition of the mandate in the spring of 1920 that brought about a massive revolt against British rule. Broken promises helped fuel the revolt. Despite British statements in favor of self-determination, the postwar government in Baghdad did not proceed toward Iraqi self-rule. Instead, the British civil commissioner, Sir Arnold Wilson, established a colonial regime in direct control of the country.[33] This provoked an anti-British uprising that Iraqis view as a popular revolution; the revolt set the stage for a long independence struggle against the British empire. After immense expenditure in money and lives to put down the uprising, the British opted for a method of indirect control, establishing a constitutional monarchy with Faysal ibn Husayn as king. Faysal had led the Arab Revolt against the Ottomans, helping Britain in its wartime struggle. The Iraqi government produced through this arrangement could not act without a British assent, and its politics have been the subject of several historical studies.[34]

The other postwar development is much less visible in history—the great flood of 1919. Both events, the imposition of the mandate and the flood, shifted the logic of the British occupation and altered the trajectory of Britain's administration in Iraq. The revolt of 1920 implicated Britain's plans for governance of Iraq's human beings, while the flood of 1919 was a struggle against the nonhuman, a force that could not be bombed into submission. Just as the 1920 Iraqi revolt drove the British toward a method of indirect control in their mandate, so too did the rivers' flood propel new understandings of how to manage water in Iraq. Moreover, the disaster produced political linkages between Iraq's rivers and its people, connecting environmental and political control in ways that would have future consequences for both.

A year before the 1919 flood, and about nine months before the Armistice of Mudros ended the Ottoman war effort, the British military established an Irrigation Directorate in Iraq to centralize irrigation and flood control efforts, which until then had been the purview of appointed provincial Political Officers. The shift from Political Officers to a centralized system was significant because the new Irrigation Directorate included very few Iraqis on its staff, even though many of the Political Officers had relied on local expertise on irrigation matters. Instead, the department imported expertise, guidance, and labor from India. In March 1918 the Inspector-General of Irrigation in India, T. R. J. Ward, appeared in Baghdad to assist in the planning of works to increase cultivation.[35] Within a few months of Ward's visits, the department doubled its professional staff, mostly with British officers. By the end of December 1918, the Irrigation Directorate had recruited another 1,000 workers, almost entirely from India, to fill all subordinate positions.[36]

In composition and scope, then, the Mesopotamian Irrigation Directorate mirrored British efforts to remake the river valleys of India. As David Gilmartin has noted in his study of the Indus River Basin, British officials in India viewed irrigated agriculture as "a civilizing instrument that could be critically important in stabilizing the frontier" and as having a kind of "moral power" over nomadic groups.[37] The full extent of British-Indian influence in Iraq remains to be investigated. However, the faith in the reformative powers of irrigation infrastructure displayed in the Indian case ultimately played a significant role in debates over water in Iraq, as we will soon see.

The primary focus of the Irrigation Directorate in its first year of existence was to respond to wartime food shortages. Thus, the Directorate sought to obtain as much grain from Mesopotamia in the shortest amount of time possible.[38] To that end, irrigation officers and their retinue of Indian subordinates fanned out across the country to survey land, dredge canals, and reinforce flood embankments. They encountered many obstacles. Everywhere local laborers were difficult to find, and the local population was more dispersed than originally imagined. A small dam built annually on the Diyala River to feed irrigation canals could not be reconstructed because Ottoman and British soldiers had cut down all the nearby trees during the conflict. Infighting over land ownership ended the building of new canal works on the right bank of the Diyala.[39]

The Directorate had some success in bringing more land into cultivation, but overall the results were dismal. Rather than the normal yield of 90,000 tons, rice production in Iraq's al-Shamiyya district was some 600 tons gross in 1917, while in 1918 it was still less than a quarter of the normal amount.[40] It seems that the Irrigation Directorate, a product of British-Indian ingenuity, expected to find in Iraq a small version of India and instead discovered a country apparently denuded of trees, people, and government. The land between the two rivers did not fit the models of the Indus or of the Nile that the British staff had carried with them to Iraq.

This misperception of the Mesopotamian environment quickly collided against the real thing. In the winter of 1919, the engineers at the Directorate encountered a very early and significant flood. On the Tigris the flood peaked on February 15. The rivers' onslaught collapsed even the newest and strongest flood embankments. By pressing military brigades into service, the Irrigation Directorate maintained the dikes protecting the military camps near Baghdad, but breaches developed above and below the city, inundating over 5,000 acres of cropland. Damage on the Euphrates was even greater. The river flooded much of the town of Tuwairij (today's al-Hindiyya), damaged several important irrigation canals, and inundated 60,000 acres of crops in the Middle Euphrates region alone. Had the partially functional Euphrates flood escape at al-Habbaniyya not been used at the proper time, officials warned that "the whole of the country downstream of the [Hindiyya] Barrage would have been flooded."[41] Even still, the flood came close to overwhelming the five-year-old barrage before the waters abated.

In the months following the flood, British officials attempted to reconcile the flood as a physical phenomenon with their notion of what a properly irrigated Mesopotamian landscape ought to accomplish. First, they had to figure out (and agree on) why the flood had been such a disaster and why the newest embankments had failed. Per Willcocks's report to the Ottoman government, the problem had to do with the rivers' topography. In Willcocks's view, the two rivers entered their delta above Baghdad, and this deeply informed his ideas about how to control them:

> The Tigris-Euphrates delta is strangely flat. Baghdad, removed some 500 miles from the sea is only 120 feet above sea-level . . . a very serious breach on the Tigris or the Euphrates has been followed by the river completely leaving its channel and forming a new one miles away, after inundating the whole country. Such was Noah's flood in the early days of the World's history.[42]

Willcocks maintained that the only way to provide security from flooding in the relatively featureless terrain of the delta was to divert flood water into nearby depressions. For this reason, he proposed a number of escape channels through the rivers' flood embankments that could be opened when needed.

Baghdad's Deputy Director of Irrigation, H. Walton, thought Willcocks was wrong. After all, Willcocks had practiced in India and Egypt, and practically everything similarly educated engineers had tried in Iraq had failed. The policy of building flood escapes, he wrote, was "nothing less than a most dangerous delusion."[43] Escapes such as those proposed by Willcocks treated only a symptom of the rivers' overall disease. According to Walton, the floodwater lost to these escapes reduced the volume of water passing downstream, which meant that the river remained narrow and shallow in its lower reaches, threatening navigation, complicating drainage, and exacerbating the flooding when high water arrived. Since Willcocks's plan was "nothing more than 'delta' treatment," Walton thought it would be a mistake to implement it. To Walton, the main problem with managing the rivers was not topography, but demography:

> For the last 6,000 years—or may be more—since irrigation of any sort was practised in the country the hand of man has been applied, not scientifically but very unscientifically, to the rivers. . . . It has been the unwillingness to admit failure and the steady pursuit of success over centuries that is the root of the evil; and such success as was attained—being at the expense of the rivers—has made subsequent success a steadily diminishing quantity up to the present time.[44]

Walton here emphasizes a different history, not one of Biblical floods but of failed irrigation engineering. Whereas Willcocks focused on the flatness and variability of the rivers, Walton argues that the real issue had always been how human societies modified the rivers. Human tampering with the rivers' true nature had resulted in such damage that the Tigris and Euphrates had over time become menaces to life and property and no longer useful and productive. The flood disaster was thus a problem of people, not of nature.

To bring the Mesopotamian rivers back into balance, Walton suggested that the Irrigation Directorate follow a program involving "the regeneration of the rivers." Under his proposal, the Irrigation Directorate should focus on confining the rivers to their banks by closing breaches and other unmonitored watercourses. Only then would the rivers "sink themselves into their beds" and act as true rivers rather than marsh makers.[45] In discarding Willcocks's surveys and observations, Walton refuted the Indian-Egyptian model for irrigation development that Willcocks represented, a model focused on measurement and analysis of the rivers' physical characteristics. Instead, Walton suggested a model he thought would work better in Iraq, one centered around the consideration and mitigation of supposed human interference. Walton's arguments convinced his superiors. They had witnessed not only the problems created in that year's floods but also how the Ottomans had used water to slow the British advance. The rivers' variability and winding course—now explained by Walton as subject to human agency—had created navigation problems for the flotilla of British river craft.[46]

Not everyone was so sanguine about Walton's ideas, however. Baghdad's policy nettled Willcocks's colleagues at the Egyptian Irrigation Department.[47] Engineers there agreed with Willcocks that the two rivers were in their delta for most of their journey through Iraq and therefore could not be controlled as envisioned.[48] British engineers in Cairo thought "Walton's ideas to be impracticable" and declared "that regeneration and rubbish are synonymous terms."[49] Back in Baghdad, Walton retorted that the rivers exhibited "deltaic" features, acknowledging Willcocks's findings regarding the topography of southern Iraq, but he continued to argue that the rivers should not actually be in their delta until Basra.

After more than six months of discussion, in late 1919 Evelyn Howell, the Revenue Secretary and assistant to Civil Commissioner Arnold Wilson, responded to the arguments. Howell regarded the dispute as "the prime riddle of Mesopotamian irrigation and on the solution to it in large measure depends the future prosperity of the country."[50] At the heart of this was a problem of agency. Were these flood disasters brought about by geography and Ottoman idleness in executing Willcocks's plans, or by something in the relationship of Iraq's people to their government and environment? Howell went with Walton; the culprit was not nature or a sluggish Ottoman bureaucracy but an Iraqi people driven by greed:

> If, as was the fact, the Turks could not prevent the Arab tribesmen from murdering their provincial governors, when they chose, what chance had they of successful interference with the same tribesmens' [sic] methods of cultivation? Consequently man, armed with the little knowledge that is so proverbially dangerous, and animated by true Semitic desire for the immediately profitable, no matter at what cost to his neighbor or to posterity, had his wicked will of the river.[51]

Whereas Willcocks and the Ottomans had seen the rivers' floods as something that could be mitigated with man-made engineering works, Walton and Howell set forth a different interpretation, one laced with racism and a belief in the superior civilizing capacity of imperial engineering. For them, the best method for managing Mesopotamia's rivers would combine river control with sociopolitical control.[52]

By entangling river action with human action in the "regeneration" scheme, Howell combined human intentions and natural forces to produce a vision of state power that solved two related predicaments. First, the regeneration plan answered the question of why British models and works had failed so miserably. It was not that British engineering could not comprehend this new environment, but that engineers were confronting a problem that was equally a social one and hence beyond their purview. Second, Walton's ideas justified a massive expansion of state power as both a political and a technical necessity. Rather than Willcocks's precise intervention in the rivers' flow, which would have centralized control only over excess floodwater, Walton proposed regulating the entirety of both rivers, giving the state control over all forms of water use. Irrigation officers would be charged with monitoring and maintaining every canal, ditch, and embankment along the rivers' course.

In defining and tightening the relationship between Iraq's river system and its sociopolitical system, the Irrigation Directorate acquired a new mission and purpose—to protect the river from its users so that not only nature but also society could take their proper course. Wartime propaganda had depicted Iraq as a stagnant, neglected polity waiting to be restored.[53] In this scheme,

Iraqi society became something even worse—a parasitic body feasting on the lifeblood of the country. Only through comprehensive management could the Iraqi people's "wicked will" be restrained.

By 1920, that "wicked will" was not only directed at disrupting rivers, but also at disrupting British rule. Peaceful protests against the British occupation erupted into a violent uprising among the Shi'i tribes located along the Euphrates River. After the arrest of a Shi'i cleric's son, leaders in Najaf and Karbala called for independence and instructed local tribesmen to resist the British. The uprising soon spread across the entire country. By the fall, British forces reached 100,000, against what they estimated to be 131,000 Iraqis, in one of the largest anti-colonial revolts against British power in the twentieth century. During the revolt it was not only the actions of Iraqi cultivators that had to be curbed with respect to the country's rivers— any Iraqi resisting imperial occupation had to be subdued. Iraq's terrain came to figure in the latter efforts in equal measure. As Priya Satia has shown, the flat landscape of southern Iraq was seen as uniquely "barren" and open, such that the techniques of air attacks and indiscriminate bombing were thought to be especially effective.[54] Even so, only with a surge of fresh ground troops from India were the British able to force Najaf and Karbala to surrender.[55]

Still, the insurgency in Iraq had put additional strain on the British Treasury at a time when London was demobilizing from the war, so much so that British leaders considered abandoning Mesopotamia altogether.[56] In the end, the British reduced the empire's footprint through a method of indirect rule. In July 1921 British officials introduced Iraqis to Faysal ibn Husayn, the erstwhile king of Syria whose democratic government had been toppled by a French invasion in July 1920. In August, British officials placed Faysal on a newly conceived Iraqi throne. They also established a government whereby Iraq's people bore the expense of creating a dependent state that served a limited set of British interests. Britain demanded that Iraq not only pay for its share of the Ottoman debt, but also for the various "improvements" the British had built to invade the country—railroads, flood protection works, and so on.[57]

Walton's plan for controlling the rivers would no longer work under indirect rule. Changing the way political power could be exercised in Iraq required a parallel reconceptualization of water management. Walton and Howell had drawn an association between Iraq's unruly populace and a pair of damaged rivers.[58] Through the violent repression of the Iraqi revolt, that unruly populace had been subdued by other means. However, the problem of flooding remained, and now without a theory for why and how it should be remedied.

In 1923, the rivers flooded again, and although the disaster was not sufficient to produce changes in policy it did reorient the course of human governance and river control. Prolonged floods greater than in 1919 wreaked havoc along the Euphrates, with high water at Ramadi lasting over a month. The flood inundated 335 square miles in the Baghdad province alone—an area larger than all of New York City—and destroyed the canals at as-Saqlawiyya, which had to be rebuilt.[59] On the morning of March 23, 1923, a floating bridge across the Tigris at Baghdad, named for the city's "liberator" General Sir Stanley Maude, collapsed under the strain. *The Times* reported that "the steel girders were twisted all shapes, and the fifteen pontoons are now dashing downstream in the direction of Basra."[60] While floating bridges had been used in Baghdad for centuries to cope with the changing flow of the Tigris (see Figure 3), reassembling them in the 1920s used newer technology. The British sent Royal Air Force planes downriver to locate the bridge parts before they could damage steamers and other vessels.[61] As for Baghdad itself, the city was saved only because of a breach in an embankment north of the city, which flooded 200 square miles. For those whose livelihood did not depend on growing crops, the damage to agricultural lands was a terrible inconvenience as it muddied and perforated perfectly good terrain, something Gertrude Bell pointed out some six months after the flood:

> On Saturday I rode out to see the Arab Army play polo.... But it's sad to ride out over that great stretch of desert which had been converted first by our army into a wonderful farm and was then taken over by the King. The floods of last spring have sent it back to desert, the roads are blotted out, the irrigation channels half filled in and the young trees which the King planted in hundreds, all killed or uprooted. And all the desert which was under water is horrid to ride on, covered with a cracked mud surface and full of holes.[62]

One wonders how the polo ponies fared. In any event, as in 1919, the 1923 flood demonstrated the precariousness of this new British "mandate" if the rivers could obliterate prominent public infrastructure and lay waste to the king's own lands.

FIGURE 3. Pedestrians traverse one of Baghdad's pontoon bridges over the Tigris River, 1932. Library of Congress, Prints & Photographs Division, LC-DIG-matpc-13237.

An even greater disaster was to come. On April 6, 1926, Baghdad received word from Mosul that a very serious flood on the Tigris would reach the city within the next forty-eight hours. High temperatures and a southerly wind had melted snow in the mountains at the same time as heavy rains arrived in the lowlands. Irrigation officers in Iraq commenced flood preparations and ordered the closure of all irrigation culverts, the channels through the flood embankments that brought irrigation water from the river to the fields.[63] While an irrigation officer generally oversaw construction of flood protection, private individuals built and owned the culverts. According to a 1923 law passed after the flood that year, landholders were also responsible for "managing, repairing, and rebuilding embankments . . . [as they] benefit from their proper function and suffer from their disrepair or deterioration."[64] The culverts in particular posed considerable danger during a high flood, as they weakened the embankment. With enough water pressure a serious breach might occur and flood hundreds of square miles.

Two days later, on April 8, an employee of King Faysal requested permission from the British Adviser in the Irrigation Department, L. E. Bury, to open a culvert near the royal palace north of Baghdad. Permission was denied and when an irrigation officer later discovered the culvert open, he ordered it closed and directed the police to fine the culvert owner.[65] The next morning, agents working on behalf of the Administrator of the King's Estate, Tawfiq Bey, opened the culvert again, and within a few hours the embankment collapsed and the resulting breach grew so large that it could not be repaired (Figure 4).[66] Water drowned the palace, forcing the evacuation of the royal family, and rushed toward the city. The flood followed the palace road toward the next line of defense, the Sarrafiyya dike, built in 1918 by British troops, which protected the northern part of Baghdad.

On the other side of this dike stood the Baghdad North Gate Rail Station and a hospital. By 5:45 pm that evening, the water reached its full height along the Sarrafiyya, at which point a twenty-five-foot-wide breach opened literally under the feet of local workers. Floodwater then met another barrier, the main road, which the railway engineers and staff thought was solid enough to protect the rail station from any serious flooding. Rather than evacuate the station, they went about preparing the night train for Khanaqin, a city to the northeast of Baghdad on the Persian frontier.

A few hours later, with water lapping at the edges of the road, the station received a call to also ready a troop train to collect 2,000 men, 1,000 horses and 100 transport wagons from Ba'quba, a city fifty miles to the northeast along the Diyala River. These men, horses, and wagons were to be brought to

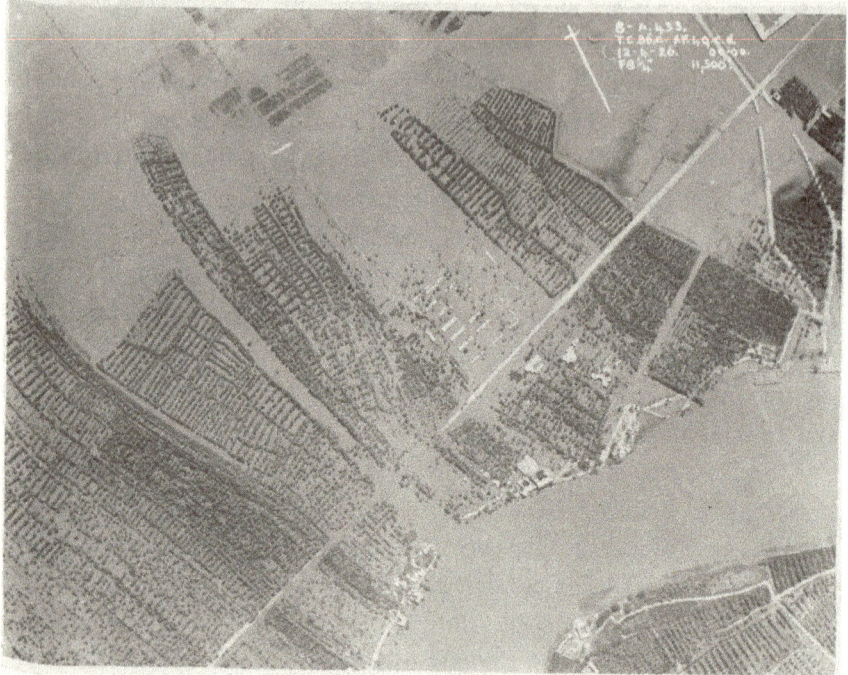

FIGURE 4. Aerial view of the main breach along the Tigris River showing flooded buildings and fields, 1926. Courtesy UK National Archives, CO 730/94.

Baghdad to help protect the city from the ever-rising Tigris. By 8:30pm, the water was rising in the goods yard of the station, and by 9:45pm the water was knee-deep. The station staff managed to evacuate the troop train, but about thirty minutes later, the floodwaters were "rising round the quarters of the staff, houses were collapsing and it became difficult for officials to keep control over the workmen who were naturally anxious for the safety of their families." By midnight the Tigris had inundated the entire rail station, destroying over 320 wagons of goods and 80,000 parcels (Figure 5).

The flooding was far from over. Several days later, Gertrude Bell reported her own experience of the flood:

> Our chief preoccupation during the past week has been water.... On Friday the Tigris dike broke on the left bank—my bank—above the King's palace which it flooded . . . from then until now we have never been sure that it would not break through and flood the low lying parts of the town, which include my quarter! I think that risk is over now, unless the Tigris again does something

FIGURE 5. Aerial view of the flooded rail station showing a breach in the bund at the lower left of the image. Courtesy UK National Archives, CO 730/94.

very perverse, but the possibility of having 6ft. of water in one's house hasn't been pleasant. How dreadfully annoyed I should have been, to be sure. It has been difficult to think of anything else. They have brought in thousands of peasants and propped the banks with reed mats and sand bags, but the worst is when the water begins to drip in through rotten places in the lower parts of the dike. They have electric light all along and people watching and looking night, and day . . .

Bagdad can never be made really safe, it lies in such low ground; but I expect that after this experience, following on that of 1923, they will do a great deal to make it safer.[67]

By the middle of April, the Euphrates began to add to the misery. The valleys of the Middle and Lower Euphrates flooded to create a river fifty miles wide, slowly working its way to the Persian Gulf.[68]

The Iraqi statesman, scholar and poet Shaykh Muhammad Rida al-Shabibi of Najaf also wrote of the flood, expressing frustration and anger at the scale of the destruction. Al-Shabibi served in Iraq's parliament through the 1940s and as Minister of Education in several cabinets. He used poetry to express support for political and social reforms. In this poem entitled "The Flood," he compares the river's destruction to a people suffering oppression:

> The valley flooded like a people with fiery spirit
> It could neither endure humiliation, nor bear it
> And when they bound it for their gain
> It refused to be bound, aiming only to burst forth
> Please, O valley, come to our aid
> And teach how we are to break free of the bonds
> Are we not a nation tired and weary
> Of the tyrants' oppression, with no release?[69]

Al-Shabibi was critical of both Ottoman and British rule in Iraq, and here equates the river's condition with that of Iraq's people, bound for colonial gain by "tyrants' oppression." In this passage, the river valley perhaps holds the answers, as if by finding harmony with the Tigris and Euphrates, Iraq's people might secure liberation—a far cry from Walton and Howell's focus on the control and coercion exercised by the colonial state.

Iraq's highest political leaders responded to the severe flood mostly by trying to avoid responsibility for the disaster. During the time of the rivers' rise, Naji Shawkat, the son of an Ottoman-era governor of Kut, was serving as the governor of Baghdad province. In his memoirs, he recalls how successive floods would isolate Baghdad as an island for "about two months, during which epidemics spread, insects multiplied, and the climate worsened." Shawkat describes how during the 1926 flood he worked with city officials and laborers to close a minor breach in the bunds protecting the city, only to learn of the collapse of the dikes safeguarding north Baghdad. The destruction there and what it could mean for the city, Shawkat writes, "robbed me of my comfort, disturbed my sleep, and brought sadness to my heart."[70]

In response, Shawkat issued orders for the police to investigate the incident, only to discover that the breach had been caused by one of the King's own. "Most of the employees of the royal court—if not all of them—were Syrians and Hijazis," Shawkat writes, "so it was difficult to arrest one of them or bring him to trial . . . but the arrest order was actually carried out." According to Shawkat, once the public's eye had turned to other issues, the perpetrator of the disaster was set free while Shawkat was dismissed from government service, only to be reappointed as the governor of Mosul by Prime Minister ʻAbd al-Muhsin al-Saʻdoun when King Faysal was out of the country.[71] While soured by the experience, Shawkat's witnessing of the 1926 flood in Baghdad colored his attitude about how the government should manage water in Iraq. His criticisms of plans in 1933 to build the Kut Barrage on the Tigris River, near the town where he was born, helped bring down a government (see Chapter 2).

As for the British experts in the Irrigation Department, responsibility for the flood disaster lay, at least in part, with the way Baghdad had been built. After the flood emergency waned, L. E. Bury submitted several prescriptions for protecting the capital. He declared that the only way to defend Baghdad from the Tigris was to move the entire city to a more secure location:

> Possibly the cheapest in the long run and certainly the safest proposal for the safety of the city would be to rebuild it on the high land near Tel Mohammed ... laying out a fine modern city ... connected to the old city and to a fresh port on the Diyala by a double or quadruple line of Tramways.
>
> It would be a city capable of being served with all the modern requirements such as roads, water, light and sewerage in a manner that it will never be possible to obtain in Baghdad ...[72]

Bury's technocratic vision imagined a new Baghdad that, instead of lurching from one perilous flood season to the next, provided every modern convenience. However, the Iraqi government, saddled by debt and blocked by various means from exploiting the country's natural resources, was in no position to act on such an expensive scheme.[73]

Moreover, what made Baghdad so difficult to defend from the Tigris had little to do with whether the city could provide "roads, water, light and sewerage." William Willcocks planned to protect the city using a new flood water escape into a depression near Samarra (Wadi al-Tharthar), but he provided no detailed construction plans.[74] He instead proposed another solution to protect Baghdad, one that was used "in ancient times" to send flood water around the city (see Figure 6).[75]

There was only one problem with Willcocks's alternative scheme: the Baghdad of 1926. The twentieth-century city differed substantially from the ancient metropolis. Postwar additions to the city occupied part of the

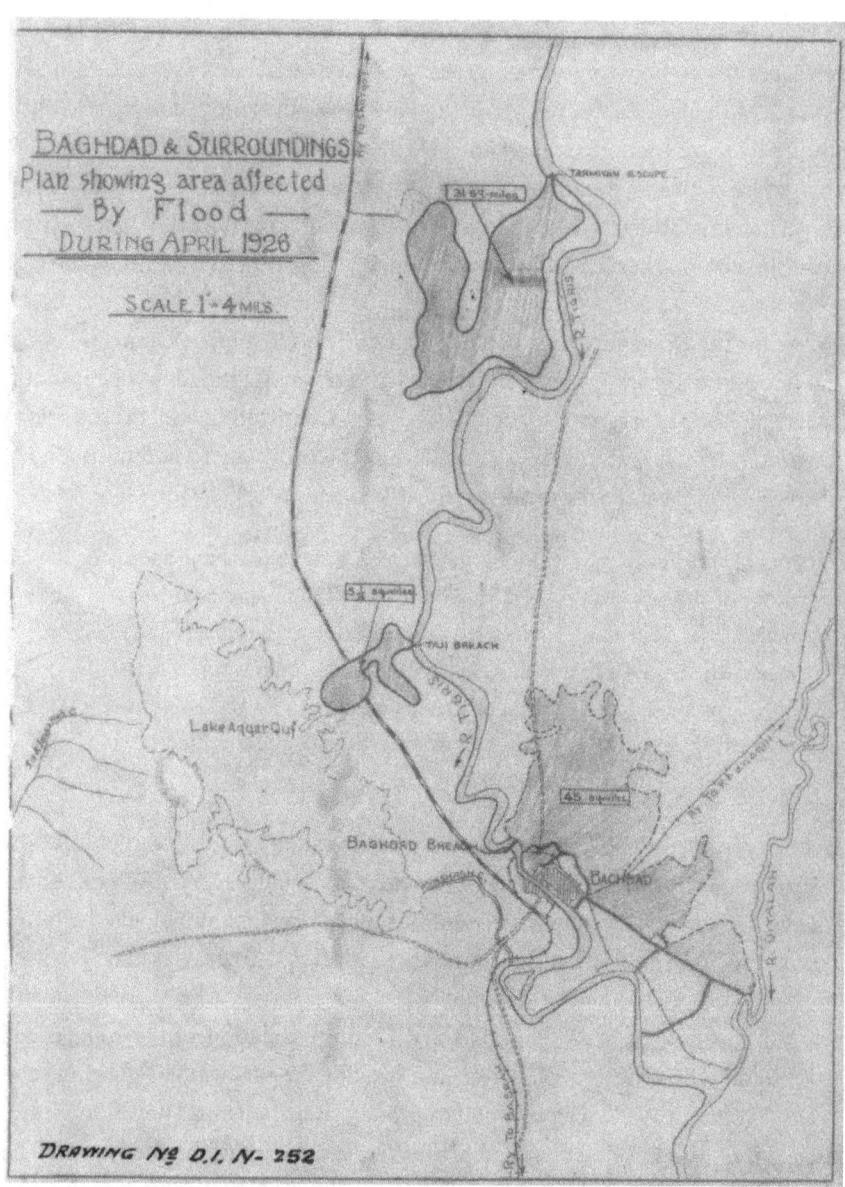

FIGURE 6. Diagram of breached bunds and flooded areas along the Tigris River north of Baghdad in 1926. Note the location of Aqar Quf to the west of the city. Courtesy UK National Archives, MPG 1/1203.

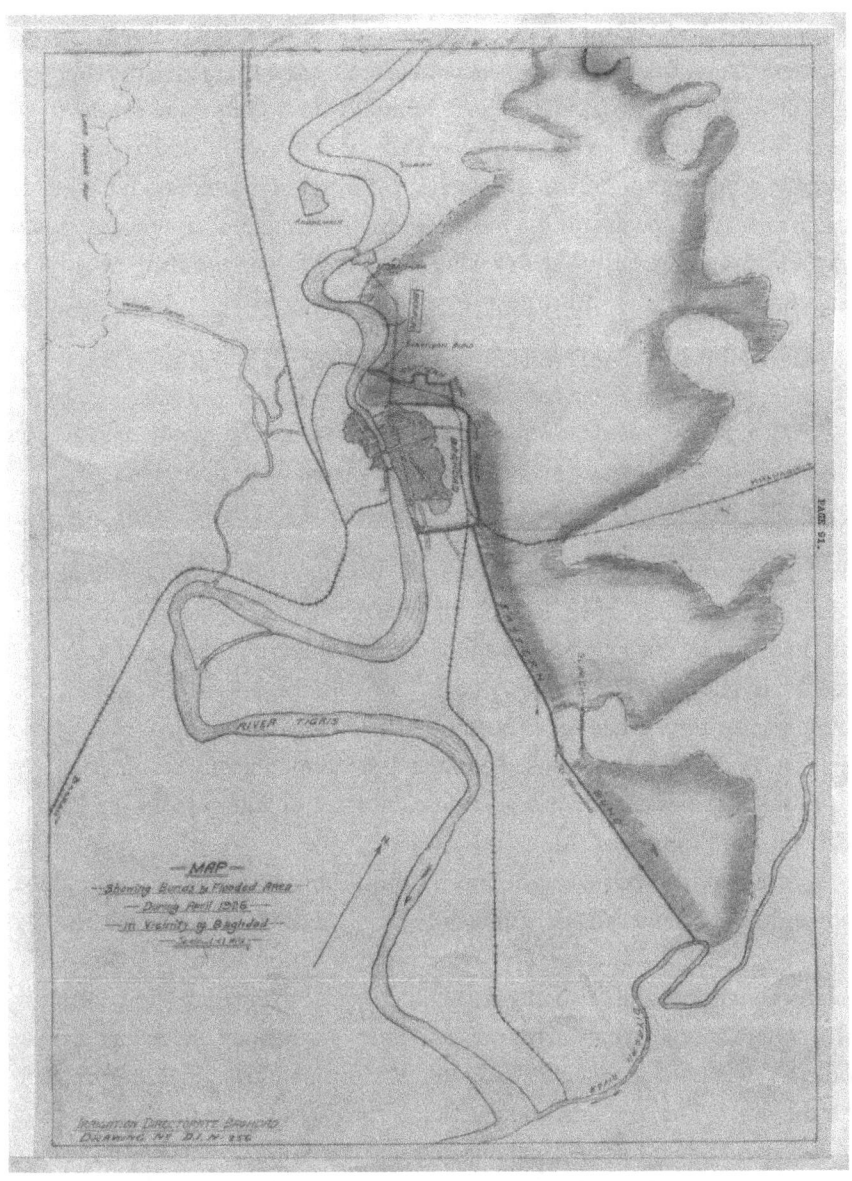

FIGURE 7. Diagram showing Baghdad's major bunds and the extent of flooded areas north and east of the city in 1926. UK National Archives, MPG 1/1203.

escape route for flood waters while British-built embankments along the river blocked access. Bury was left with just two options for managing the threat of flooding. The first involved using a potentially hazardous flood escape in an area west of Baghdad, the Aqar Quf, which if it failed would destroy the western city (the location of the British High Commission offices) and valuable cultivation. The second option was to continue current policies to strengthen the embankments on the river's edge (Figure 7), an option Bury noted was effectively a policy for future disasters:

> But, if we make the left bank safe and thereby keep Baghdad City . . . safe we still have the right bank to consider. . . . And if we make the right bank safe, we shall then have caused the whole of the flood to reach Baghdad and it will have to pass through the narrows between the two halves of the town, a thing that it has never done in history.[76]

Yet, in opting for continued disasters, the British administration potentially made itself responsible for the ensuing damage. The King's own agents, after all, had caused the 1926 breach by opening a culvert in the flood embankments. In addition, the Sarrafiyya dike, built by the British and maintained by the government, had failed easily and led to the swamping of the railway station. The disaster had caused enough destruction to commercial property to bring a formal complaint and a demand for an enquiry from the British Chamber of Commerce.[77]

Since Baghdad could not be moved from the river's edge, and the flood embankments could not be sufficiently strengthened, the British needed a way to resituate the disaster. The river's physical action had to be channeled—appropriated in some fashion—or the floods might someday wipe out more than important buildings. Arguments about the 1919 flood had helped British officials articulate a connected social and environmental engineering program for the rivers. The combined destruction of the 1923 and 1926 floods moved the government toward a different conception of the rivers' social life. In the long run, British officials needed to find a way to manage Iraq's water resources to minimize expenditure and international exposure, while still reaping the potential gains in revenue and stability from the expansion of agriculture.

British officials came to rely upon several approaches, both technological and administrative, to manage water and disaster in Iraq. Just after the flood, the Iraqi government made it possible for investors to claim public lands by building irrigation pumps and bringing new land into cultivation.[78] Until that time, state-controlled lands had been held by custom for grazing and cultivation by local tribes. The pump law meant a significant shift in the control and use of both water and land, affecting land ownership along the rivers' banks and undermining the rationale behind flood protection.[79] Moreover, the introduction of irrigation pumps on a large scale developed an important nexus of collaborations that served British interests in several ways, including reducing political pressure for greater flood protection.

Constructing an irrigation pump along the rivers required capital and access. In Iraq, this kind of capital accumulation could be found primarily with merchant interests in the country's largest cities, especially Baghdad.[80] The capitalist also required access to the rivers' banks to construct the pump and, once constructed, some means of protecting their investment, both physically and financially.[81] Such access and protection were to be found in Iraq's newly empowered tribal leaders. Through the *Tribal Criminal and Civil Disputes Regulations* of 1916, the British military occupation gave local tribal leaders wide political and economic powers, including the ability to register land in their own names, and limited peasants' means of legal recourse.[82] Those leaders continued to wield power in the time of the Mandate and, in alliance with other wealthy interests, obtained ever greater control over the land.

A good deal of scholarship has focused on these economic connections and the consequences for land distribution and tenure, particularly in relation to the Ottoman Land Code.[83] Much less attention has been paid to water—how floods shifted policies and how pumps helped produce a techno-political base for British power. A landholder's domain held little value without the ability to obtain and control water. The pump provided that means, and the capitalist provided the pump. Through special legislation the state made the pumping of irrigation water into a claim on state-owned land, creating a mechanism for the expansion of both land ownership and water use. By 1930, pump owners had used this mechanism to claim "something approaching two million acres of State land."[84]

Losing sight of material, environmental factors has led to analyses situating the production of an Iraqi landed, often absentee, elite solely within the context of culture, economics, or politics. In reality, the physical characteristics of the rivers propelled an effort to craft mechanisms for combined social and environmental control and to manage ecological risks to state legitimacy.

The movement of water was thus deeply entangled with the construction of a narrow elite that ruled Iraq through the monarchical period.

Further, by shifting the focus toward water, it becomes possible to better understand the construction of the mandate state, the later deployment of major waterworks, and the ecological problems engendered by government policy. In making water part of the mechanism for extending land ownership, the British mandate of Iraq not only assembled a techno-political apparatus necessary to sustain state power but also reduced the state's responsibility for flooding by shifting ownership of land and water from the state to private interests. The very parties best positioned to demand government action to mitigate flood damage benefited most from the pump legislation, whereby they gobbled up most of the land and water and, by extension, agricultural labor. Meanwhile, vast and ever-growing holdings reduced the probability that a single major disaster would ruin the landholder, even if it wrecked the livelihood of peasant farmers.

As agriculture at this time was the most important economic sector, the most direct way to fund water management projects in interwar Iraq would have been a tax on land or water or both. In fact, officials in the Irrigation Directorate made clear that allowing irrigation and flood control to devolve to private parties was not sound policy, partly because pumps were more expensive to operate than canals.[85] By diverting control into private hands, the British administration produced a client elite with little interest in taxing itself to pay for new river management schemes, even as the floods cost the government more and more in regular revenue.[86] Indeed, the only piece of major water infrastructure built in the interwar period, the Kut Barrage, was meant not to control flooding but to bring more land under cultivation. And, while the Kut Barrage was under construction, it came very close to being destroyed by a massive flood in November 1936.

Finally, the extension of pump irrigation on the rivers contributed to a rising ecological crisis: salinization. Far from regenerating a glorious ancient civilization, the British-dominated government produced a situation where Iraq's land and people were caught between two disasters: fast-moving floods and slow-moving salinization. The scope and implications of this double bind occupy the discussion of the following chapter.

Far from being exclusively the result of human "invention," the Iraqi state in the 1920s was also produced by the rivers' floods, the rivers' physical properties, and the ways these properties interacted with technology, policy, plants, and people. Thus, the state produced in Iraq was one entangled in and dependent on natural forces. Its actions were informed by particular ways of knowing nature, by natural events themselves, and by the conjunction of the two.

Successive natural disasters, and the government's responses to them, produced the Iraqi state through successive, related actions of reordering and effacement. The floods swept away the products of human ingenuity and natural productivity, from rail lines to date palm plantations. In response, the British reshaped Iraqi society by deliberately reorganizing social relations. Communal, customary modes of land use gave way to individual, direct ownership accomplished partly through the medium of the diesel irrigation pump. A landed elite followed, not merely the creation of British cultural frameworks or political machinations, but in response to material destruction. Where the two processes, human and natural, converged, they remade one another. Great floods drove new policies of social reordering. By connecting water management to land ownership after the 1926 disaster, Britain obtained the loyalty of powerful social interests and secured their adherence to British aims. In return landowners gained ever greater control over both water and land resources, to the detriment not only of Iraqi farmers who were largely enserfed, but also to the land itself, as we will see in the next chapter.

The loss of control over water resources meant a loss of state revenue, but it also made the state less responsible for disasters caused by water. In combination, social reshaping dictated methods of water management and vice versa. Further, Iraq's experience with disaster demonstrates how a state could secure a measure of control over land and people not through direct, physical interventions on the landscape but through administrative means, by organizing a set of disciplinary processes in response to disaster. By setting key parameters regarding the acquisition and exploitation of so-called natural resources, the British colonial state in Iraq resituated the rivers' natural action, redirecting the physical characteristics of the rivers to minimize risk and to protect its own limited interests in the country.

Two

SALT

Salt sits at the interface of water and land. Wind and water weather rocks and minerals, introducing salts to the ecosystem; water moves and concentrates them in the landscape. Surface deposits of salt are sometimes apparent to the eye, clinging to the soil in crystalline crusts. These deposits may appear in such volume as to define the landscape itself: salt pans, salt basins, salt flats. More often, salts are hidden, dissolved in groundwater and residing beneath the surface.[1]

In the Tigris-Euphrates River Basin, water moves salt both horizontally and vertically and does so at multiple scales. Flowing water dissolves salts and moves them across the landscape from higher ground to lower. This may occur on the scale of the basin, as the two rivers move salts from the mountains to the alluvial plain, and, in this case, from a wetter climate to a more arid one. It may also occur at the smaller scale of an irrigation canal, which conveys salt-laden water across a field. Meanwhile, in the other direction, capillary forces draw water—and thus salt—vertically upward from relatively deep underground—as much as twenty feet below the surface. Evaporation also influences the deposit of salt, as the transformation of water from a liquid to a gas (water vapor) pulls water into the atmosphere, leaving the dissolved salt molecules behind.

In this way, hydrological events such as floods, shifting watercourses, tidal effects in an estuary, and the release of irrigation water are equally salinity events, with the potential to increase or decrease the concentration of salts in water and land. As a result, salt is as defining a feature as water in the ecologies of the lower Tigris-Euphrates River Basin, influencing not only the life and growth of various plants, fish and other animals, but also the human societies that rely on fresh, low-salinity water for agriculture, aquaculture, drinking and other basic needs.

Crops require both water and salt to grow, but too much of either can inhibit growth or kill the plant. Plant roots draw water through their cells using osmosis. Cells in a plant's roots contain water as well as sugars and other organic compounds (solutes). When water (the solvent) is present in higher concentration in the soil, water molecules flow across the semipermeable cell walls to equalize the concentration on both sides of the membrane. Higher concentrations of salt in the water reduce the difference on the two sides of the membrane, making it increasingly more difficult for the plant to take up water, even if water (albeit "salty") is present in large quantities. Plants compensate for salinity by producing more organic compounds or by taking up the salt, but this requires expending energy below the surface instead of above. Hence, plants do not grow as well and produce less. Some cultivated plants are better than others in handling the challenges of salinity—barley, rye, and safflower are examples of crops with greater salt tolerance.

As salinization grows worse, even salt-tolerant plants cannot grow. There is a temporal component here, with salt accumulating year after year until the growing medium itself becomes difficult to cultivate. High concentrations of salts break down and disperse soil particles. Plants exhibit symptoms of extreme drought. A salt crust may begin to form on top of the soil. In some cases, soils exhibit "sodicity," which means a larger presence of sodium ions (Na^+). Sodic soils, sometimes referred to as alkali, lower the permeability of the soil to air and water, creating dense, impermeable crusts that hinder plant life. To make matters worse, the sodium in the soil reacts with other elements plants need to grow, such as calcium and magnesium; these elements become unavailable to the plant.[2]

There are not many good options for ridding the soil of salt. Any event that raises the water table—an ebbing flood or an irrigation canal or introducing more water to a field to wash the salt away—can usher saline groundwater closer to the surface, making it easier for capillary action to bring it to ground level. Waterlogging the soil with too much irrigation water might wash the salt away, but it also might connect the upper layers of soil to saline groundwater, undermining any effort to reduce salinity. In arid regions, saline water thus transported to the soil surface evaporates rapidly, leaving those salts behind in the soil level critical to agriculture.[3] Even if the "soil washing" is effective, the salt must go somewhere and may merely be relocated from one plot of land to a neighboring plot. In effect, the only good place for cultivators to dump the lower basin's subsurface salt is the Persian Gulf.

Sodic soils require another kind of treatment—adding elements such as gypsum or calcium chloride to alter the soil's composition. Cultivators without access to such amendments must play a long game instead, growing sodic-tolerant crops such as alfalfa and wheatgrass and over time incorporating leftover organic matter. Finally, the accumulation of salts on the soil surface usually indicates other problems. High levels of salinity can affect human needs and enterprises by degrading ecosystems supporting important food sources such as freshwater fish, by corroding machinery and infrastructure, and by increasing the cost of water treatment and the availability of drinking water.[4]

Due to these factors, dams and irrigation alone are not enough to raise a crop on Mesopotamian soils. Further, the supposed benefits of farming in the lower basin—abundant sunshine, a long growing season, and two great rivers—are each dangers as well. Heat and aridity lead to evaporation, the long growing season is only useful if you raise crops in summer heat, and, as this chapter will demonstrate, the consequences of putting river water onto the land have never been straightforward, nor has the proposition been a simple one.

Ancient societies in the lower basin confronted the puzzle of soil salinity, and archaeological studies suggest that salinization plagued the civilizations of Mesopotamia, forcing cultivators to abandon fields and villages. The blighting of agricultural lands had large-scale effects, and one study suggests that "growing soil salinity played an important part in the breakup of Sumerian civilization."[5] Even so, cultivators in ancient times learned ways of coping with saline soils. They avoided overwatering and grew crops that could tolerate some salinity. They also fallowed fields for extended periods before excessive salinization took place; salt-tolerant plants in the fallowed areas could reestablish a dry zone between the saline water table and the upper layer of soil. These antisalinization methods worked just as well in the twentieth century

CE as they did in the twentieth century BCE. However, by their nature, this style of agriculture involved a lower intensity use of land and water.

Many of these salt-adaptive agricultural practices ran counter to the development discourses of the post-World War II period and the ideas of the so-called "Green Revolution." The Green Revolution came about through the union of two spheres. The first involved scientific advances in the use of fertilizers, genetically modified seeds, and water infrastructure. These technological breakthroughs in combination with the extension of liberal economic policies emphasizing private ownership and free enterprise for a time brought about a significant increase in food production in different parts of the world.[6] However, as Vandana Shiva writes, these high-intensity uses of the land and changes in sociopolitical structures "turned out to be conflict-producing instead of conflict reducing [sic]."[7] Centralizing control over agriculture, whether by the state or a private entity, stripped and simplified complex ecosystems and substituted short-term technological gains for long-term stewardship.

The Green Revolution coincided with, and often relied on, a revolution in energy production and use. Petroleum's share in energy consumption skyrocketed after World War II, underpinning a significant economic expansion in North America and Europe and increasing the demand for oil. Meanwhile, many of the industrial inputs facilitating the Green Revolution—equipment such as tractors, fertilizers such as ammonium nitrate and superphosphate, and synthetic pesticides—required or were derived from fossil fuels.[8]

The large-scale postwar revolutions in energy use and industrialized agriculture met in a distinctive way in the lower Tigris-Euphrates River Basin. There, within the borders of the state of Iraq, the raw materials for these two connected revolutions could be found—vast, underground reservoirs of oil, and a perception of an underdeveloped agriculture. Yet, accompanying those underground reservoirs of oil were significant reservoirs of salt. Both the salt and the oil might have plausibly remained underground, undisturbed, and having little effect on agriculture. However, if brought to the surface by an agent or actor with the power to conduct them, via irrigation canals or the application of fertilizer, into the root zone of plant life, their effects could compound. Indeed, a major part of the twentieth-century ecological history of the lower basin and its two great rivers is a story of the entangling of these two subsurface reservoirs, one of oil and one of salt.

A local historical process—another revolution in fact—may be traced to this entangling. In 1958, a military coup d'état brought an end to Iraq's Hashemite monarchy and, with it, overbearing British influence in the country. Hanna Batatu's magisterial book on the 1958 revolution in Iraq analyzes both the long- and short-term social and political factors resulting in the monarchy's overthrow and underscores an important irony in postwar state building. Batatu argues that the Iraqi state, through its investments in social and physical infrastructure, produced the very social classes that overthrew it.[9] These social classes depended on ecologies—a set of human-nonhuman relationships—underpinning economic production and ways of life. Of these ecologies, the ones surrounding oil and water (and thus also salt) were the most important. So, in several vital ways we might consider that the state also helped produce the *ecologies* that overthrew it, for the natural processes that shaped the Iraqi economy and state-driven infrastructure projects during the 1940s and 1950s were not outside of or merely incidental to political life.

The ultimate ecological result of the entangling of oil and water in the river basin—and of the economic, technological, and political revolutions of the twentieth century—was not what politicians, development planners, and irrigation engineers of the postwar period intended, as neither a verdant, resuscitated Fertile Crescent nor an industrial powerhouse emerged in the lower basin. Instead, something more complex came about. The result was a political ecology of salt, by which I mean a social and political situation defined by a slow-moving ecological threat, a precarity if you will, based in the protracted and increasing salinization of water and land, a situation that had its origins as much in historical and sociopolitical change as in environmental transformation.

This chapter explores the origins and outcomes of the political ecology of salt by following the history of how water, land, oil, and salt were known, connected, and manipulated along the Tigris and Euphrates Rivers. The discussion begins by analyzing what was known about salinization in Iraq during the British Mandate period and investigates the concurrent development of oil and salt reservoirs during the 1920s and 1930s. The efforts during that time to bring oil and salt to the surface—one intentional and one less so—met in the late 1930s at Kut, a town located on the Tigris River between Basra and Baghdad. In the planning and construction of the Kut Barrage, the four ecological factors—water, land, oil, and salt—became entangled in politics, development ideologies, settlement policies, and agricultural practices to the detriment of Iraqi settlers and farmers. This entanglement only intensified in the post–World War II period, and the remainder of the chapter contends with the implications of that intensification. As oil revenues accrued to the Iraqi government in the 1940s and 1950s, so did salt accrue in Iraq's lands. The regime sought to construct additional water engineering, but the ensuing plans largely failed to address the degradation of lands. Criticism of the engineering, along with growing social unrest in the late 1940s, pushed the Iraqi government to make increasingly grandiose development promises. Yet, critically, the engineering plans for transforming the rivers remained unchanged—with dire consequences for Iraq's people and for the ecological, political, and economic systems on which their lives depended.

Advances in scientific knowledge during the twentieth century helped soil experts in different countries better understand the action of salt in the environment, with substantial work done by both Russian and American experts in the more arid regions of those countries.[10] While this scientific knowledge may not have reached those governing Iraq in the early part of the twentieth century, both Ottoman and British officials were aware of salt's implications for agriculture. Ottoman officials appointed to positions in the lower basin knew of salinization issues; indeed, the problems for agriculture were apparent on the landscape. William Willcocks, in his surveys before World War I, notes the extent of salinized land south of Hillah: "As one goes south, the salted land increases in area, and then the marshes begin with their stretches of rice."[11] Willcocks did not include any salt remediation in his survey, even though his proposed works would later add considerably to the salinization predicament.[12]

British irrigation engineers across the empire knew of the issue of saline soils prior to the invasion of Ottoman Iraq from their experiences in other river basins. In 1912, the Government of India—responsible for training many of the irrigation engineers later sent to Iraq—began studying the effects of waterlogging in the Indus River Basin, shifting its focus to drainage by 1917. In his study of that basin, David Gilmartin notes, however, that "waterlogging was not generally approached before partition [1946] through the same *systematic* science that shaped the hydraulics of surface-flow integration..."[13]

A pattern like the Indian case would later play out in the studies of salinization and drainage in the Iraqi context.

British irrigation officers and soil chemists began studying and cataloguing the effects of salt accumulation in Iraq's soils at the start of the occupation. The scientific staff accompanying the British administration began studies of what they referred to as the "alkali lands" of Iraq in the early months of 1920. J. F. Webster, the Deputy Director of Agriculture, and B. Viswanath, the Assistant Agricultural Chemist, oversaw a study of approximately 400,000 acres of land between the Tigris and Diyala Rivers running roughly from Baqubah to Baghdad. The soil survey involved several tours across the landscape, facilitated by district irrigation officers. The 1920 Iraqi revolt, which began in May, curtailed the two chemists' journeys into the countryside, but they went on to conduct a series of laboratory experiments later in the year. The survey and experiments provided the government not only with a notion of how soluble salts could affect agriculture in Iraq but also supported policies for land and water use with implications for the entire country.[14]

The soil survey along the right bank of the Diyala River gave Webster the data he needed to sound a general alarm about government plans to expand and intensify irrigation and cultivation in Iraq. In his policy recommendations Webster contended that Iraq's agricultural lands were at a tipping point and salinity would soon cause a crisis that could threaten the soil's fertility. Natural forms of drainage benefited some Iraqi cultivators in the irrigation zone south of Baghdad. In some areas, an elevation difference between rivers, from the Diyala to the Tigris, for instance, allowed salt-laden irrigation water to flow out of the fields and into another waterway. In the Shatt al-Arab estuary, tidal movements pushed freshwater into fields, and then, as the tide receded, allowed for that water to drain, taking salts with it. Other areas in the irrigation zone benefited from the natural drainage provided by nearby depressions, where salty water flowed after passing through the fields. Engineering works and other modern forms of water management tampered with these modes of natural drainage; this was already proving to be disastrous. An attempt in 1918 to use pumps to irrigate land near Basra outside the effects of the estuary had caused rapid salinization.[15]

It was likely, Webster argued, that most of the prime agricultural land with natural drainage features was already under cultivation since salinization typically worsened over time. As a result, any planned expansion or intensification of agriculture would occur on lands without drainage and would thus encounter progressive salinization. Webster and Viswanath's experiments suggested the best way to deal with such troubles. The two scientists argued that it was more effective and economical to preserve soil fertility from the start than to depend on slow and costly reclamation efforts after salinization had occurred. Webster concluded, "The problem is not one of reclamation but of preservation, and unless preventative methods are adopted the situation will rapidly become serious."[16]

Preservation meant careful attention to the ways water moved across the landscape. According to Webster and Viswanath's surveys, in most areas salts arrived on the soil through irrigation. The drawing of salts up through the soil from the water table was only a worry in places where the soil was exposed to water for long periods of time, namely along canals. Summer weather in Iraq, when temperatures could rise above 120 degrees Fahrenheit (50 degrees Celsius), compounded the effects caused by irrigation because the high temperatures increased evaporation. If irrigation was undertaken only during the winter, when water supplies were relatively plentiful and temperatures were low, then salt accumulations might never reach problematic levels. This was how Iraqis in many parts of the country practiced irrigation—fields lay fallow during the summer months. However, government plans for intensification of agriculture and the requisite use of summer irrigation would accelerate the accumulation of salts, such that a plot of land in less than a decade might become too salinized for cultivation.

Webster reasoned that the reclamation of salinized lands should be a second-order priority because plenty of irrigable land was already available. It was less costly to safeguard those lands from salinization than to reclaim degraded lands. To accomplish this safeguarding, Webster advised in his policy document that the government proceed quickly to designate lands for a drainage system: "... it can only be urged that even now it is not too early to establish drainage reservations, so that the necessity of buying developed land from private owners for drainage purposes, may not arise in the future."[17]

In his warnings to the British administration Webster recognized salinization as a long-term feature of Iraq's agricultural landscape that modern irrigation engineering would only exacerbate. In fact, he already had an example. The region of Iraq dealing with the most significant salinization at the start of the Mandate also happened to be the one with the greatest piece of modern water engineering—the Hindiyya Barrage (Figure 8).[18] The dam had increased the provision of water to many areas, but that had meant increasing the degree of salt as well. By 1920, seven years after the dam's construction, salinization in the district had increased beyond even Willcocks's own observations, and the government's own scientists pointed to the barrage as a leading factor. The Hindiyya Barrage provided a clear lesson for agriculture south of Baghdad; it was not enough to get the water to the fields—the water (and the salt) also had to be removed from the fields.[19]

It is here that our story of two soil chemists meets the narrative about the British Mandate administration introduced in the preceding chapter, demonstrating again how expedient measures to preserve colonial control compounded environmental degradation. Due to the change of administration stemming from the 1920 Iraqi revolt against British occupation, Webster and Viswanath's detailed work on soil chemistry was not continued and a planned countrywide soil survey was postponed.[20] Changes made to agricultural policies during the 1920s and 1930s mainly aspired to avoid the sharp pain of a sudden flood rather than the chronic, slow ache of salinization. To improve the country's economic situation, the Mandate government undertook to increase cultivation, but these policies paid little attention to the environmental factors the chemists had outlined in their studies. These legal measures encouraged the use of lift irrigation; that is, the placement of diesel pumps on the rivers' banks to bring irrigation water on to the land.

In addition to the political benefits discussed in the preceding chapter, installing additional pumps on the rivers gave landholders control of state land and reduced the state's exposure to the risk of flooding. Unfortunately, in applying more water to the land, the pump law's expansion of cultivation also increased salinization and the consequent degradation of Iraq's cultivable lands.

And, in a perverse sort of way, progressive salinization made installing diesel irrigation pumps on the rivers an economic *and* ecological imperative. Economic analyses have tended to view the expansion of pump irrigation as a land grab when it would be more accurate to call it an *unsalinized* land grab. Without adequate drainage, increased irrigation almost always caused salinization. So long as more land could be brought under cultivation, the lands becoming less and less productive due to salinization could be abandoned. This created a cycle incentivizing the expansion of irrigation, and thus salinization, as the irrigation pumps spread the problem to more and more lands. So, even as the irrigated land area in Iraq quadrupled from the start of the British occupation in 1918 to 1942, an environmental crisis began to build just below the surface.

Economic and agricultural analyses of this time have tended to treat salinization as one of several economic impediments to intensifying agricultural production rather than an ecological feature of the basin exacerbated by social, legal, and industrial practices. For instance, in a study of agriculture in Iraq, Kamil Mahdi argues that landholders had little reason to intensify agricultural production so long as they could simply capture more land. Mahdi notes that any agricultural production "growth under this system depended upon continued expansion of cultivation."[21] Put another way, there was no need to adopt new agricultural practices, such as the expensive techniques of the Green Revolution, when cultivation (and control of land) could be expanded merely by installing a pump. After all, diesel irrigation pumps required fuel that could be refined just downriver at the refinery on the Shatt al-Arab at Abadan, Iran—a fuel that Iraq might eventually produce itself.[22] Thus, experts at the time, and economic analysts later, have blamed the lack of intensification for Iraq's stagnant agricultural fortunes, even when salinization has been recognized as a factor; why invest in new agricultural techniques and infrastructure when there was more land to claim and exploit?[23] This view has tended to minimize or even ignore the environmental imperative to obtain additional lands. Moreover, these studies generally leave aside how the degradation of cultivable land grew entangled with other social and political processes, a story to which we will soon turn.

The advent and spread of diesel irrigation pumps on the two rivers paralleled the rise and dramatic expansion of oil exploitation in Iraq, setting the stage for growing connections between these two subterranean forces. In 1925, the League of Nations awarded the Mosul province to the British Mandate of Iraq, ending Turkish claims to the area.[24] European and American petroleum geologists then proceeded with systematic study and exploration of oil reservoirs in Iraq's north.[25] Just as local cultivators had long knowledge of salinization in Iraq's south, Kurdish families in northern Iraq had been exploiting oil from seepages and using primitive bitumen wells throughout the Ottoman period (Figure 9). As Arbella Bet-Shlimon details, the Naftchizada family held rights to seepages in the Kirkuk area. The family's name indicates

FIGURE 8. Two men paddle a *quffa*, the traditional Mesopotamian round boat made of tree fiber and tar, near the spillway of the Hindiyya Barrage, 1930s. Library of Congress, Prints & Photographs Division, LC-DIG-matpc-16185.

a generations-long engagement with the oil business (*naft* means petroleum in Arabic and Persian). The Naftchizada struggled with the Baghdad government throughout the 1920s to enforce the Ottoman decree that had granted those rights. Local claims were largely set aside at the behest of Western powers vying for control of Iraq's petroleum resources.[26]

In the autumn of 1925, European and American geologists arrived in northern Iraq to conduct a geological survey. They gathered at al-Fatha, a town near the confluence of the Tigris and Lower Zab Rivers, where the seasonal low flow of the Tigris revealed oil seeping into the water. The survey, a joint expedition of the era's major oil companies—including Anglo-Persian, Royal Dutch, and American firms—led to a series of experimental wells.[27] Two years later, in October 1927, drilling at Baba Gurgur—where gas had vented from the earth since the time of Nebuchadnezzar—unleashed a gusher. Oil, nearly 95,000 gallons per day, poured out onto the desert and began flowing down the valley. 2,000 Iraqi workers blockaded the wadi, stopping the oil

FIGURE 9. A man stands before a bituminous well near al-Qayyara, a town on the Tigris River south of Mosul, 1930s. Bitumen, a viscous component of petroleum, was traded for centuries; Ibn Battuta visited the wells of al-Qayyara in the early fourteenth century. Bitumen had many uses in antiquity, from waterproofing boats to embalming the dead. Library of Congress, Prints & Photographs Division, LC-DIG-matpc-16196.

from reaching the river, while others sought to bring the well under control. Two American workers suffocated from gas exposure, and three Iraqis trying to rescue them died in the effort. Had the oil reached the Tigris, "it would have caused an environmental catastrophe far beyond Kirkuk."[28] It is telling of later history that the most important dam constructed in Iraq in the 1920s was built not to stop a river of water but a river of oil.[29]

Once contained by new infrastructure the flow of oil from the wells of Kirkuk provided the revenue for the Iraqi government to build other dams. Money accruing to the government increased in the 1930s, though only after the Baghdad government in 1931 expanded the Iraq Petroleum Company's (IPC) concession from 192 square miles of Iraqi territory to 35,000 square miles.[30] The new agreement granted the IPC rights to exploit all the oil fields in the Baghdad and Mosul provinces east of the Tigris River.[31] In 1935, the IPC opened a pipeline connecting its oil fields in Iraq's north to two ports on the Mediterranean Sea. One line traveled through Transjordan to Haifa in Palestine, the other through Syria to Tripoli. A British Petroleum film released in 1955 chronicles the lines' construction, noting how the pipelines carried "a river of wealth, a third river" that flowed west to the Mediterranean instead of south to the Persian Gulf.[32] The pipelines accommodated more production, which brought additional revenue to the government.

The interwar period, then, was a time when both oil and water brought increasing concentrations of wealth. Oil revenues flowed into the coffers of foreign petroleum companies, first and foremost, and then into the national treasury. Spreading water on the land facilitated expanded land ownership, mostly in large tracts consolidated by the few. And, both the oil and the water produced creeping ecological crises—one in the climate and the other in the land.

It was only a matter of time before these two material processes met. They did so at Kut, a town on the Tigris River about 300 kilometers northwest of Basra. Like the Euphrates River, the Tigris had once taken a different path to the sea, making its turn south toward the Persian Gulf at Kut instead of nearly 100 kilometers farther east. The channel through which the river had flowed in the past was known as Shatt al-Gharraf. By the 1930s Shatt al-Gharraf had silted up and only received water from the river during times of flood.[33] King Faysal and his cabinet believed the oil revenues from the 1931 expansion of the IPC concession could be spent to build a barrage at Kut on the Tigris River. The Kut Barrage would distribute some of the river's flow to its former channel in the same way that the Hindiyya Barrage on the Euphrates had provided water to the Hillah Canal.[34]

Increasing oil revenue did not, however, bring about increasingly stable governments, especially after King Faysal's death in September 1933. Governments in Iraq changed rapidly with twelve different cabinets in the period from 1932 to 1939.[35] Independence, too, had not secured political stability. While before his death Faysal had negotiated an end to the League of Nations Mandate, the far-reaching oil concession in 1931 pointed to the quasi nature of Iraq's independence. The country in 1932 had won only a limited sovereignty, one curtailed by a treaty that codified British military and political interference.[36] Ensuing "independent" governments were shaky at best, a situation historians have ascribed to several factors, from venal elites to admiration for Turkish leader Mustafa Kemal's authoritarian style, to the international environment of the 1930s.[37] At least one additional reason must be added—struggles over the mixing of oil and water. Shortly after the King's death, the Kut Barrage and Gharraf Project led to the collapse of one of Iraq's governments.

In November 1933, Jamil al-Midfa'i formed a cabinet that included Rustam Haydar as Minister of Economy and Transport. Al-Midfa'i had served as an aide to Faysal when he was King of Syria. Haydar had served as the late king's chamberlain. Both men, then, had reason to admire and to secure the king's legacy. As scholar Najda Fathi Safwa reports, al-Midfa'i brought Haydar with his "modern and scientific manner" of work into the cabinet specifically to execute the Kut Dam and Gharraf Project.[38] Once appointed, Haydar moved quickly to issue the international tender and commence design and construction.

This act surprised the Minister of Finance, Nasrat al-Farsi, who oversaw funds in the Iraqi Treasury. Al-Farsi believed the monies would be better allocated to the Iraqi Army. Naji Shawkat, at that time the Minister of Interior, had witnessed and suffered politically because of the 1926 flood on the Tigris River (see Chapter 1) and was eager to see the capital city protected. He argued that no works on the Tigris should be built unless they were "to protect Baghdad from the flooding that it is exposed to every year or two."[39] Haydar replied that King Faysal had told him that "there should be no project before the Gharraf," to which Shawkat replied that he didn't think the king had ever said such a thing nor had the sayings of a monarch the sanctity of religious texts.[40] To Shawkat, the Gharraf project would only "achieve feudal goals" [*ahdāf aqṭāʿiyya*] by helping four or five large Shiʿi landowners, who were also leaders of major tribes. The cabinet split with Nuri al-Said, Minister of Foreign Affairs, siding with Shawkat. Salih Jabr, the Minister of Education, joined Haydar.

Shawkat's comments and press attacks on Rustum Haydar's Lebanese origin emphasized sectarianism in the division over the Gharraf project.[41] Yet, the political struggle was equally about how oil revenues should reshape flows of water. Should the government prioritize flood control, which the Kut Barrage did not provide, or the extension of irrigation, which it did? Jamil al-Midfaʿi sought assistance from Faysal's son King Ghazi, but the new sovereign held little sway. Al-Midfaʿi resigned in February 1934 amidst the turmoil but then reconstituted a new government without any of the previous members save Jamal Baban, the Kurdish Minister of Justice. This cabinet lasted only seven months but succeeded in moving forward the Kut Barrage and Gharraf Project. Criticism of the government's intentions had forced something of a compromise. This second government also committed to building Willcocks's planned escape at al-Habbaniyya on the Euphrates River, to be completed after the Kut Barrage.[42]

Along with showing how water projects exacerbated preexisting social rifts, the government's prioritization of irrigation extensions over flood control—at a time when irrigation pumps had already expanded both cultivation and salinization—reveals a critical facet of the political ecology of salt: Iraq's elite privileged gaining control of land and water over safeguarding their ecological and economic viability and over protecting settlements regularly affected by flooding. Moreover, the government knew that the Kut Barrage's supposed improvement of lands would generate controversy. So, the plans allocated a relatively small portion of the newly irrigated land for a settlement project, which it was hoped would inoculate the government from the charge of favoring only the landed elite.[43]

Construction on the Kut Barrage and Gharraf Project began two years later, employing over 2,500 Iraqis and requiring the movement of more than 1.6 million cubic yards of earth. The huge edifice stretched a half kilometer across the river. Fifty-six gates, each twenty feet wide, controlled the flow of water, funneling a regular supply down the Gharraf channel for distribution across the dry plains separating the Tigris and Euphrates Rivers between Kut and al-Nasiriyya. The Gharraf Project allowed the construction of additional canals and the irrigation of 900,000 acres.[44] Oil royalties helped cover the £1.12 million price tag. The dam became operational in 1939; the March 28 ceremony opening the Kut Barrage was King Ghazi's last official act as monarch before his death in a car accident on the night of April 3.[45]

As always in the lower basin, the movement of water onto new lands brought with it the agricultural nemesis of salinization. Tucked between Shatt al-Gharraf and the Kut Barrage, the headworks of another canal were constructed, known as the Dujayla. The canal diverted Tigris River waters into an area of "potentially productive desert" east of Kut.[46] Giving credence to Shawkat's complaints about the project's benefits, large landowners claimed 76,000 of the 164,000 acres the canal could irrigate. The government allocated the remaining land to the Dujayla Land Settlement Project, a scheme that established landless families on state-owned lands. War interrupted work on the project; the first families were settled in the area in 1945.

Historian Sara Pursley has studied the Dujayla Project and notes in her analysis how salinization affected the project, such that the "deterioration of the land . . . block[ed] even a modest local future" for settler families.[47] Despite land surveys that would have allowed the building of drainage, none was constructed. No provision was made for clean water, leaving the irrigation canal as the only source of water for crops and drinking, and for carrying away human and animal waste. Purpose-built latrines failed because the canal and subsequent irrigation raised the local water table. Water-borne diseases, such as bilharzia, trachoma, and malaria, spread rapidly. In the end, the Dujayla settlement scheme was an abject failure. By 1970 the area had reverted to its original place in local economic life: as land for raising sheep, not families. In fact, a 2013 salinity study in Iraq showed that Dujayla still suffered from salinization, with only 42% of the irrigable area under cultivation.[48]

Pursley's diagnosis of the failure focuses on the "new kinds of spatial immobilities, especially for peasants, nomads, women, and the rural nonhuman world" that the Dujayla scheme brought about. The Dujayla Project worked by fixing land, water and farms to a grid, and like other modernizing projects enforced a "conceptual separation of people from their environment."[49] Settlers were affixed to the land by debt and forced to grow unsuitable crops.

They were sent to schools where they were put through patriarchal education programs and forbidden to congregate in village centers.

Yet, when understood as part of a growing political ecology of salt, the Dujayla Project becomes less about immobility and fixing to grids and more about the mobilization and movement of nonhuman forces, with some acting in unintended ways. The Dujayla settlement came about because of the movement of land to build the Kut Barrage and of water through the canals and on to fields. Salt moved accordingly, drawn up from the saline water table into the newly saturated layers of soil. Barrage and canals came about through the transportation of oil across the Syrian desert to the Mediterranean Sea in exchange for materials and development experts flowing the opposite direction. And, so, while a social view of the environment at Dujayla demonstrates the "complex hierarchies of human and nonhuman kinds" produced by developmentalism, an analytical knot emerges from thinking through such hierarchies, namely putting human intention first and conceiving of nature as fixed or stable. Water and salt moved and interacted at Dujayla as they had in the alluvial plain for millennia; the failure was in imagining somehow that they wouldn't, that oil revenue manifested in a dam might suspend not only what was happening to the lands of the lower basin but also what was known to have happened—and then subjecting hundreds of Iraq's poorest families to the consequences.

These consequences were in many ways "unintended," in that development experts and engineers did not want the Dujayla Project to fail. That their intentions took little account of Iraq's environment is fairly clear. However, in all the ways Pursley describes, and in all the actions during the 1930s that entangled water, land, oil, and salt, the consequences were manifold and expansive. So, instead of "unintended consequences," a concept that seems to have risen to a law in some fields of study, this history offers a different explanation: that of the unequal distribution of consequences. This, too, is a feature of the political ecology of salt, for salinization happened without intention and against intention, but only some suffered the results.

Finally, at the Kut Barrage, water, land, oil, and salt converged in ways that revealed the emergent contours of the twentieth century—economic, environmental, political, and social. In the fluid and mobile postwar carbon economy, oil revenue made it possible for the state to draw these natural forces together. With petroleum wealth, the Iraqi government generated new flows of water; that water drew forth salt from the land. And, as that salt rose, slowly and inexorably degrading the soil over the course of the 1930s and 1940s, Iraq's oil offered another promise: the flow of land itself. What oil had enabled it could now attempt to replace. Oil revenue thus promised not only new supplies of water, but also of land, fueling a new and precarious dependence on international markets to secure the crops, materials, technologies, and capital needed to sustain Iraq's people.

In the years following World War II the connections wrought at Kut intensified, not only at Dujayla but throughout Iraq. The war confirmed fossil oil as an indisputable economic and strategic asset. British investments at Abadan in Iran and increasing oil production in Iraq helped make the head of the Persian Gulf, where the British Empire predominated and the two rivers met the sea, critical to geopolitics.[50] Cultivation in Iraq expanded and salinization accelerated because of Mandate-era water management policies. More and more land degraded, leading farmers to adopt more salt-tolerant crops, while salinization left some lands unusable. The Tigris and Euphrates Rivers wreaked havoc in the Iraqi countryside in eight significant floods from 1940–1954. And the development expertise that started remaking Dujayla in 1945 came to play an ever more prominent role in Iraqi politics and economic life.

The concurrence of overflowing oil wells and riverbanks, of lands overburdened with water and salt, led to extravagant plans to reshape the rivers and, through them, aspects of Iraqi society and government. The basic premise of these plans was to harness Iraq's abundant water for agriculture while safeguarding the country against the ravages of flooding, an accomplishment that would lead to widespread prosperity and the glorification of Iraq's monarchy and its imperial patron. The floods posed real peril for Iraq's people, of course; poetry and memoirs from this period reveal the struggle and grief prompted by the near-annual torrent. But, as with the Kut Barrage, what the Iraqi government planned and built after 1945 did little to mitigate the hazards producing the political ecology of salt—misapplication of irrigation water, consolidation of land ownership, political infighting, and progressively worsening salinization. These were left mostly untouched by Iraq's grand plans. Moreover, the intensifying entanglement of water, land, oil, and salt underpinning this political ecology caused wide social, cultural, and economic effects that could be seen in such disparate realms as local diets and internal migration.

It is uncertain how much land grew salinized during the war, but agricultural statistics show an increase in barley cultivation over wheat.[51] Barley is more salt tolerant than wheat, indicating that increased salinization may have been driving a shift in cultivation. Such a change, along with wartime shortages, chiefly benefited the large landholders, who produced more for export. Sharecroppers and smaller landholders could not as easily absorb the vicissitudes of cultivation in the lower basin: floods, low summer water supplies, salinization, and wartime inflation. These factors in turn increased the large landholders' sway over labor, land, and water.[52]

At the same time, the interwar model of constant, low-intensity expansion of cultivation could not continue. The additional growth in agricultural output had depended not on an increase in productivity, but on the relative freedom to claim available land and water. In the decade after World War II, the practice and potential of grabbing excess land, water, and labor by exploiting the pump law and other legal means had exhausted the ready supply of land and repressed the labor. Supposedly, only intensification, including a more "efficient" use of water, and the technological replacement of labor could produce further gains.[53]

As a result of these developments, when it became possible after the war to invest in water infrastructure, the issue of salinization occupied Iraqi leaders. In 1945, the Iraqi Prime Minister Hamdi al-Pachachi requested help from the British to create a new survey of the rivers. The last such survey by British engineer William Willcocks, undertaken at the behest of the Ottoman government, was woefully outdated. At a scheduled meeting on May 1, the day after Adolf Hitler's suicide in a Berlin bunker, the British Ambassador to Iraq, Hugh Stonehewer Bird, met with the Prime Minister. Bird reported to London on the meeting:

> What this country needed, said Saiyid Hamdi al Pachachi, was sound internal development. Since the country's livelihood depended primarily on agriculture, this development must depend on irrigation. . . . Another aspect of the scheme on which the Prime Minister very rightly dwelt is the importance of thorough drainage being undertaken simultaneously with irrigation. He said that no less than 30 per cent of the land of Iraq was "dead" owing to salinity.[54]

What should not be missed here is the scope of salinization—nearly one-third of Iraq's agricultural land had already succumbed by 1945—and the way water, land and salt all appear in conjunction, as connected environmental factors. That these connections were known makes what happened after this meeting that much more of a tragedy. Moreover, the prime minister wanted to surpass the Ottoman development program in the early part of the century. Hamdi al-Pachachi had indicated that "the poverty-stricken Turkish Government . . . had been able to spend thousands of pounds on providing Sir William Wilcox [sic] with a staff of 40 experts. The Iraqi government, with its healthier finances proposed . . . to do even better."[55] After the ambassador's meeting the British Embassy in Baghdad immediately sought an expert who would meet Iraq's needs, cabling messages across the empire to London, Egypt, and India.

The greatest concern for the British at first was not how to accomplish what the prime minister wanted but to outflank the United States and capture Iraq's development contracts for British interests. Worries about rising American influence and competition in the region drove a sense of urgency. Ambassador Bird warned:

> If we miss this opportunity there is little doubt that the Iraq Government will turn to America and even less that, if they do, an American expert will certainly be found. It would be a fine advertisement for American engineering to be able to re-make the irrigation system of Iraq, and might profoundly affect, to our detriment, the whole balance of British and American relationships in the Middle East. This, in fact, is exactly the type of capital development in which United States interests would wish to participate.[56]

American interference became an even greater worry when the King's regent, 'Abd al-Ilah, visited the United States in May 1945 and toured the country, making a stop at the Boulder (later Hoover) dam on the Arizona-Nevada border.[57] As Joseph Sassoon argues, economic competition between Britain and the United States led each country's representatives to encourage the Iraqis toward grand projects—they then jockeyed for the ensuing lucrative contracts.[58] The incentive, then, in international development was not to respond to Iraq's peculiar ecological situation, but to build as big as possible.

The task of finding the required British expert took many months, and it was October before one was recommended to the Baghdad government:

Frank Fraser Haigh, the Chief Engineer of the Punjab Irrigation Department. Punjab, as the site of the Canal Colonies and extensive works on the Indus River, was one of the places in the British Empire where an irrigation engineer could gain considerable hands-on experience. At that time, Haigh led one of the largest irrigation departments in the world. He had been Chief Engineer since 1941 but planned to retire in a few months after twenty-eight years in the Indian Public Works Department.[59]

Haigh arrived in Iraq four months later, in February 1946, to head the new Irrigation Development Commission. He reached Baghdad just in time to witness a massive flood. The flood that began in March 1946 was the highest in forty years and required opening breaches in the Tigris floodwalls north of Baghdad. The deluge followed its usual path around the city to the east. Then, in mid-March, the Diyala River began to rise. Breaches in the levies on the right side of the river allowed the Diyala's flood to combine with the waters from the Tigris, leading to the collapse of embankments protecting al-Rashid military camp. "One of the most unusual aspects of this flood," Iraqi engineer Ahmad Sousa reports, "was the continued rise of the Tigris River until the end of May." Thus, the flood lasted more than two months, with the river falling somewhat only to rise again to serious flood levels five successive times. Baghdad became an island. The eastern part of the city was caught between the normal Tigris riverbed, swollen with floodwater, and the ten feet of water covering the flooded lands to the east. Muddy river water oozed under the embankments, as if coming from the ground itself, and submerged the streets and the lower floors of buildings. It seemed as if the flood sought to erase the familiar foundations of the city.[60]

For Frank Haigh, the 1946 flood served as a potent reminder of the destruction the rivers could cause. For engineers, then, the event was one to be measured and controlled. However, there were other ways to understand the flood and human relationships to the river—ways endangered by the technical remaking of the rivers. For poets like Nazik al-Mala'ika, the 1946 flood was a moment of rupture, one that implicated Baghgdadis' memories and the city's permanence. Al-Mala'ika wrote her poem after hearing a story about a graveyard that had been inundated. In "The Drowned Cemetery," she evokes the strangeness of a gravesite lost to "the terror of the raging storm." She describes human remains uncovered so that they "floated, bewildered, over the water's face." Then, after these images of terror and loss, the narrator addresses the Tigris River:

> O river, do not be harsh on the dead
> Enough with the misery you have caused!
> Enough with mourners you have displaced!
> Be gentle with the blameless dwellers of the earth.[61]

Al-Mala'ika's poem serves as both a lament and a reminder of how individual tragedies may be lost within the larger context of a collective disaster. So many "blameless dwellers," both above and below the surface of the earth, had suffered in the long-lasting 1946 flood. Moreover, al-Mala'ika's poem is not only about the flood's destruction—it is about the fragility of memory itself. Just as the river unearthed graves, it exposed the ways in which human lives and histories can be washed away, whether by rising waters or by neglect.

Neglect for the specificities of Iraq's ecology and society characterizes much of what Frank Haigh's Irrigation Development Commission produced. In the wake of the massive flood, Haigh assembled a team of six British engineers and twenty-five Indian surveyors and draftsmen—no Iraqis were part of the core staff. In November, the commission submitted a preliminary plan for the Iraqi government's approval. The plan followed almost exactly Haigh's fellow India-trained engineer William Willcocks's original ideas for water control in the country—a continuity that underscored how an imperial form of hydraulic thinking still shaped Iraq's water policies.[62] Some cabinet members voiced concerns about the lack of innovation but failed to address the deeper issue: even if the models imported from India and Egypt had worked in those countries (an open question, even then), they still had not been designed with Iraq's ecological realities and variabilities in mind.[63]

Haigh's schematic and technical treatment faithfully exhibits each river as a "water resource" subject to management and engineering, and his plans for flood escapes on each river offered hope for ending the devastation of the annual floods.[64] Yet, time and again, salt disrupted Haigh's vision, exposing the limits of an engineering mindset that saw rivers as manageable resources rather than living, unpredictable systems entangled in a wide web of relations. The reality of salinization revealed how Iraq's rivers refused to conform to the neat calculations of hydraulic planners.

Haigh's proposed escapes transferred excess flood water from the rivers into low-lying depressions— these were intended to become reservoirs for irrigation. Unfortunately, the depressions were already reservoirs of salt. On the Tigris River, Haigh proposed building a barrage at Samarra with a channel that would divert excess flood waters into Wadi al-Tharthar, a depression 100 kilometers northwest of Baghdad. A seasonal stream already drained from the Jebel Hamrin into Wadi al-Tharthar. As a result, water sometimes covered fifty square kilometers of the depression during the winter and spring. Summer heat evaporated the water in the depression, leaving behind salt. Wadi al-Tharthar's crystalline encrustations of salt were evident at the time and would become even more evident to engineers when the construction company utilized aerial photographs to determine the basin's features.[65]

On the Euphrates River, Haigh proposed to redesign and complete a project that had been started and stopped several times during the twentieth century—a flood escape at al-Habbaniyya. The Habbaniyya depression near Ramadi had functioned as a spillover for Euphrates River floods for many years, sometimes regardless of human intention. The river had simply spilled over its banks into the depression. Constructing a flood escape there had been attempted multiple times in the first half of the twentieth century, with each subsequent effort halted by war.[66] However, beyond the bad luck of wartime interruptions, Buhayrat al-Habbaniyya was not a perfect solution to Euphrates River floods. The depression was too small to offload enough floodwater, and there were concerns that improperly locating the channel could flood Ramadi. In addition, the proposed outlet channel, which would direct stored waters back into the river during low-water periods, imperiled a British Royal Air Force Base. Unfortunately, Haigh's solution to al-Habbaniyya's limitations held major implications for the Euphrates River south of Ramadi.

To augment al-Habbaniyya's storage, Haigh proposed building a channel between the Habbaniyya Depression and one farther south. This low-lying area was known as Abu Dibs, but it also went by another name—Bahr al-Milḥ, which in Arabic translates as "sea of salt." One can imagine, then, what this area of desert looked like before it was flooded by river water. In fact, the Commission's survey of the Bahr al-Milḥ suggested the depression contained 44 million metric tons of salt. Such a massive quantity meant that the reservoir would carry water of 2150 ppm salinity, well beyond a suitable concentration for irrigation (<120 ppm is ideal). Haigh thus proposed leaching the salts from

FIGURE 10. Area of waterlogged land with little vegetation near Hilla on the Euphrates River, 1952–1953. Salt has precipitated from the water to form white crusts on the soil. Robert T. Hatt, photographer. Image courtesy of Cranbrook Archives, Cranbrook Center for Collections and Research.

the Bahr al-Milḥ into the Euphrates River, which would require not only the construction of a more formidable outlet channel but also four years to complete and many billions of cubic meters of water.[67] That saline water would be dumped into the river at a point above Hillah, below which were some of the most heavily salinized lands in the lower basin (Figure 10).

Moreover, the relocation of so much water and salt required a substantial revision to Iraq's irrigation areas and southern marshlands. Opening large areas of new irrigation along the Euphrates River entailed halting flows elsewhere, a course of action that would result in substantial dispossession of "blameless dwellers." The plan called for a closure of the "right bank of the River downstream Shinafiyah" with "the cultivators settled thereon [to be] compensated with new land in the extension areas."[68] The report suggests two disastrous options for these downstream areas: either devastating salinization or the elimination of Iraq's southern marshes. In either case, the same engineering solutions meant to ensure Iraq's prosperity threatened to undermine the very foundations of the environment and society in the southern part of the country.

By the time all the additions to irrigation were added, Haigh notes, "the only supply in the [Euphrates] river south of the Kifl Barrage will be drainage and seepage water unfit for irrigation use, . . ." while other irrigation extensions required "the drying up of the marshes which will result from keeping the River to the defined channel proposed . . ."[69] The marshes of southern Iraq were one of the most important wetlands of southwest Asia and inhabited by the Maʿdān, so thousands, if not millions, of lives, both human and nonhuman, depended on suitable water supplies in these areas. It appeared from the report that ending the floods required fomenting a new tragedy—the destruction of the marshes and the ways of life they supported. Protecting Iraq from flood disasters, in Haigh's vision, came at the cost of salinizing land, draining marshes, and dispossessing villagers.

In these ways, the Irrigation Development Commission became equally a Salt Development Commission, as each step in altering the rivers' flow implicated, and in many cases intensified, the process of salinization. The Commission's meager plans for drainage stand in contrast to its detailed surveys and analyses of flood control and irrigation, suggesting that the process of dumping excess flood water into salty depressions was the more straightforward part of the plan. Rather than integrating drainage with irrigation and flood control as Hamdi al-Pachachi had indicated, direct remedies for salinization appear near the end of the Commission's report, an appended afterthought to all the grand (and expensive) plans for high-profile infrastructure. The problem itself is clearly expressed. Haigh notes that "on the average 60% of the flow irrigated area [of cultivated land] is affected" by salinization, some of it so severe that lands were abandoned.[70] Haigh's report diagnoses the mechanism of salt transport—a high and saline water table—and the process of reclaiming salinized lands by a method of leeching.

Yet, the solutions the Irrigation Development Commission offered were few and tentative. Haigh attributes this to a lack of good information—the Commission had developed only a few small studies and the "surveys which have been made of salt affected lands are limited in extent." A survey of the water table had been undertaken but was imprecise. Iraq lacked the system of wells to monitor the water table, a program that had been undertaken in the Punjab where Haigh had worked. However, it's also clear that resolving the pervasiveness of salinization would require engaging the broader scope of the rivers' ecological connections, especially those supporting the entrenched political power of large landholders. The report intimates that both the government and segments of the public might need some persuasion concerning the utility of drainage and the costs and sacrifices involved. One proposed land reclamation project was to be "situated in the estate of a large landowner whose interest and help, it was thought, would be useful on the experimental side." Meanwhile, the other proposed drainage project would be "very suitable for demonstrative purposes owing to its proximity to the Hilla Kerbala road."[71]

Without precise figures on the height of the water table and additional experiments and demonstrations, Haigh's commission struggled with how to effectively address the central conundrum of Iraqi agriculture, despite its stated intention to extend irrigation to millions of acres of additional land. At one point Haigh wonders on the page about the eventual effect of his proposed works: "The question arises, is the present salt affected area permanent or is it likely to increase or decrease?"[72] In the end, the Commission decided to recommend an ad hoc fix: "Drainage should be introduced whenever ... the condition of the crops show it to be necessary."[73] In other words, build the irrigation system first, watch what happens, and then fix it. This strategy was the opposite of what Webster and Viswanath had proposed in 1921 and ran counter to centuries of agricultural practice in the Iraqi countryside. Moreover, reclamation was a good deal more expensive and time-consuming than proactively protecting the land from salinization. Thus, by prioritizing irrigation expansion over drainage, Haigh's commission departed from both historical precedent and established best practices. More than that, the approach reflected a deeper assumption that water could be channeled and controlled without fully investigating and understanding the social and ecological consequences.

Like the Kut Barrage and Gharraf Project, the infrastructure Haigh designed to expand Iraq's agricultural economy rested on fragile assumptions about the movement of water in the alluvial plain. Further, while Iraq's concentrations of oil had reshaped the country's economic outlook and possibilities, using that revenue for Haigh's water projects supposed the desirability of other kinds of concentration: of water in depressions, of salt in irrigation water, and, ultimately, of power in the state and its agents. Haigh and his engineers, together with Iraqi leaders, embraced a postwar faith in technocratic solutions and imagined a future of stability and abundance. Yet the very plans they advanced ensured that Iraq's environmental dilemmas would not only endure, but grow more severe.

As Haigh's plans took shape, critiques emerged—from engineers and scientists at first, but increasingly from other quarters—challenging the assumptions that underpinned Iraq's proposed hydraulic transformation. In December 1946 Haigh's irrigation plan found its way to the desks of experts in the newly founded British Middle East Office (BMEO), which had taken over the remaining activities of the wartime Middle East Supply Centre.[74] Officials at the BMEO raised several objections to Haigh's scheme. Herbert Stewart, who had formerly served as the Agricultural Adviser in India, noted that Haigh's job was to supply water "where it was needed, [but] to other scientists must be assigned the equally important task of determining how best and where the water may be used."[75] The Director of the BMEO, Arnold Overton, worried about the singular focus on irrigation and flood protection schemes:

> It seems to me important that when detailed schemes are worked out some sort of development commission or planning committee is set up so that irrigation, agriculture, public health and social services can be planned together, and so that all the ramification [sic] of that plan will be ready before any water is put on the ground.[76]

While the Director insisted on a coordinating body, the technical experts at the BMEO focused on three main issues: salt and soils, agriculture and animals, and disease. Stewart suggested that the Iraqi government hire a soil chemist who would be charged with determining "the suitability of soils . . . which will be commanded by irrigation," as well as "the question of drainage." He also called for an agriculturalist to assist with crop choices as Iraq might "become the wheat granary of the Middle East" by virtue of the new irrigation works, conjuring again the idea of an especially fertile Fertile Crescent. Meanwhile, E. D. Pridie, the former Director of the Sudan Medical Service, insisted that "a British health expert should be appointed to the Iraq Ministry of Health at once" to address the rise in cases of bilharziasis and malaria that would accompany the extension of irrigation.[77] The criticisms of the BMEO experts not only demonstrated the growing power of technical expertise at this time, but also that the intricacies of irrigation expansion were well known.

The BMEO's criticisms reached the highest levels of the British and Iraqi governments, but the solution the British proposed was not to rethink the infrastructure but to expand its scope. In February 1947 Britain's Foreign Secretary Ernest Bevin met the Iraqi Minister of Foreign Affairs, Muhammad Fadhil al-Jamali, and proposed that the Baghdad government set up a "Development Planning Committee" that would "consider the interdependence of irrigation, agriculture, public health, labor policy and social services."[78] Water management could produce a new kind of political legitimacy in Iraq, it was thought, one that would extend the government's longevity and, for Bevin, preserve British power in a key petroleum-producing area:

> [T]he energetic development of Iraq's great agricultural potentialities on broad democratic lines should provide a powerful answer to subversive criticism of the present regime. It will give Iraq the basis for that economic prosperity that will assist her to achieve real political stability and progress.[79]

Bevin framed Haigh's scheme to expand cultivation as a vehicle for a "modern democratic form of land tenure," but in doing so he conveniently ignored Britain's long-standing role in perpetuating Iraq's deeply unequal land tenure system and its previous failed settlement projects. When Salih Jabr took over the Iraqi prime ministry in March 1947, he mostly agreed with Bevin's diagnosis, noting that "something practical must be done to raise the general standard of living, for speeches and threats would not check the danger" to the regime posed by inequality and poor living conditions.[80]

But, rather than correcting these inequalities, all signs pointed to the proposed "Development Planning Committee" entrenching them. With nationalist activism and regional unrest rising in the late 1940s, both British and Iraqi leaders turned to environmental engineering as a tool for political containment. The idea of a development planning committee was attractive not because it could resolve the ecological challenges at the heart of Haigh's proposals, but because it provided an illusion of control—over land, water, and people—despite its glaring flaws. So, even as Jabr's fears about dangers to the regime proved correct, the government's response—after repression—amid a succession of political crises in the late 1940s was ultimately to sustain and expand a flawed vision for remaking the country's rivers.

The troubles began in early 1948 when public outrage over the Portsmouth Treaty, which codified continuing British influence in Iraq, led to mass demonstrations, the killings of hundreds of protestors, and the collapse of Jabr's government.[81] Events outside of Iraq then added fuel for social unrest. In late 1947, the United Nations approved a partition plan for Palestine. When the plan took effect in May 1948 the Iraqi government declared martial law to suppress ongoing anti-treaty protests, which had expanded to include labor strikes and Palestine solidarity demonstrations.[82] The government then dispatched troops to join the Arab offensive against the newly declared Israeli state and shut down the Kirkuk-Haifa oil pipeline, cutting off a major source of revenue. With war expenses and no pipeline income, Iraq faced a financial crisis. The Iraq Petroleum Company (IPC) refused to extend a loan against future royalties until the pipeline reopened. This further strained state resources, forcing a halt to Haigh's Irrigation Development Commission and jeopardizing Iraq's applications for international development aid.[83]

The protests intensified the government's search for legitimacy through planned development, even as Iraq's financial crisis and stalled aid requests highlighted the country's vulnerabilities in the international system. When in early 1950 Prime Minister Tawfiq al-Suwaydi's government passed the Development Board Act as a prerequisite to obtaining foreign aid, the moment wasn't about "exemplary seriousness, organization, and accomplishment" as al-Suwaydi professed in his memoirs, but about contending with an entangled political, economic, and ecological crisis.[84] Far from being a product of farsighted vision or democratic opening, the Development Board was a product of unrest, which not only included political factors such as unequal treaties, but also the slower rhythms of ecological change. And, because of these intertwined crises, the Iraqi government shunted nearly all its oil revenue into the new Board, which included two foreign members—one British and one American.[85]

The great irony of the Development Board's creation in 1950 is that even though it emerged from critiques of Haigh's plan—especially its neglect of salinization—the Board ultimately adopted Haigh's blueprint, embedding its ecological ramifications into an even more powerful institutional framework. Haigh's scheme, which treated irrigation as an unqualified good, was the only plan available when the Board came about. So, rather than mitigating salinization, the Board used its vast authority to extend it, and with its control over oil revenue and public works, made salinity a secondary concern to Haigh's grand vision of hydraulic modernization. What had begun as a plan to expand agricultural production instead magnified the political ecology of salt that threatened it. The Board also showed how a state may double down on environmentally destructive practices when they have already been woven into extant economic and political structures.

So, even as the Development Board made headway in building the schools, roads, and hospitals to address some social and economic factors of discontent, its water engineering plans exacerbated the ecological processes sustaining or even causing social and economic distress. In addition, before the vaunted flood control could be built, Iraq experienced one of the worst floods of the century in 1954. Political narratives of the 1950s view unrest in Iraq as linked to a series of events—anger at Western influence, especially after the 1948 war in Palestine; rising Arab nationalism, particularly after the 1952 Egyptian Free Officers coup in Cairo; and frustration at military alliances with European powers against other Arab states, such as the Baghdad Pact in 1955. These, when added to longer-term social and economic studies illuminating inequality, unemployment, poor living conditions, and the rise of new social classes appear sufficient to explain the contestation—and violent repression—of social movements in the 1950s, which eventually led to the violent overthrow of the Iraqi government in 1958.

However, by thinking through ecological factors, we may see how floods and salinization involved rhythms of change different than the ones animating the political narratives of the period: the suddenness of rising waters, the shriveled bleakness of dying plants, the crusts of salt rising like ghosts from the soil—the rich, black oil dug up and sent west. By considering the movement of water and salt, and how petroleum drove salinization during this period by fueling pumps and financing dams and canals, we may move these entangled social and natural processes from the background of history to the foreground, from the offhand mention to the center of a story about a political community under duress and embedded within an interconnected and changing ecology.

And by doing so it becomes clear that salinization factors into nearly all the most frequently cited causes of social, political, and economic distress in Iraq during the 1950s: declining agricultural yields, indebtedness, dispossession, urban migration, poor health outcomes and starvation. Studies of economic policy, land tenure, and agricultural growth in the 1950s help to illuminate the Iraqi farmer's abysmal economic condition, while later studies connect the Iraqi farmer's economic condition to the similarly abysmal condition of the land.[86] However, the mechanism of salinization—meaning not just the presence of salinity but how its long-term unfolding had radiated outward into myriad sociopolitical processes—is left out in these studies. Yet in 1951, a report estimated that one-third of Iraq's cultivated land had been abandoned while "on a large part of the remaining land yields have declined by 20 to 50 percent and even more."[87] This is a description of devastation in the agricultural sector, the sector providing sustenance and employment to more than two-thirds of Iraq's people.[88] Thus, salinization was not simply an agricultural problem—entangled as it was with petroleum and irrigation, salinization was an ecological force shaping Iraqi society.

Indebtedness and lack of ownership were among the major reasons Iraqi farmers suffered such grievously low standards of living, but these difficulties were not merely the products of social class or legal decisions. Ecology played a significant role, too. The British Mandate government had produced a land registration system that abolished customary rights and favored large landowners and wealthy urban dwellers, forcing many cultivators into a state of tenancy. Tenancy terms in Iraq were onerous, with the landowner and his agents in many cases taking as much as 70 to 80 percent of the harvested crop as the cultivator's rent. The remaining crop did not provide enough for both subsistence and for the next year's equipment and seed. The shortage obliged the cultivator to take on debt with the landholder, and a 1933 law required the cultivator to remain on their landlord's holding until that debt was repaid.

Several studies at the time, and since, have suggested that a more equitable distribution of the crop would have afforded the tenant farmer subsistence and continued production, alleviating the debt burden.[89] However, the legal framework enserfing Iraqi cultivators had more than a class-based bias—it had an ecological and thus a temporal underpinning. Salinization happened by degrees. Salt-affected plants did not necessarily drop dead or fail entirely; sometimes the consequence was merely reduced yields. Absent a comprehensive and well-maintained drainage system, remedies were labor-intensive and reduced the size of the crop. The measures Iraqi farmers used to mitigate salinization—widely spaced crops in built-up rows—meant that a given plot of land produced less. As salinization progressed yields declined, but the effects could take time. According to a study by Iraqi economist Muhammad Salman Hasan, agricultural "productivity declined from an average of 375 kg per acre in 1920 to 238 kg per acre in 1953–58 . . ."[90] Figures of this sort often appear in economic and agricultural studies, but the more important point here is not that there was a decline of about one-third over three decades. Rather, what matters is that a share of produce from a less salinized plot might have covered a farmer's expenses at first, but, as time went by, that share became increasingly less valuable. The inflexibility of the land tenure system, then, meant that indebtedness would surely follow, while the ecological factor of salinization meant that such inflexibility would become progressively more vicious and punitive.

Debt and the inability to provide for a family fueled internal migration and urbanization. Salt, too, figures into this social change, and to wider effect, as salinization impacted even those who owned land outright. By virtue of the size of their holdings smallholders who faced salinization had fewer options in how to deal with it. A farmer could only fallow so much of their land, hoping that a lack of irrigation and deep-rooted fodder crops and weeds would dry it and lower the water table, thereby preparing the field for another crop. Again, there was a temporal component to this entangled socio-environmental process. As was true with sharecroppers, plots that had once supported a certain family size could no longer do so as families grew. In surveys of Iraqis moving from the countryside to the city, Atheel al-Jomard found that "the rural factor," meaning rural push factors that, he explains, involved "floods, soil salinity, and fluctuation in weather conditions," almost fully explained the movement of people from the countryside to the city.[91] Another study, by Doris G. Phillips in 1957, found it difficult to untangle the answers given in interviews of 259 migrant families, for "'hunger' could be caused by a rapacious sheikh or by soil salination or both."[92] Rather than a quandary about variables, Phillips's observation underscores the essential condition of the political ecology of salt—rapaciousness and salinization *were* the socio-ecological system. Moreover, ameliorating the plight of the fellah—the peasant laborer—could not be achieved solely with land tenure policy because it wasn't caused solely by those policies; it also required adaptation to ecological factors.

Beyond impoverishing and pushing farmers into the cities, salinization affected Iraq's urban centers by shifting diets and influencing trade policies. One response to progressive salinization was to grow a different crop.

As noted earlier, barley is significantly more salt-tolerant than wheat and, over time, barley came to occupy a larger and larger share of Iraq's total grain production. After World War II, barley became Iraq's most important agricultural export, behind dates—both were also important for subsistence.[93] At the same time, to accommodate increased consumption in urban areas, the Iraqi government was forced to prohibit the export of wheat. Far from becoming the "wheat granary of the Middle East" as predicted by the British Middle East Office, by 1961, Iraq had become a net importer of wheat—and had stopped exporting barley, too.[94] So, continuing salinization not only reduced yields and helped to drive both smallholders and sharecroppers toward the cities but also rippled into dietary changes and then into national imports and exports.

In studies of water security, importing food is understood as a way to import water. Buying wheat or other crops means acquiring the water used to grow them. Geographer Tony Allan in 1993 introduced the concept of virtual or embodied water, arguing that many countries in the Middle East secured their food supply by substituting "oil revenues to purchase food which cannot be produced at home because of water shortages."[95] But Iraq's case was different. Iraq did not lack water—at times it had way too much water. Iraq lacked arable land. The problem was not water scarcity but land scarcity, and this in a country the size of California with less than seven million inhabitants.[96]

Oil revenues allowed Iraq to import not just food, but the land its farmers could no longer cultivate due to salinization. The country's oil exports funded the purchase of wheat and other staples, effectively outsourcing the ecological costs of food production to other regions. Scholars have long noted how fossil energy enables societies to exceed previous ecological limits—moving people and goods farther and faster, growing more food, and even escaping Earth's gravity.[97] But, just as crucially, fossil energy allows societies to abandon damaged ecosystems and sustain themselves on the productivity of distant lands. (One of the complications of global warming is not its effects in one place but its effects in all places.) Iraq's oil wealth made land and water fungible, allowing the government to circumvent rather than address salinization. At the same time, this fungibility relied on an international system of trade and exchange that could move food as a replacement for salinized land around the globe. In the end, then, oil wealth helped Iraq's government not only to sidestep the ecological consequences of salinization but also to build a hydraulic system that would exacerbate the trouble. As a result, salinization made Iraqi society ever more dependent on fossil energy and on the markets that allowed the country to trade it for substitute land (and in the twenty-first century for water, too).

So, far from being only an agricultural issue, this history shows that salinization intensified the cycle of debt and dispossession, drove the displacement of farmers, fueled the growth of Iraq's urban poor, and increased Iraq's dependency on its oil reserves and access to international trade. These forms of economic and social distress were thus not merely the result of a one-off policy failure—a bad law passed in 1933, say—but the cumulative outcome of decades of change, by which the political ecology of salt had become pervasive by the 1950s. The Development Board could not resolve the political ecology of salt because it was part of it; by expanding irrigation without suitable drainage, the Board facilitated the salinization that was the environmental foundation of political and economic inequality. The Board's ambitious public works projects on the rivers were thus not just hydraulic works, they were mechanisms for perpetuating an unsustainable system that would lodge land degradation and economic vulnerability into Iraqi history for the next several decades.

Then came the flood. Amidst the depredations of the political ecology of salt, a flood to eclipse its twentieth century kin arrived in March 1954. On the one hand, the scale of the flood's destruction appeared to justify the Development Board's intensified push for flood control.[98] After all, why build anything if a natural disaster could come along and wreck it all? On the other hand, the government's impotence in the face of cataclysm put the Development Board's promises in stark relief: why build if the promised constructions sought only to preserve the status quo and extend the political ecology of salt, wrecking the land and destroying the southern wetlands? By some coincidences in timing, the latter question appeared the more salient one in March 1954.

Things were not going so well for the Iraqi government in the months leading up to the flood. In 1952, Iraqi agriculture still could not produce enough food, high prices persisted, and the government imported twenty thousand tons of Canadian wheat. Frustration with the government grew; strikes and political agitation increased in frequency and ferocity later that year.[99] Meanwhile, the political class sniped at each other and made empty promises. Nuri al-Said led several of the governments during the 1950s, and Salih Jabr, as founder of the opposition Socialist Nation Party, criticized those governments.[100] One can see in a statement from his party about boycotting the 1954 elections just how much economic development activities had obtained the quality of empty rhetoric:

> [T]here is no one who does not wish to enhance the national spirit, or to review tax laws, achieve social justice, utilize the country's agricultural and mineral resources, care for the Iraqi countryside, promote modern villages, revive state lands, expand small ownership, attend to labor matters, improve their professional and living standards, raise the standard of living, and provide essential goods to the Iraqi people at the lowest prices. All these are obvious matters that any responsible government must pay great attention to and are not reasonably or conceivably subjects for a referendum or for seeking the people's opinion.[101]

In setting aside any questions about how, why, and at what pace these items should be accomplished, Jabr's statement emphasizes not only the government's lack of progress but also how little would change no matter the person in charge.

As a further example of how few answers the regime had to its own predicament, Baghdad officials acted as they had in the late 1940s and represented yet another report on economic development as "change already in hand."[102] They called for an outside expert who would analyze the development program and "give first-class advice on the direction in which they should proceed."[103] In one of those uncanny ironies of history, the expert hired to fix the problems of the Development Board was named Lord Salter.

Like Frank Fraser Haigh, Lord Salter was a technocrat whose career had been shaped by colonial development projects, though Salter's career had also involved international economic planning. He had served as the head of the economic section at the League of Nations secretariat and as a British Cabinet Minister. In March 1954, Lord Salter arrived in Baghdad, and again like Haigh, his visit to Iraq's capital coincided with one of the largest floods in recent memory. Salter's presence in Baghdad might have encouraged British and Iraqi officials that they were taking necessary action, but his mission was part of a long-standing pattern of submitting Iraqi ecologies and communities to the gaze of British experts.

The 1954 flood arrived to remind Iraqis of both the limits of expertise and the hollowness of government promises. The flood was particularly severe because the Diyala River, a tributary of the Tigris that joined its parent just south of Baghdad, rose in flood at the same time. On March 25, the government opened breaches in the Tigris River bunds north of Baghdad, reducing the amount of water trying to make its way between the two halves of the city. The water spread into the surrounding countryside and pressed against the eastern levies protecting the city. Meanwhile, as the Diyala River swelled, the government evacuated fifty thousand people in the low-lying areas around the confluence so that breaches could be opened in both rivers to reduce pressure on more densely populated areas. Without such actions, the swollen Tigris would have risen above the embankments and flooded the capital.

The flood was not over, though. The Tigris's other tributaries, the Greater and Lesser Zab Rivers, swelled as well, dumping their excess water into the Tigris north of Baghdad. Adding to the hydrological danger was the huge load of sediment carried by the Tigris. Opening breaches in the banks had slowed the river, allowing suspended sediment to settle. Additional sediment in the riverbed reduced the river's capacity to carry water. So, as the Tigris's other tributaries added to the deluge, the water levels rose and pressure increased on the city's eastern defenses even as the bloated river passing between Baghdad's two halves became less capable of carrying its portion of the flood.[104]

The city's residents, the army, police, and college students banded together to save the city. Nasir al-Chadirchi, the son of prominent politician Kamil al-Chadirchi, was an employee of the Ministry of Transportation during the 1954 flood and recounted his experience in his memoirs:

> I had turned our house into a "crisis rally point" (*khaliyya 'azma*), making my room a gathering place for the youth of the area.... At the beginning of each meeting, we listened to the latest flood news broadcast over the radio, then we would go out at night to the levy, urging people along the way to work with us to save their capital, and this or that person would join us until we reached our destination with a larger group than we had started with.... We also saw hundreds of college students working diligently amid patriotic cheers and songs.... We worked earnestly by placing earth on the dam to contain the water and prevent it from crossing into the capital, and each of us was energetically shoveling dirt onto the dam until our strength waned...[105]

Despite the efforts of citizens, water seeped into houses and filled public spaces, turning Sarai Square into a lake. Officials decided to sacrifice al-Rashid military base and New Baghdad—these areas were completely flooded, "leaving nothing but the brick kilns' chimneys intact."[106] Government officials called on all companies with earth-moving equipment to bring their machines to the eastern side of the city to help raise the embankments there, the city's last line of defense.

On the evening of March 29, amid concerns that easterly winds were putting too much pressure on those levies, the Iraqi cabinet decided to abandon the eastern half of Baghdad. To evacuate that part of the city, nearly a half million people living in harm's way would have to make a treacherous journey across one of Baghdad's two bridges spanning the Tigris, enormous in flood. These crossings, already a choke point in the best of times, threatened to become sites of chaos and calamity. The Minister of the Interior, Sa'id Qazzaz, refused to implement the order, arguing that a worse tragedy would unfold if a panicked populace was forced across those bridges. Instead, he urged continued vigilance at the embankments and exhorted "every citizen to promptly assist the nearest group to him by all possible means . . ."[107] The next day, the Tigris flood began to recede; the city was not evacuated nor was it inundated. Most of Baghdad survived, but the same could not be said for the rest of the country. In April, the Euphrates River rose as well. The use of the Habbaniyya and Abu Dibs depressions saved Ramadi but the collected waters of all the rivers met in southern Iraq, where they inundated thousands of square miles of winter crops.[108]

Rough estimates of the flood's financial cost put damages at over 40 million dinars or $112 million—a sum that would have absorbed nearly sixty percent of that year's oil revenues. The financial total, while substantial, does not adequately describe the damage. Nearly twenty percent of the country's irrigated agriculture was destroyed; 110,000 acres of orchards were inundated. One-third of the country's livestock was killed. So, just as salinization had over time forced greater and greater food imports, the flood devastated wide swathes of Iraq's agricultural production. Just as salinization had slowly displaced farmers, the flood rapidly displaced thousands of city dwellers and encouraged many in the ravaged countryside toward urban areas.[109] The flood involved a staggering loss of wealth and livelihoods, one that could only have added to political discontent.[110]

For the government, the 1954 flood was an unfortunate deviation—a challenge to be solved through engineering already underway. But for many

Iraqis, it revealed yet again the emptiness of the state's promises of progress. Iraqi poet Muhammad Bahjat Al-Athari captured this frustration in his poem, "Baghdad and the Flood," in which he casts the capital as adrift, its leaders failing to steer a course through the rising waters. "O Noah, rise, for the times have circled back," al-Athari writes, "... where is your ark, / O Noah, towards which humans flee?" The mood of public frustration is palpable in the poem:

> They claimed progress, but did not understand it, nor did they fulfill it.
> They were enamored with the grandeur of descriptions, and for them,
> In virtuous deeds—if ever recalled—they have no place.
> They are like your people in misguidance, but,
> They were oblivious—despite the knowledge of the times—at their peril.

For al-Athari, the disaster of the flood revealed not only the platitudinous quality of development promises, "the grandeur of descriptions," but also how little Iraq's leaders seemed to care, "despite the knowledge of the times." Those in charge had left the capital city rudderless in an endless sea, "as if Baghdad in its midst, / Is a ship, but without a captain." And, perhaps, like the story of Noah, the flood offered an opportunity for something new, for "the land has turned into a raging sea, / Like a people whose oppression has ignited their fury."[111] The flood, like the salt rising in Iraq's fields, was more than an environmental disaster—it was a political reckoning. Within a few years the people's fury was unleashed on Iraq's monarchy, its political leadership, and the British empire builders who in their hubris believed they had invented Iraq.

The growing connections between water, land, oil, and salt ecologies in the 1950s produced a political ecology of dam-building and indebted peasants, export controls and abandoned countryside, imported land and migrating families, saline irrigation water and rapid urbanization. The entanglement of ecological with social, political, and economic factors did not simply shape the country's urban and rural landscapes—it underpinned the very structures of governance and economy, driving political discontent that would culminate in a revolution. The state's hydraulic ambitions, embodied in the Development Board's large-scale infrastructure projects, were meant to secure Iraq's future by controlling the rivers' water. Yet, by prioritizing flood protection over drainage, reclamation, and other forms of resilience, these projects deepened the country's dependence on imported food, fossil energy, and an international economic system that left Iraq vulnerable to shocks.

The oil fields of Iraq had long promised national prosperity, but their profits instead financed a hydraulic vision that—like oil extraction's own use of water, land, and labor—drained one resource to exploit another. The Kut Barrage, built in the 1930s, had already demonstrated the long-term consequences of mismanaged irrigation expansion, which accelerated the spread of salinity across the lower basin while enriching the landowning class. By the 1950s, Iraq's leaders were repeating the same pattern on a much larger scale: channeling oil wealth into waterworks that prioritized visible, politically expedient development over long-term ecological and social viability.

The 1954 flood, with its staggering destruction of wealth and livelihoods, was a turning point—not just for the government's infrastructural ambitions, but for the entire political order. For many Iraqis, the promises of development had long since given way to disillusionment, and in its timing, its destructive effects and its promotion of solidarity amongst different groups, the flood acted as punctuation to the long sentence of salinization.[112] When the monarchy fell in 1958, then, it was not only the result of political currents and social unrest, but also of the entangling of environmental change, economic mismanagement and dependency, and related socioeconomic stressors, which had been building over decades like sediment along the banks of the Tigris. And yet, the forces that helped to shape the 1950s—oil extraction, water engineering, and the slow violence of salinization—did not stop with the 1958 revolution. Iraq was not the only state in the river basin keen to build huge dams and reservoirs. When political and engineering logics like those that had brought about Iraq's water infrastructure began to be applied throughout the basin, the political ecology of salt became international, transnational, and began to surround Iraq, creeping in from all sides.

If the pipelines carrying Iraq's oil wealth constituted a third river, then the country's eventual response to salinization—the Main Outfall Drain—was its fourth, and yet another attempt to resolve a naturally occurring feature of the basin with massive infrastructure.[113] Frank Haigh's proposal for embankments and canals in the lower reaches of the Euphrates provided the initial basis for a canal dedicated to reclamation and drainage. For thirty years, the Iraqi government worked on the project, which runs hundreds of kilometers from south of Baghdad to the Persian Gulf. The canal was equipped with pumping stations and reservoirs into which cultivators could dump salinized wastewater. Soviet engineers and construction companies carried out the first major section from 1973 to 1977, while German and Brazilian companies worked on the later stages. The drainage canal offered the promise of desalinating Iraqi agricultural lands, and in many places, it worked to that end—transporting millions of tons of salt to the sea.[114]

Yet, as with previous efforts to control water, the effort to control salt through engineering reshaped Iraq's landscapes in harmful ways, for dams, drains, and dikes could be employed less to remediate environmental ills than to extend them. As Haigh had envisioned in 1946, the same infrastructure built to purge salt from Iraq's fields helped to drain the southern marshes. These wetlands supported a great deal of life and had given rise to a distinct, aqueous way of living (Figure 11). As Ariel I. Ahram writes, by the late twentieth century, the marsh dwellers were defined less by specific ethnographic characteristics than by the way they had adapted to life in the wetlands while avoiding and resisting state intervention. By the 1980s, the marshes were under pressure not only from drainage schemes but also from upstream dams, with river flows having fallen by forty percent. Then, during the Iraq-Iran War of the 1980s, the marshes became a war zone, with Iraqi forces laying mines and diverting water to defensive moats.[115]

After the 1991 Gulf War and an aborted Shi'i uprising against the Saddam Hussein regime, the marshes became a haven for deserting soldiers and enemies of the state. So, in 1993 the Baghdad government launched another hydraulic drain project—the Saddam River. Ostensibly built for land reclamation, the Saddam River was in reality an effort to drain and destroy the wetlands. Due to the regime's efforts and changes upstream, by the early 2000s, both the wetlands and the population of marsh dwellers had declined by ninety percent. The United Nations has called the deliberate destruction of Iraq's marshes, "one of the world's greatest environmental disasters."[116] Further, the marshlands have become salt lands. The same report notes that the "[c]entral and Al Hammar marshlands have completely collapsed with ... their land cover transformed into bare land and salt crusts."[117]

FIGURE 11. Boats on a channel in Iraq's southern marshes near a village with *sarifa* (reed houses). Photo by Kamil Chadirji (1897–1968). Kamil and Rifat Chadirji Photographic Archive, courtesy of Aga Khan Documentation Center, MIT Libraries (AKDC@MIT).

At the same time as the marshes were being drained, Iraq's reliance on imported food and international markets resulted in "a grave humanitarian tragedy," as post-Gulf War sanctions severed Iraq's access to international trade. The Food and Agriculture Organization in 1993 noted that sanctions had "virtually paralyzed the whole economy and generated persistent deprivation, chronic hunger, endemic undernutrition, massive unemployment and widespread human suffering." Without the possibility of turning oil into unsalinized land and water, Iraq in the 1990s faced many of the same problems as it had faced in the 1940s and 1950s. Indeed, per capita income in 1990s Iraq was lower than in 1950.[118] In response to criticism of the sanctions, the United Nations formalized the use of oil wealth to feed a hungry Iraqi populace in an "Oil-for-Food Programme." The program provided essential aid, but like the Development Board before it, the program was criticized as insufficient to the need.[119]

The political ecology of salt has returned in other ways, partly as a result of the global carbon economy. After the Iraqi state constructed the hydraulic works proposed by Willcocks and Haigh, the depressions running north and south of Ramadi were turned into reservoirs. The Bahr al-Milḥ, the Sea of Salt, became an artificial lake and acquired a new name: Lake ar-Razzaza (*Buhayrat ar-Razzaza*). The lake supported wildlife, including fish, and recreation. The fish supported fishermen, who made a livelihood on the artificial lake's bounty.

In the third decade of the twenty-first century, however, Lake ar-Razzaza is disappearing. Upstream dams, built in part to provide cheap energy, have curtailed the Euphrates River floods that once replenished the lake's waters. At the same time an intense, multi-year drought has tightened its grip on the basin.[120] As the lake recedes, it slowly reveals the landscape it once submerged—the Sea of Salt. What had been a white crust on the desert floor dissolved into the water, hidden from sight and perhaps also from memory by virtue of a new name. But now, the salt reasserts itself as an immediate and visible consequence. As salinity rises, fish die en masse, washing ashore in a stark and smelly testament to the lake's decline. Without fish, the fishermen suffer; some have turned to tourism, ferrying visitors from nearby Karbala across waters that will not last. Without regular infusions of water, the lake will continue to shrink until it vanishes entirely. And when it does, the Sea of Salt will once again reclaim its place in the desert.[121]

The other depressions-turned-reservoirs face a similar fate, their waters growing saltier as fresh supplies dwindle. Even the groundwater deep beneath the basin is no longer immune to salinization. Lake Habbaniyya and Wadi al-Tharthar, once crucial components of Iraq's hydraulic infrastructure, now suffer from such high salinity that hydrologists have recommended discontinuing Wadi al-Tharthar's use for irrigation.[122] Meanwhile, in Erbil, the capital city of Iraqi Kurdistan, wells drilled nearly 600 feet below the surface now yield only brackish water, unfit for drinking or agriculture.[123] The salt is no longer confined to depressions or the water table—it is rising like oil from the depths.

The forces that once promised control over Iraq's water—petroleum wealth, engineering ambition, and hydraulic expansion—have now amplified the threats of salinity. The oil that financed Iraq's dams and reservoirs, that saved Baghdad from rising floods and helped construct the Main Outfall Drain, has also increased the greenhouse gas content of the atmosphere, driving climate change and increasing the hazards of salinity. Rising seas now push higher tides into the Shatt al-Arab estuary, threatening the fertile landscapes around Basra. This climate-related factor has combined with lower river flows and pollution to affect the region's date palms, which are suffering under the weight of progressive salinization—an ominous sign for food supplies and Iraq's export trade.[124] The demise of the date palm plantations has led to other effects. As the trees die and become firewood, the lack of vegetation transforms orchards into desert.[125] Moreover, without trees and other plants to hold the soil, the rich sediment deposited by the Tigris and Euphrates Rivers contributes to sand and dust storms. The storms at one time only happened during certain seasons but now occur throughout the year.[126]

Meanwhile, in the upper basin, the building of new dams and irrigation schemes injects more salt into the basin's hydrology. Irrigation water leaches salt from the land and washes it into the rivers, while increased evaporation from large reservoirs and irrigation networks concentrates salts in the water. Fertilizers meant to sustain exhausted fields only add to the salt burden. As a result, the water flowing into the lower basin, from upstream and downstream, is now saltier than ever.[127] Further, without significant changes to water use and storage practices in the rivers' basin, the rivers could dry up by 2040, with no water reaching the Shatt al-Arab and the Persian Gulf.[128]

And so, as Lake ar-Razzaza recedes and the desert reclaims land once vegetated or submerged, the Sea of Salt takes on a new meaning. As a result of the rivers' ecological transformation, the Sea of Salt is no longer just a geographic feature but an expanding reality—an ecological condition that now stretches across Iraq, north to south. The political ecology of salt has become a more encompassing condition, making Iraq into a land where salt is difficult if not impossible to escape and where any solution must contend with entangled histories.[129]

Three

ROCK

The terrain of the upper basin is a rocky one, the opposite of the lower basin's flat alluvial plain. In the lands that became part of Turkey and Syria after World War I, the two rivers run between high bluffs or towering mountains. Here, wind and water carved canyons as much as floodplains. This rocky topography offered an opportunity less plausible further south: to move that rock around in the building of great dams. While human designs on water in the lower basin entailed dealing with massive floods or seemingly endless quantities of salt, water management in the upper basin has entailed working with rock.

Histories of dam building often focus on how human groups move rock around. Yet, the rock in the upper Tigris-Euphrates River Basin also moves on its own. The Taurus Mountains were formed through orogeny, a mountain building process involving the collision of continental plates. Volcanic mountains also dot the region, including the famous Ağrı Dağı, known in English as Mount Ararat and in Armenian as Masis. The upper basin remains tectonically active, as the tragic and destructive 2023 earthquake centered north of Gaziantep confirmed. Two fault zones converge in the region of Eastern Anatolia where the two rivers begin, subjecting the area to considerable tectonic forces (see Figure 12). The Arabian continental plate, comprising the Arabian Peninsula, Iraq, and the Levant, is moving northward and colliding with the

Anatolian plate, comprising much of the Anatolian peninsula. The resulting East Anatolian fault zone runs in a diagonal line from İskenderun on the Mediterranean coast through Malatya and Elazığ. Meanwhile, the Anatolian plate is slowly turning counterclockwise because its northern motion is blocked by the large Eurasian plate to the north. Where the Anatolian and Eurasian plates meet, the North Anatolian fault zone may be found; this fault zone runs across northern Turkey and through the Marmara and Aegean Seas, passing mere kilometers south of Istanbul. The East and North Anatolian fault zones meet in the Bingöl province of eastern Anatolia at the Karlıova Triple Junction. This confluence of fault zones sits 200 kilometers from where the Turkish government built its first major dam on the Euphrates River near Keban, itself the site of another confluence, that of the Murat and Karasu Rivers.

Large-scale, long-duration tectonic movement is not the only way the rocks of the basin shift and change. Wind and water erode particles of rock, transforming the landscape. The Euphrates River has sculpted kilometers-long canyons through the Taurus Mountains, some with cliff walls hundreds of meters tall. The Tigris has done the same, though for less of its length. The rocks the rivers flow through were once the bed of a prehistoric ocean, the Tethys, which separated the ancient continents of Laurasia and Gondwanaland. The oldest rock, or "basement," of the Taurus Mountains is igneous and

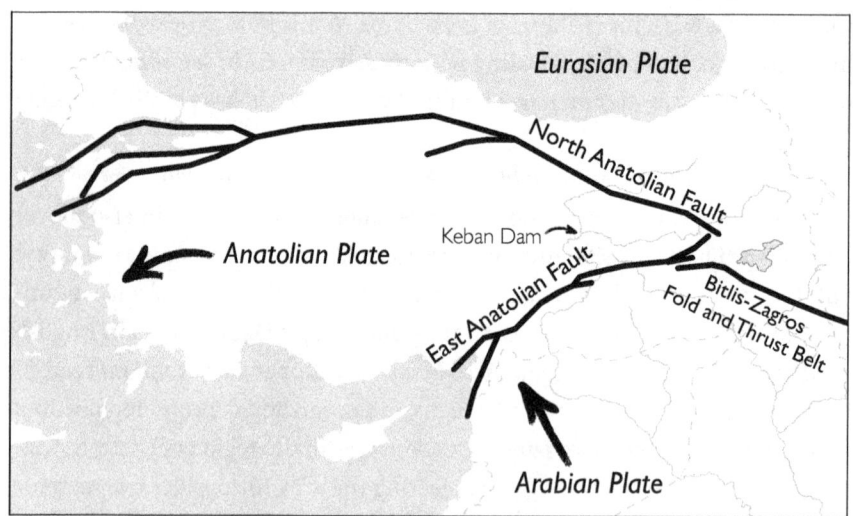

FIGURE 12. The major fault zones in Eastern Anatolia. Adapted from "Anatolian Plate" by Mike Norton, licensed under CC BY-SA 3.0.

metamorphic. However, because of the land's oceanic origin, the next layers are limestone, a rock formed from the accumulation and sedimentation of marine debris, and the precipitation of calcium carbonate, an inorganic salt, from seawater.

In addition to forming seashells and limestone, calcium carbonate is a useful treatment for heartburn and indigestion because of the way it interacts with an acid. Rainwater is such an acid, as the water contained within clouds reacts with carbon dioxide in the atmosphere to produce carbonic acid. Just as tap water dissolves a calcium carbonate tablet, weak acid rain (~5.6 pH) dissolves exposed limestone. Faults and fissures in the rock produced by plate tectonics allow water to penetrate beneath the surface and dissolve the rock underground. The resulting landscape is known as karst and is characterized by ridges, sinkholes, caves, and towers. Forty percent of Turkey is a karst landscape, with some of the most famous in western Anatolia at Pamukkale. The Tigris and Euphrates Rivers flow through karst landscapes at several points in their journey to the sea.[1]

Along the way the rivers pick up a lot of fine rock particles known as silt. Prior to the advent of large water control structures, the Tigris and Euphrates Rivers carried several times more silt than the Nile River per unit of water. The rivers carried these tiny bits of rock toward the sea, sometimes depositing them along riverbanks, over marshes, in irrigation canals, or today, behind dams. The structures that trap or divert flood water now suffer from siltation. At the same time, the dams add to the accumulation of sediment along the rivers' course by reducing flow and thus the rivers' capacity to carry these grains of rock. Such accumulations of silt in the riverbed and along the rivers' banks have ironically made areas of some cities such as Baghdad more susceptible to flooding, not less.[2] So, the vaunted flood and water control structures have neither fully controlled the floods nor have they held back the salt, all while they fill inexorably with dirt.

Starting in the mid-1960s the Turkish and Syrian governments used the rock of the upper basin to each build a great dam on the Euphrates River at the same time (see Figure 13). The installations, which became operational in the mid-1970s, are immense. The dam in Turkey at Keban is 207 meters (680 ft) high, about the height of the Metropolitan Life tower in Manhattan. If you enjoy running a 5k road race, you may have a visceral sense of the size of the dam in Syria at al-Tabqa, which is 4.5 kilometers (2.8 miles) long. The Keban

Dam in Turkey is a composite, part rockfill and part concrete. Al-Tabqa Dam in Syria is built mostly of earth fill, namely gravel rubble and soil. Both are huge constructions built on rock, out of rock, and because of rock.

As a result, they appear stable. However, as noted above, rock can have faults. It can be permeable, and it can dissolve in the presence of water. Some kinds of rock wear easily, translating into more sediment. Since rocks in all these ways change over time, they have a history. But that history is equally unstable and uncertain. At one time the main stem of the Euphrates River began at the confluence of two of its branches, the Karasu and the Murat.[3] The Karasu and Murat each flowed nearly 1,000 kilometers through deep canyons in the mountains of eastern Anatolia before entering an open and level area of land characterized by their confluence. At this spot the river, formally known as the Euphrates, began. Below this confluence, the Euphrates River carried the combined waters of the Karasu and Murat south through the Taurus Mountains, carving its own long S-shaped canyon. Where the mountains ended the Euphrates River continued, entering the broad plains of the Syrian steppe and then on to the desert and alluvial plain of the lower basin.

Today it's not clear where the Euphrates River begins. Great cliffs once towered just below the confluence of the Karasu and Murat, and between those slabs of rock the Turkish state built the Keban Dam, which produced a reservoir 675 square kilometers in area—about the size of Singapore—with a volume of 30,600 cubic hectometers.[4] A hectometer is 100 meters, so the place where the Euphrates River once began is now somewhere beneath a volume equal to 30,600,000,000—thirty billion six hundred million—cubic meters of water. That is the stated capacity of the reservoir, but the actual volume of water can differ at any given time because of droughts, floods, seasonal changes or climatic shifts. Not only then is the start of the Euphrates River unclear, so is the size of the reservoir hiding it. Once the reservoir filled it became Turkey's third-largest body of water after Lake Van and Lake Tuz. As it grew, the reservoir displaced as many as 40,000 people, drowning ninety-four inhabited villages and an unknown number of archaeological sites.[5] Hence, the confluence that once marked the start of the Euphrates is lost, and the people who could have pointed to where it was are long dead or dispersed.

The nature of rock, then, has affected a range of other histories. Historian Nancy Reynolds has considered some of the ways this happens. In her studies of the Aswan Dam, Reynolds reveals the "rockscape surrounding Aswan" and

FIGURE 13. Euphrates River dams in Turkey and Syria with opening dates.

how it "shaped the structure of the dam and the ways in which the cultural, political, and environmental narratives about it could unfold."[6] The concept of the rockscape provides a method for showing how rock influences human communities. In addition to structuring narratives as Reynolds suggests, however, rocks do things all on their own as material, earthly and unstable.

To understand human action in building the dams, we must accomplish three connected analyses. First, we must see how nature—rock, in this case—entered human representation—as narrative, or rockscape—and, especially, how these narrative representations expressed a transformed nature. Second, we must contend with a nature that at the same time exceeded both human imagination and human intention, acting in unexpected ways. This earthy unintention shaped multiple contexts: the engineering of the dam itself, the action of the river, and the social transformations brought about by dam building. And, finally, rock, river, dam, and reservoir—an aggregate of human and nonhuman, a jumble of human intention and rocky recalcitrance—continued to affect those contexts, acquiring and carrying with it an accumulated history. The mass of rock and representation that became the dams at Keban and al-Tabqa gave rise to additional dams and further representation, all the while contending with the mixture, with what had been gathered and combined. Later history, then, could not help but be shaped by wildly entangled agencies, no matter how stories might tend to separate them. The next two chapters contend with the intertwined histories of the upper basin; this chapter focuses on the rock and the next chapter on the reservoir.

So, at Keban, old certainties have been drowned by vast volumes of water: the confluence, the canyons, and even measures of capacity that can change by the day. To build the dam, new certainties had to be created, and in the process, a new Euphrates River came into being, both conceptually and materially. Some of these concepts were cultural: new metaphors and stories were required to explain with authority why the river's confluence should be disappeared. In this poetic engineering of a new Euphrates River, poets and storytellers constructed the river around Keban as an environmental and historical problem that only a vibrant nation-state could resolve. Other concepts were technical, told in the language of engineering and written by engineers for state and international institutions. Despite the technical quality of this work, engineers assessing the plans for Keban employed similar linguistic techniques to the poets and storytellers to present an imagined dam and a changed river. Together, poets, storytellers, engineers and other experts produced the rockscape, the distinct narratives about the river and the dam that expressed multiple "environmental imaginaries," ways of seeing and interpreting the environment.[7]

It may seem odd to place engineering alongside poetry and stories as works of imagination. However, this chapter will show how the engineers assessing plans for the Keban Dam and later projects employed linguistic techniques, some quite similar to the poets and storytellers, to convince their audiences. Technical documents of this kind adhere to specific formal structures and stylistic conventions, making it possible to conceptualize a poetics of engineering—a techno-poetics, if you will.[8] In this way, technical writing operates as a genre: a socially embedded mode of communication governed by norms intended to produce particular effects. And, at Keban, we may see the two forms of engineering at work: poetry as a genre of engineering and engineering as a literary genre. This approach offers new ways of understanding the activity of writing about nature, whether as a poet or as a technical expert, and helps to recast concepts like legibility, fact-making, and techno-politics.

Still, just as the interactions of water and land in Iraq subverted human intentions to extend agriculture, so too did a water-rock interface at Keban foil human intentions to build a dam. The rising reservoir revealed the instability of rock, and, in so doing, uncovered the instability of history by unsettling modes of material construction and cultural interpretation. In this narrative, what poetry, politics, and engineering promised at Keban comes into question, as material changes operating outside of human design, and at times human notice, confounded notions of cultural and social progress. The agentic qualities of nature joined with two genres to engineer a new Euphrates River, whose point of origin came to matter much less than other points of origin. This new Euphrates River was both poetically and technically engineered; this chapter begins with the poets, then moves to the engineers, and finally to the rocks themselves.

The town of Keban in eastern Turkey is a small hamlet tucked against a ridge line at a bend in the Euphrates River canyon. The river itself flows some 100 meters below the town. Just upriver from Keban, cultural formations and rock formations joined forces to construct a dam. From the early 1960s, local periodicals began to fill with reports, stories, and poetry about the river and the dam. Much of this writing was overtly nationalistic and propagandistic, extolling the great project, the great men who would bring it about, and the great future for the nation that the dam would facilitate. But other works resonated more as forms of reconciliation, as attempts to make sense of the human place within an environment about to be transformed on an inhuman scale.

There were two main venues for this cultural production. A local newspaper in the nearby provincial center of Elazığ, the *Elazığ Gazetesi*, published journalistic accounts, stories, editorials, and poetry about the dam project. The editor of the newspaper, Necip Bingöl, ran the printing house that produced the periodical, first published in 1950, and his son Sadık later served as a journalist.[9] Then, in 1962, the arts and culture journal *Yeni Fırat* (*New Euphrates*) began publication in Elazığ. Founded by Fikret Memişoğlu, a lawyer by training but with a strong interest in poetry, folklore, and folk songs, *Yeni Fırat* was meant "to benefit from written work in order to learn about history, geography and folklore." The editors chose the title, *Yeni Fırat*, for two reasons. They wanted to acknowledge the journal's debt to another publication, called simply *Fırat*, which began its short-lived publication history in 1918. And the journal was intended "to describe the effulgence that remains today of the metaphor of the Euphrates, the source of life for our environment and perhaps for all of Turkey."[10] According to the editors, the river bore cultural meaning alongside the life-sustaining waters it delivered to the mountains and arid plains of eastern Anatolia. A third reason for the journal and its title was not directly articulated: *Yeni Fırat* was also about the making of a new Euphrates. One might imagine, then, the journal's poets and storytellers as engineers of another kind, making not machines but artful structures out of language, intended to shift cultural meanings much as a dam shifts water and the river itself.

Some of the poets and storytellers published in the *Elazığ Gazetesi* and *Yeni Fırat* had achieved national stature by the time their work was reproduced in these local publications. All had strong ties to eastern Anatolia, while evincing a commitment to Turkish nationalism. Arif Nihat Asya was born in 1904 in the Çatalca district of Istanbul. In 1928, he moved to Adana and spent the next fourteen years there as a teacher and administrator. In the 1950 elections, Adana elected Asya to parliament where he served one term. Asya published dozens of books and trained many students but rose to prominence with his most famous poem, *"Bayrak"* ("Flag"), which he delivered in Adana in 1940 on the anniversary of the 1922 French withdrawal from the city after World War I.[11] Asya noted how he wished the poem to capture the excitement in Turkey about the 1939 transfer of Hatay (Alexandretta) from French-ruled Syria to Turkey. The narrator in the poem speaks to the flag, declaring their devotion: "What happens if there is no morning or sunrise; / The light of the moon and star is enough for our country." (*Sabah olmasın, günler doğmasın ne çıkar; / Yurda ay-yıldızının ışığı yeter.*)[12]

A more radical and irredentist Turkish nationalism may be found in the works of Niyazi Yıldırım Gençosmanoğlu, who served as a columnist at the *Elazığ Gazetesi* and published works on the dam project. Gençosmanoğlu was born in 1929 in the Ağın district of Elazığ province. After the rise of the Keban Dam reservoir, for forty years, Ağın could only be reached from the provincial center by ferry. Gençosmanoğlu styled himself an epic poet (*destan şairi*), and his first book of poetry, published in 1952, *The Spirit of the Gray Wolves* (*Bozkurtların Ruhu*), contains a 6,176-verse poem based on a 1946 novel by Hüseyin Nihal Atsız, *The Death of the Gray Wolves* (*Bozkurtların Ölümü*). Atsız advocated a muscular and racist pan-Turkism—the idea that all Turks should be united in a single state—and labeled the Turks a "master race," drawing from Nazi ideologies of the 1930s and 1940s. He even sported an Adolf Hitler-style haircut.[13] Set in Central Asia in the seventh century, Atsız's novel depicts the Chinese subjugation of Turkic tribes and a bloody, failed revolt against the Tang dynasty. The novel presents "the utopia of revival" where "the Turkish race . . . steadfastly preserves its own culture and characteristics, refusing to be a servant to other races."[14] While Gençosmanoğlu worked mainly as a teacher and civil servant, his later career revealed a commitment to pan-Turkism. In the decade before his death in 1992, he worked for the East Turkestan Foundation (*Doğu Türkistan Vakfı Müdürlüğü*), which advocates for the independence of Turkic areas under control of the People's Republic of China, and launched its official publication, *Voice of East Turkestan*.[15]

The contrasting themes and tone of Asya and Gençosmanoğlu's poetry reveals how a new Euphrates River came into existence over the course of the 1960s in the pages of local publications. Over the course of a few years, the river was transformed from a symbol of nature's beauty and bounty in Asya's rendering to a destructive phenomenon that required human mastery and guidance in the work of Gençosmanoğlu and others. In Asya's poem, "Euphrates" ("*Fırat*"), printed by the Elazığ newspaper, the narrator describes a river in its natural and cultural setting, flowing through the rocky uplands of "blue mountains far away" and giving its name to a beloved folk song. The river "flows into the flavor of the song (*türkü*)," becoming a national icon. Asya's narrator goes on to express his life's hope, crying out at the end, "*Ve senden doğacak kızımın / Adı 'Fırat' olsun!*" ("And my daughter who will be born from you / Let her name be 'Euphrates!'").[16] A similar image evoking the river's life-giving capacity appears in the first issue of *Yeni Fırat*, such that the river's name befits a newborn child. At the same time, the poem marks the river as feminine and, with its nationalist overtones, makes the Euphrates a participant in giving life to the nation.[17]

Later poems, from roughly 1963 forward, published in the newspaper and *Yeni Fırat* describe a very different river. Yıldırım Gençosmanoğlu's "Reckoning with the Euphrates" ("*Fırat'la hesaplaşma*"), takes the river to task, evoking images of the river as deluge, inundating the land. "You split my valleys, you swallowed my plains," the narrator accuses, "You scattered the ashes of many hearths; you gave them to the wind!" The river is now a destructive and unfriendly force: "You raged and frothed, you gave my country to salt / Neither a road for my caravans nor a ford for travelers / We depended on you; you gave us suffering." In this rendition the Euphrates is a harmful and resentful presence, devastating human constructions and thwarting human designs. The poem's title suggests something of what Gençosmanoğlu believes should happen to the menacing river. The root word, "*hesap*," suggests accounting, calculating, and planning—the very processes crucial to dam building.[18]

Gençosmanoğlu's poem continues with imagery connecting the construction of a modern nation to the taming of a wild river, crashing down from rocky uplands. "Oh lion who roars down from the Palandöken [mountains] / We will put a chain around your mane / And forget the past; and you, me, shoulder to shoulder, / We will found a nation that rises with prosperity." The personal, and personified, quality of Arif Nihat Asya's Euphrates defers in Gençosmanoğlu's version to collective demands. The river's past identity and character—the narrator likens the Euphrates to an oppressive sultan—can be and must be forgotten. In this new, democratic Turkish Republic the river could no longer act like a despot. Such a sentiment against Ottoman despotism, directed not at rivers per se but at other features of Ottoman life, was longstanding both within Turkey and beyond.[19]

A poet of more local renown, Cenani Dökmeci adopts a different style, one with historical overtones. Dökmeci was born in Elazığ in 1927 and at first followed his father into coppersmithing. However, with the encouragement of Fikret Memişoğlu, Dökmeci pursued his love of poetry and joined Memişoğlu in publishing *Yeni Fırat*. Dökmeci wrote 144 poems but never published a book of his poetry; his work appears mostly in literary journals and newspapers.[20] In one of his poems published in *Yeni Fırat* in 1966, Dökmeci follows Asya's lead in personifying the river, while adopting a more measured tone than Gençosmanoğlu about controlling it. Dökmeci attempts a "Dialogue with the Euphrates" ("*Fırat'la söyleşme*"), reaching even further back in the history of the river to make his claims. Dökmeci's narrator beseeches the river, "Stop Euphrates! ... Don't flow and burst forth into the deserts, stop, no more." The river should instead become "an inland sea in the nation's breast." As a reservoir rather than a river, the Euphrates will be more productive. "Urfa Elaziz is very thirsty," the narrator notes, "... let your shore, your contour be painted in greens." Rather than watering empty wastelands, the poem conjures a dammed river bringing relief to the pilgrims of Urfa and promoting new agriculture.[21]

In this later poem, Dökmeci mentions the dam site specifically, connecting the infrastructure and the river to events nearly a millennium past. "Tell me of Keban," Dökmeci writes, asking the river to slow down and learn about the human communities along its banks. The poet presents the dam as "a new saga" (*yeni bir destan*) that may be told by the ancient river, an epic as important to Turks as Seljuq sultan Alp Arslan's defeat of the Byzantine Empire at the Battle of Manzikert (*Malazgirt*) in 1071. Just as Alp Arslan's conquests ushered in an Anatolia dominated by Turkish tribes, in this poem the Keban Dam proclaims the construction of "an enlightened nation" capable of dominating Anatolia in an entirely new way.

In addition to poetry, writers told stories to imagine a new Euphrates and shape new conceptions of the environment around Keban. One story by Bahattin Senemoğlu tells of an American who arrived in Elazığ shortly after the founding of the Republic. One day the American "went down to the shores of the Euphrates." After wandering a bit in the canyon, he asks a villager about the river, "Does it always flow like this?" The villager confirms the American's observation but thinks nothing of it. Senemoğlu then claims that "the American had discovered a great treasure imprisoned in the river," a treasure that would not be recognized by Turks until much later. Only when "Turkey entered the age of planned development" did the people of Elazığ recognize what lay hidden. With the coming of the Keban project, "the eyes of the entire Turkish nation are turned toward . . . the promise of the shining sun that will rise with the construction of the dam." Though an American technical presence was not as pervasive in the Middle East until after World War II, Senemoğlu projects this influence into the interwar period and credits American expertise for the discovery of the river's potential. It is also worth noting that the story conveys none of the tragedy of that American presence, as would appear two decades later in 'Abd al-Rahman Munif's *Cities of Salt* (1984), for instance. Nor does the story betray any of the characteristic tension and contestation between American and Turkish modernizers.[22]

Instead, these works of poetry and storytelling engineered a new cultural signification around the river and dam that foregrounded nation-building, technical assistance and prowess, and a range of positive social outcomes. The river, in a state of nature, was rendered anew in these works as a problem rather than a source of beauty or life. Powerful and wild, bursting and roaring, the river could only be tamed by the dam, a project worthy of the modern nation. With the building of the dam, the river's strength—dissipated, wasted, and destructive in floods—might be harnessed for the nation itself. While these writings clearly sought to reshape cultural meanings to justify the construction of the dam, these efforts convey something more than simply a public relations effort. The Euphrates was not just *any* river. It was the river that had watered some of the first human agriculture and fed the canals of great ancient civilizations. The Bible has the Euphrates emanating from the Garden of Eden; the river is also mentioned in the Qur'an.[23] Changing the river's cultural meaning entailed producing new contexts, new historical connections, and a new vocabulary for imagining and understanding the human engineering of the environment around Keban.

In this new cultural context, the river was represented differently than in hydrological treatises, and in ways that often complicated the binaries of human and nonhuman or human and nature inherent to technical inquiry. In this cultural invention of the new Euphrates, the river became anthropomorphic and zoomorphic, represented as both human and animal. In addition, the above writers, all male, used gender and ethnicity to anthropomorphize the river, linking it to Turkish and pan-Turkish national projects and using metaphors and historical references familiar to those ideological referents. Though forms of scientific calculation were more directly responsible for producing the dam itself, poetic rhythms and clever storytelling helped produce the dam as a *cultural* object worthy of construction in the first place.

The construction of the dam as a material object required poetics of a different kind. While poets and storytellers in the region around Elazığ used image, metaphor, and historical allegory to produce a new environmental imaginary, Turkish and American engineers deployed technical knowledge and description to organize institutions, construction, and finance. Though different in content, engineers nevertheless produced their own environmental imaginaries and told their own stories using language, structure, and form. Moreover, poets, storytellers, and engineers shared the same goal—the damming of the Euphrates River. Keeping this shared purpose in mind, consideration of a poetics of engineering provides another view into the rockscape, the narratives structured by the rearrangement of rock to build the dam.

While poetry is the application of linguistic forms to the problems of existence, engineering is the application of scientific knowledge to real-world design problems. Engineering emerged as a distinct vocation with the advent of industrial technologies in the late eighteenth century, while engineers as practitioners evolved into a distinct profession as those technologies proliferated. The process of professionalization happened fairly quickly. In 1816, estimates suggest that in the United States there were only a few dozen engineers, and many who did engineering work did not call themselves by the title of "engineer." In 1850, the country's census included the professional category for the first time and found 2,500 civil engineers, mostly working on railroad and canal building as technical staff.[24] In countries like the United States and Britain, engineers were most often affiliated with business, while in colonial situations they were more associated with government, as army engineers or as staff in public works departments charged with designing and building transportation and other works, such as dams and irrigation.[25]

As a result, analyzing the spread of engineers, engineering knowledge, and engineering institutions is one way to understand the connections between technology, imperialism, and expertise. While some scholars in the 1970s and 1980s still referred to a "designer-less" technology as aiding imperial ambitions, in the decades since engineers have found their way into studies, alongside their designs, as agents of empire.[26] The environmental imaginary is one framework for understanding the work of engineers. The imaginary explores how engineers perceive natural processes. However, the study of environmental imaginaries has mostly focused on analyzing the content of *what* has been imagined rather than *how* it has been imagined and communicated. Considering engineers' technical documents from the perspective of poetics targets the forms and structures of language at work in shaping the process of imagination, thus not only evaluating the content of the imaginary but also the context and means of its production. In philosophy or anthropology, this might be referred to as "the linguistic act." Word choice, syntax, context, purpose—these features work together to give meaning to a linguistic construction. Since engineers write texts in advance of production or building, techno-poetics offers a method for assessing how these texts communicate, how they intend to affect or convince the reader, and, ultimately, how they participate in the work of altering the material world.

Like poetry, a techno-poetics of engineering operates in three important registers: as an imitation of nature, as an expression of universal (physical) experience, and as an evocation of the future. Technical documents are, at root, an example of what Aristotle refers to as mimesis—an imperfect imitation of the real world—in technical form. Poets across the centuries have imitated nature, using language to evoke natural phenomena from rivers to birds to snowy evenings. Engineers use methods in addition to language—models, drawings, charts, graphs, et cetera—to imitate nature and express what it "ought to be."[27] A motive of material change, then, variously referred to as improvement, development, remediation, et cetera, is the purpose of a techno-poetics.

Engineering studies contend, according to specific methodologies, with what was and is, but are most concerned with what may happen. Thus, engineers evoke a possible future according to universal laws (of nature and humanity) as they understand them at the time and in response to specified goals and needs. Notably, language and engineering operate according to an essential principle—the predictability of the world. In considering an "anthropology beyond the human," Eduardo Kohn notes, "it is only because the world has some semblance of regularity that it can be represented" through language. Further, it is only because the world has some semblance of regularity that it can be engineered.[28]

Techno-poetics offers another way to understand the role of engineers in histories of science, technology, and the environment. Analyses of these actors have become more sophisticated in recent years. For instance, in *Science in Action*, Bruno Latour considers how scientists and other technical professionals produce the authority of facts. Latour demonstrates how, in scientific literature, statements become fact through argument, assertions of credibility, reference to other literature, and increasingly technical descriptions.[29] These are all important methods for establishing "fact-ness," but they are not the only means by which one may be convinced.[30] Indeed, while engineering studies work as documents asserting the so-called facts, they often work in other ways: as visions, portrayals, and proposals.

A techno-poetics places less emphasis on engineering documents standardizing, simplifying, and depoliticizing, and more on the many ways a technical document "evokes a concentrated imaginative awareness of experience ... through language chosen and arranged."[31] Instead of noting how these documents simplify nature, making it "legible" to government bureaucrats,[32] we look into the complexity of narrative structure and technique as they are combined and associated with diagrams, maps, and other graphical information. Techno-poetics asks how technical documents employ linguistic structures to evoke contingencies, possibilities, and the subjunctive—not "the" reality or "all" of reality, but things that may happen and may be experienced. Moreover, taking a cue from new materialism, techno-poetics recognizes that one of the vital contexts shaping linguistic forms in engineering is the intended outcome—material change. Like techno-politics, then, a poetics of engineering acknowledges the purpose of engineering: to create an imitation of a putatively universal reality so that the real world in all its specificity and variability may be acted upon by forms of power.[33] But, it asks a distinct question—how, through choices in detail, language, and structure, do engineers imagine, evoke, and imitate?

Finally, a poetics of engineering offers another way to historically contextualize technical documents, one less dependent on orienting texts to social or political theory or to conditions—high modernism, for example—that have their own genealogies. Instead, techno-poetics offers a way to examine how a technical document accomplishes its linguistic purposes within its own historical context, including its material one. Put another way, rather than seeing engineering as a set of techniques coming from outside, we may explore the ways engineering directs attention specifically and locally. A poetics of engineering allows us to see how technical documents construct, simulate, and represent on the page a localized material reality—a reality very often destined for radical change at the hands of the engineers themselves through their documents. Furthermore, placing aspects of artistic and scientific production in the same conversation generates possibilities for comparison; for example, we can see how very different genres of knowledge production (a poem and a technical document) can nonetheless have similar approaches to the nonhuman, or how their approaches influence and infiltrate each other.

One of the most important engineering documents written for the Keban Dam project required engineers to transliterate technical designs—the size, shape, and components of the dam—into a script that could secure the international finance needed to build the project. This form of transliteration is known as the third-party feasibility report, and, like a volume of sonnets or of free verse poetry, the report is representative of one of the forms within the literary genre of engineering. A feasibility report outlines how a large-scale infrastructure project will work in the context of a national economy, so that creditors can assess whether the investment risk is worthwhile. At root, the report evaluates the viability of a given project for given aims. Production of the feasibility report for the Keban Dam was a turning point in the project's history, as the report was an essential part of the "paper technologies" that helped enlist political and financial institutions to undertake the dam.

On its surface, the feasibility report is about how the dam would function as a power plant—large sections of the report detail electricity demand, usage, load, transmission, and distribution—within the context of the Turkish economy. The report discusses the material aspects of construction with diagrams and close detail. For instance, descriptions of the exact weight of cable accompany two-dimensional renderings of electrical transmission towers. This is the work of mimesis; these diagrams portray material facets of reality that evoke an awareness of the dam as it will be. However, equally important to the report's findings are representations of cultural knowledge, of "things as they are said or thought to be."³⁴ (I often think of this aspect of techno-poetics like the -mış or reported tense in the Turkish language—things that have been heard but not witnessed.) Using these two modes of description in tandem—the imitation of material reality and representations of cultural assumptions—the engineers writing the feasibility report generate two allegories, sometimes in tension. The first portrays the rock, earth, concrete, and steel of the dam as social order. As the following discussion will demonstrate, the Keban Dam project became a key component in rectifying supposed "disorder" and "imbalance" in eastern Anatolia. Thus, the engineers had to render the material facets of electrical production, from transmission wires to turbines, as capable of bringing about social reform, even as their own rules for assessing the dam's feasibility set aside social and cultural criteria.

The second allegory portrays the dam as an economic object. For the dam to produce sufficient revenue to repay creditors, the engineers had to fit the project into the Turkish economy. The dam had to be shown to enhance industry, further growth, and raise incomes. For, even in this ostensibly technical document, the dam itself—the rock to be rearranged, the concrete to be poured, the large pipes (penstocks) to be fitted to the powerhouse—is only important for what it will do. As a result, in conveying the material realities of dam construction, the engineers also expressed assumptions about what an economy is, how it works, and what features of economic life are deemed most significant.

Dams were seen as methods of social reform and industrialization the world over, but understanding the meaning of these two allegories requires specific knowledge of how the report came into being and the local and global histories it implicated. To that end, the following discussion first traces the intersection of development planning with ideas of social order in eastern Anatolia. Kurdish resistance to physical and cultural erasure exposed fractures in a supposedly homogeneous and united national self-concept. Like the poets analyzed earlier, proponents of the dam positioned the project as vital to "integrating" eastern Anatolia into the national whole; engineers later recapitulated those ideas within their own genre. The argument then turns to how engineers produced the two allegories mentioned above using thematic, topical and linguistic devices. In addition, the engineers adopted American pretensions of an apolitical and acultural assessment of infrastructure, and this analysis demonstrates how the report ultimately undermines that pretension.

May 1960 is the origin of the feasibility report, when a military coup in Turkey overthrew the democratically-elected government of Adnan Menderes. About five months after the coup, the military junta established the State Planning Organization (*Devlet Planlama Teşkilatı* or DPT). The law creating the DPT required the organization to "evaluate thoroughly the natural, human, and economic resources and potentialities of the country" and in doing so prepare the first Five-Year Development Plan.[35] The plan is chiefly a set of directives for the public sector, including government budgets and state enterprises, and a package of incentives and disincentives meant to guide the evolution of the private sector.

Turkey's adoption of the five-year planning model may be traced to the Soviet Union after the First World War. The USSR became the premier site for development planning in the twentieth century, as the country sought to recover from internal and external conflicts.[36] The Soviets adopted their first five-year plan in 1929 and within four decades produced the world's second largest industrial economy. This success provided a model to several countries, most notably India and the Philippines. Both countries began planning processes while still under foreign rule. Development plans proliferated in country after country after World War II. Colonial tones also colored post-1945 planning, which was often driven by the pressure or direction of an outside power, as in the case of Iraq, or as a prerequisite for World Bank or other forms of foreign aid.[37]

Modernization theory also played a role, serving partly as a framework for economic progress but more so "a political project . . . to prevent the inclusion of the underdeveloped countries in the Soviet-type development model."[38] So, while the Soviets provided an example for planning, American and other European governments and institutions encouraged modernization theory with its anti-communist rhetoric and policies as a means for maintaining influence and economic access. Turkey was no exception to these trends. Soviet advisors had assisted the Ankara government in developing a five-year industrial plan in 1933 and, upon its completion, provided economic assistance.[39] However, with Turkey joining the North Atlantic Treaty Organization in 1952, the country's post-war and post-coup Development Plan, like Iraq's Development Board, was in part a method for suppressing forms of communist and anti-imperial dissent.[40]

The Keban Dam entered Turkey's Development Plan as a project to further the central government's aims in controlling economic life and dominating the geography of eastern Anatolia. Moreover, the political struggle over the project shows the influence of the industrial and colonial genealogies of planning. First, the effort to insert the dam project into the plan required expressing the government's motives in specific language—order, balance, and justice, to name a few—that inflected the project with cultural meanings engineers would later have to represent. Second, the political contestation in the government over the Keban Dam was not about its location or purpose in eastern Anatolia, but about who would benefit from its construction and when. In other words, the question was not about whether damming a river would dislodge dissenters or produce colonial-style economic benefits from the country's eastern hinterlands, but about which river should receive such treatment. Finally, due to previous studies, the Keban Dam was known as a keystone project that would if built engender much greater investment along the Euphrates River. Building the dam would have a domino effect and direct government attention and funds for years to come. For all these reasons, the wrangling in Turkey's parliament over the dam's place in the Development Plan was not merely over whether to pay for a feasibility study.

The political battle over the Keban Dam began in 1961, at the same time as the Development Plan was being debated in parliament. A group of members submitted a proposal to add five million Turkish lira to the budget of the Electric Works Study Administration (*Elektrik İşleri Etüt İdaresi* or EİE). The legislators wanted EİE to pay a foreign firm to prepare a feasibility report, which would be the first step in obtaining foreign credit. While a paltry sum in comparison to the dam's total cost, the act of paying for the report signaled a larger political commitment to the Keban Dam project.[41]

The parliamentary Budget and Plan Committee, which considered the proposal first, dissolved into acrimonious debate. Fethi Çelikbaş, under whose Ministry of Industry the EİE operated, declared that the money was not required and that he would not use it, even if it were appropriated. Another member of the committee, Dr. Suphi Baykam, MP from Istanbul, resigned in protest. The proposal's opponents felt the vast sums required to build the Keban Dam would be better spent elsewhere on smaller projects that could be more rapidly realized—projects perhaps in their own jurisdictions—and resisted funding a study for a project that could take a decade or more to realize. The constant infighting over plans largely divorced from the conditions of communities eventually led an MP representing Urfa,

Kadri Eroğan, to complain loudly on the parliament floor: "The people want rice, not a plan!"[42]

Despite the acrimony, the committee passed a proposal for consideration by parliament but, as parliaments are wont to do, another committee was formed to study the first committee's proposal. The National Assembly charged this new committee with determining whether the Keban Dam project fulfilled the ideals of the recently promulgated Five-Year Development Plan. The dam project's feasibility was moot if it didn't fit into the state's overall plans. The architects of Turkey's post-coup constitution of 1961 intended the Development Plan to guide the efficient allocation of government funds and reduce political infighting. The struggle over the Keban Dam suggested that the Plan might accomplish the former goal, if not the latter. Moreover, the constitution spelled out the purpose of development planning:

> Economic and social life shall be regulated in a manner consistent with justice, and the principle of full employment, with the objective of assuring for everyone a standard of living befitting human dignity. It is the duty of the State ... to draw up development projects.[43]

Due to constitutional directives, affiliating the feasibility report with the aims of the Development Plan became the only way for the new committee to move forward with the dam project. Further, the Development Plan's authors had associated the Plan with the constitution's language and extended its overarching goal: "[T]he achievement of a high growth rate and the consequent rise in incomes is not the ultimate aim. The real aim is to promote social welfare ..." Moreover, the Plan had targeted specifically the eastern half of the country for development to promote "the establishment of a balanced social order."[44]

The terms "social welfare" and "balanced social order" in this context were euphemisms for efforts to eliminate political and physical resistance to the central government in eastern Anatolia. As Begüm Adalet has argued, the supposedly "modernizing" reforms of the latter half of the twentieth century were part of a longer "historical lineage" of Turkish nation-building reaching back to the late nineteenth century.[45] In the 1890s, Russian ambitions in the Ottoman east led Sultan Abdülhamid II to empower local Kurdish militias, the Hamidiye Alayları or "Hamidian Regiments." These militias went on to commit massacres against Armenian Christians; Armenian groups retaliated with acts of terror aimed at Ottoman institutions and officials. As the empire progressively lost territory in the Balkan Peninsula and the Caucasus region in the years leading up to the First World War, the Istanbul government confiscated Armenian lands, dispossessed Armenian communities, and distributed the spoils to Muslim and Turkish refugees arriving from the western parts of the empire. During World War I, these policies intensified to mass murder and dislocation, resulting in the Armenian Genocide.[46]

Government attitudes toward eastern Anatolia hardly shifted after the war and eastern Anatolia remained an internal frontier. With nearly the entirety of eastern Anatolia's Armenian population deported or destroyed, the homogenizing aims of nation-builders, now based in Ankara instead of Istanbul, focused on Kurdish communities. In the name of integrating eastern Anatolia into the national economy, of "modernizing" rural agriculture through land reform, and of reducing the appeal of communism, Turkish governments designated Kurdish communities as backward and in need of civilizing reforms. Government policies were directed at both cultural and physical erasure, including the banning of the Kurdish language, land confiscation, deportation and forced relocation, and increasing military surveillance and repression.[47]

Kurdish groups resisted these measures in several rebellions over the course of the twentieth century, many of which began in the region around Elazığ where the Keban Dam would be built. In 1925, Shaykh Said of the Sufi Naqshbandi launched a rebellion to protest Kemalist restrictions on religious practices and on Kurdish language, culture, and economic autonomy. The Turkish military crushed the rebellion, while the central government adopted additional measures aimed at the Turkification of eastern Anatolia.[48] A 1934 resettlement law played a role in instigating a rebellion in the Dersim region later that decade. The law enabled the confiscation of Kurdish lands, which were handed over to Muslim migrants from the Balkans who it was supposed "share the Turkish culture."[49] Resistance to the law's implementation, along with a general refusal to adhere to state regulations, led the Turkish government to place Dersim under martial law in 1936. In March 1937, a bridge was burned and telephone lines were cut. This minor act, which may not have been directed at the government, led to a massive military campaign aimed at "pacifying" the countryside. Tens of villages were destroyed by the Turkish military, either by ground forces or by bombing campaigns, with as many as 7,000 people killed.[50] The sum of these acts against Kurdish lives, lands, and culture led Turkish sociologist İsmail in 1969 to declare Kurdistan as subject to a form of colonial rule.⁵¹

The Keban Dam project, framed as a material intervention for "social balance," added another dimension to the social conflicts in eastern Anatolia, amplifying the region from a cultural and colonial frontier into a techno-political one as well.⁵² In other words, if legal and other physical measures could not "rebalance" eastern Anatolia in the way government officials wished, then perhaps massive technical infrastructure and environmental engineering could accomplish the government's goals. Key to this amplification was the dam project's inclusion in the Turkish government's Five-Year Development Plan, which, as a response to the "imbalance" and "disorder" of eastern Anatolia, coded the "Kurdish question" as a problem of colonial economic consolidation and integration.⁵³

During the Keban Dam committee deliberations, the dam's proponents affiliated the project with this promise and demonstrated how the installation would fulfill it. The project's detractors offered less expensive alternative options that could accomplish similar aims. A member of the committee, Mehmet Turgut, notes how the representative of the State Hydraulic Works (*Devlet Su İşleri* or DSİ), Senior Engineer Selahattin Kılıç, kept the committee busy by showing one day a set of proposals for the Seyhan River, then insisting the next day that the Ceyhan River might be a better possibility.⁵⁴ The interchangeability of river projects supports Christopher Sneddon's concept of the "concrete revolution" and reveals that the political contest was more about who would benefit than whether a dam could in fact ameliorate social conflict.⁵⁵ Only after a leading lawmaker grew tired of the delays did the committee coalesce around pursuing the project. In its final report in support of the Keban Dam project, the committee declared:

> We earnestly desire the benefit to our country of this project by its quick realization. This project will play a large role in the social and economic development of Southeastern and Eastern Anatolia and as large a role in agricultural and industrial growth as in the question of our country's energy.⁵⁶

It is significant that growth in energy, industry, and agriculture is associated here with "social and economic development," for these associations both reflected and help set the parameters for engineers' assessment of the project's effects. The commission then recommended that the dam be included in the first Five-Year Development Plan—the only project to enter the plan with its own title—and concluded that the Keban Dam was the "key project" for improving conditions in the "undeveloped" part of the country.

This official designation elevated the project's significance within Turkish planning frameworks—but it was American engineering expertise and US federal policies that gave that vision its initial technical form. With the money now allocated for the feasibility study, EBASCO Services Inc., an engineering firm headquartered in lower Manhattan, signed a contract in December 1962 to produce the feasibility report. As a result, American histories of damming rivers, engineering the environment, and professionalizing and training engineers played a role in bringing the Keban Dam into material existence. These histories defined the genre of the feasibility report and shaped how engineers used their professional and technical training to assess the dam project. So, along with the terms, definitions, and meanings established by the histories of development planning and of eastern Anatolia, and by the political struggle over the dam project, some American histories governed the report's efforts to intertwine economy, society, and culture with river, rock, and canyon.

To understand how these American frameworks took material form on the ground, it helps to look more closely at the company tasked with producing the feasibility report. EBASCO Services began as part of the Electric Bond and Share Company, a utility holding company founded by the General Electric Company in 1905. The EBASCO unit provided services to the holding company's various gas and electric companies. By the time EBASCO engineers arrived in Turkey in February 1963, the company had designed several power plants outside the United States, most notably for the Soviets and Chinese.[57] That international experience not only reflected the transnational reach of US engineering firms during the Cold War but also signaled a broader professional confidence in the global applicability of American technical methods. By May of that year, the engineers had investigated the dam site and some of the surrounding area. An American representative of the company announced to local reporters: "There is no reason or question to prevent the building of the dam at Keban. This place is one of the most important dam sites in the world."[58] By the engineers' eyes, the river and rock were all in the right place to be remade.

To arrive at that conclusion, EBASCO engineers applied a method for calculating the costs and benefits of water projects that adhered to American water management practices. Earlier federal analyses of American water projects included "social objectives," but by the early 1960s, the preferred method evaluated "all inputs and outputs of projects in economic terms to the extent possible and ... all nonmonetary effects as 'intangible' benefits and costs."[59] Later in the 1960s, other objectives, most notably environmental ones, were added to US federal policy. Yet, American engineers were expected to avoid assessments deemed political or social.[60] In the case of the Keban Dam, federal policy helped EBASCO's engineers define the boundaries of their inquiry.

From a techno-poetic perspective, we can see these policy constraints not only as administrative rules but also as conventions of genre. Just as science fiction typically explores alternative futures (or pasts, as the case may be), the engineering genre here confined itself, supposedly, to technical and economic futures. As the engineers write in the report's summary, the Keban Dam "has demonstrated its worth on the basis of power benefits. Other benefits which accrue to the Project such as flood control, irrigation, and so forth, are recognized but no dollar value assigned."[61] With this assertion, the engineering approach appears to exclude a direct discussion of the project's role in social reform. From a genre standpoint, the engineers are articulating the formula they were trained to follow—and that their institutional audience expected to see.

Yet, the engineers clearly understood the project as a component of the development plan—there is a section on the plan in their report—and social questions found expression in the feasibility report despite the neat delineation of technical and economic issues. Just as authors of magical realism and fantasy slide into philosophy or political commentary—Marquez and the subjectivity of experience or Tolkien and the corrupting influence of power—so too did the EBASCO engineers find ways to represent a dam as economically beneficial and socially reformative. So, just as in a poem, the layout of the report, the arrangement of technical detail, the choice of topics, and the language employed were critical to formulating an imaginative awareness of the dam's presence in the landscape and its impact on the Turkish economy and Anatolian society. Further, by applying American policies to articulate the dam's utility, EBASCO engineers not only "proved" the viability of the dam to investors but also gave credence to another of İsmail Beşikçi's theses, that of the "ortak sömürge" (cooperative colony) of Kurdistan. The report's mobilization of foreign capital, material, and expertise supports Beşikçi's concept wherein different states benefit from a common extractive approach to the territory of eastern Anatolia, its people and environment.[62]

What emerges, then, in the American engineers' writing is not merely a technical study, but a narrative device—one that encodes both economic aspirations and visions of social transformation. As noted above, the report accomplishes its purposes by producing two allegories—the dam as economic object and the dam as social order. It accomplishes this feat through narrative techniques such as juxtaposition, imagery, and figurative language. It may seem odd to apply these literary analysis techniques to an engineering document, for surely the EBASCO engineers were not trained to write in these ways nor did they intend to produce literary structures. They used language, though, and narrative: choosing topics and describing aspects of the world in close detail. And as we will see, using and choosing language produces effects, intentional or not, and in the case of a report such as this, the effects radiated widely into the material world. Further, it is only through a techno-poetical examination of *how* engineers imagine social order and a national economy that we can understand the construction of environmental imaginaries and connect engineering work to other imaginative arts.

To construct the dam as an economic allegory, the EBASCO engineers use a narrative that brings the economy into existence as a set of discrete components.[63] They then make the dam into one of those components through juxtaposition and imagery. In the report's second section, the engineers consider economic sectors such as mining, timber, fishing, agriculture, and tourism, each in turn and each connected to the whole only by virtue of being included in the list. One might expect that these sector discussions would detail the projected impact of the dam on the industry in question, but that is not the case. Instead, the report simply notes the location of Turkey's logging industry along the Black Sea coast some 300 miles north of Keban. With respect to the fishing industry, the authors point out that mackerel and Atlantic bonito, both saltwater fishes, represent 70 percent of the yearly catch for an expanding fish-canning industry based in Istanbul, a coastal city 1,000 kilometers from the dam site.

What does the timber industry on the Black Sea or the fishing industry on the Sea of Marmara have to do with building a dam on the Euphrates River in the middle of the Taurus Mountains? Perhaps the new reservoir behind the dam could become a new locus for canning fish? That seemed unlikely, even to the engineers, who noted, "The fish-canning segment of the fishing industry is negligible, amounting to only 2 percent of total production."[64] It could be that the engineers thought the dam's electricity production might revolutionize those industries, but if that were the case they offer little to no evidence. In fact, the dam itself is almost entirely absent in this chapter. Only in the report's appendices, nearly 250 pages later, do we find a faint connection. In a list of expected large electrical loads, the report includes thirty textile plants, sixteen water pumping facilities, and fifteen cement factories, but only one factory processing fish: the existing Meat and Fish Association Refrigeration Plant in Istanbul. The plant's electricity demand was in fact expected to decrease with only a modest increase in consumption.[65] Only two of the 113 prospective large energy consumers had anything to do directly with forest products, and timber as an essential construction material was not explicitly detailed.

EBASCO engineers imagine the economy, then, through a narrative series of economic images. That imagery is disjointed, characterized by disparate pieces, but this disparity has a purpose. If a national economy is merely some industrial sectors within defined political boundaries, then the dam may be placed easily within that schema as yet another economic image. To be sure, such a characterization of the dam's economic position grossly simplifies and abstracts the myriad processes of production and exchange taking place within Turkey's borders. However, the narrative effect of the report's juxtapositions of economic imagery is to produce the dam as a puzzle piece—like the forests and the fish-canning industry—that may be chosen and fitted into an economic panorama. Through this stylized montage of economic activity, the engineers produce not the economy itself, but a coherent-enough fiction in which the dam finds its place.

To describe the dam as an object within that imitation and to connect it to the other industrial sectors required another step. The fitting of the dam puzzle piece comes about by embedding within this economic imagery several discussions of debt. Topically, the second section of the report progresses in this fashion: geography, population, mining, forestry, fishing, debt, agriculture, industry, more debt, communication (roads, railways, airports, et cetera), tourism, the development plan, even more debt. Tellingly, the primary section on debt is labeled "economy," as if a bit of math were all there was, while the other discussions of debt come under "economic factors."[66] The debt sections note how the Turkish economy is "plagued" and "troubled" due to inflation, "unbalanced" due to foreign trade deficits, "mismanaged," and "burdened by a large annual debt payment."[67] Framed through the lens of debt, the dam is not merely a development project but a device of economic reassurance—crafted to appeal to creditors, justify state intervention, and naturalize foreign investment.

Moreover, debt is a summing up of the processes of exchange, a totaling of transactions within a given economic sphere. Money, as Michel Callon argues, is "a trail, a wake, a visible, materializable, traceable trajectory."[68] If so, then debt is a way to calculate the multiple trajectories, the circulation, of funds and trace the ways those monies materialize. As a linguistic device, debt thus becomes a metaphor not only for the economic life and labors of millions of people but also for the forests, fishes, and waters upon which that economic life depends. Its characterization, and the multiple positions it occupies within the engineers' economic panorama, stitches together fishermen and miners with dam builders, farmers, and lumberjacks. Through descriptions and explanations of obligations, the various sectors of the economic panorama may be grasped as a unified whole, linked by a debt-calculating process that entangles social, economic, political, and natural relations. In this framework, EBASCO engineers unified disparate economic activities, transforming the dam from an isolated infrastructure project into an integral component of Turkey's economy—an economy imagined through fragmented yet strategically assembled images.

Through these descriptions of the plague, trouble, and burden of poor economic performance, the Keban Dam moves from a material project being tested for its economic feasibility to a metonym for a specific socioeconomic order. For debt is not merely a way to place the dam piece within the economic puzzle; it can also serve to define and enforce social order. As Begüm Adalet argues, "Biopolitical infrastructures sutured the expansion of global capitalism with projects of developmentalist state formation, consolidating the power of global markets over local producers and Turks over minorities."[69] If the infrastructure is the suture, then the EBASCO engineers show how debt is one of the threads tying the wound together. Trade deficits provided a reason to build the dam: its electricity would reduce Turkey's dependence on imported fossil energy and would enhance industrial capacity to produce for domestic and foreign consumption. Those forms of "debt relief" were meant to flow in a particular direction to particular places as we will see. Debt was also a reason for the United States and other Western nations to fund the dam's construction, a factor that will be discussed in greater detail in the next chapter, though it is useful to note here that monies lent for dam construction largely cycled back to the donor countries as credits for foreign labor and technology. And indebtedness was a selective mechanism for social control, as in the case of Iraq, where loan obligations fastened peasants to relations of socioeconomic domination.[70]

The second allegory of the dam as social order, and thus an instrument of governance, becomes ever more apparent as the engineers' turn toward the provision of electricity. In fact, the most technical chapters on electricity providers and electrical loads are the clearest windows into the meaning of the dam as social order. Here, for the first time, the report uses the terms "social reforms" and "social progress" in relation to the state's goal of developing electricity resources. A reader learns, for instance, that "foreign financial firms" were offered "franchises . . . to construct and operate electric systems" because of the government's pursuit of "social and industrial reforms."[71] The engineers list the five major organizations in the Turkish bureaucracy pursuing electrification and note:

> It is apparent from the number of agencies and organizations involved with electric power that the government recognizes the importance of electric power in furthering its social and economic progress.[72]

Syntactically, this declaration seems to suggest the government's progress rather than that of the society being governed, which one could be forgiven for assuming is the case considering the detail used to describe the five bureaucracies involved in electric power.[73] While the engineers at the start of the report eschewed an interest in anything other than the dam's "power benefits," the feasibility study readily links social progress and the maintenance of the economic order to the regulation and distribution of electricity.

Moreover, though the report presents electrification as a driver of "social progress," its geographical exclusions expose a deeper reality: the Keban Dam's power was meant to serve western cities, while eastern provinces, and particularly Kurdish regions, remained sites of resource extraction rather than development. In a chapter on electrical load forecasts, growth rates, and power stations, the report announces that engineers surveyed electrical resources in "45 of Turkey's 67 provinces," which covered "78% of the total population."[74] Of the forty-five provinces chosen by the engineers for examination, only one—Diyarbakır—lies east of Keban. The engineers ignored several cities in eastern Anatolia, such as Kars, Erzurum, Van, and Mardin. Despite the parliamentary commission asserting that the Keban Dam was "the key project" for the development of Eastern Anatolia, electrical power and the social progress built upon it were not meant to touch those areas.[75] Of the fourteen provinces making up the Eastern Anatolia region, only two were part of the study area, while just half of the Southeastern Anatolia region was included. Moreover, the study area included only five of the seventeen provinces considered to form Northern Kurdistan. According to the report, nearly all the electricity (and, thus, social progress) produced by the dam was to flow from Keban to cities in the western two-thirds of the country, as "all of the Keban power is usable in the Ankara and Istanbul areas..."[76] The socioeconomic order produced by the Keban Dam, then, was one of colonial extraction. The river was to be transformed, and Kurdish villages were to be flooded, to provide electricity to the cities of western Anatolia.

In the penultimate chapter of the report the two allegories—dam as economic component and dam as social order—meet a description of the dam's material construction: its size, shape, features, and positioning. But the engineers' feasibility report does not merely describe the dam; it presents an indisputable future where construction is complete and transformation inevitable. Yet, this engineered certainty is a fiction—one that abstracts time, process, and contingency to create an illusion of control over a world that, in reality, resists such simplifications.

In this section of the report, language and diagram work in concert to conjure the dam into imaginative existence. A topographical map shows the narrows where the dam will be built and the fault lines striating the rock, while another shows all of the geological soundings. Elevation diagrams depict the dam in multiple cross-sections. Several floor plans show the installed generators in the powerhouse. The language accompanying these graphical illustrations is one depicting a future beyond doubt. Unlike the economic and governmental sections of the report, one can grasp the engineers in their element as they narrate the construction of the dam. "The powerhouse will be located on the left bank." "Water discharged through the spillway openings will flow downslope in a concrete lined channel..." "The rockfill shells of the dam will be placed on the firmly consolidated alluvium in the river channel."[77]

There is no should, could, or maybe in these sentences. Such a huge, costly installation could not be justified based on suppositions or assumptions. The engineers bring to life their dam on paper with simple, declarative statements of a future that will be.

The report's certainty extends into the foundations of the earth, as the engineers portray the movement of vast quantities of "material... obtained from quarries in the limestone-marble on both banks" and "from proven deposits on river terraces along the river in the reservoir area." Any "minor faults" in the rock will be filled with concrete according to a "program of grouting." There is a nod to some uncertainty in the modeling of the reservoir, but even here a remedy has been devised in the form of an emergency spillway that "would permit passage of an even greater flood should such ever occur."[78]

The report's treatment of time reinforces its claims to certainty about the future. Rather than discussing construction or process, the American engineers focus instead on the imagined moment when the dam already exists. There is almost no discussion of the building of the dam, the methods of moving the rock, or how much time or labor might be needed to build each component. There are no specific instructions about construction beyond noting the existence of various parts of the installation and where they will go. In this poetics of engineering, nothing happens. Everything will already have happened. The homes and mosques will have flooded, the people by then will have been relocated, the natural world already will be altered. In such a narrative scheme, the human expert's design dominates—not just because the human is the only actor imagined, but because the world is depicted as already transformed, leaving no trace of the processes that led to that outcome.

Many of the indisputable statements about rock and river in this part of the report were, in fact, disputed by natural processes. During the material construction of the dam—not its linguistic one—the powerhouse had to be moved because the consolidated alluvium wasn't very consolidated, and it turned out the faults were not so minor. During the dam's actual operation, the spillway's concrete failed. As anyone involved in construction will tell you, these issues are common. Paper plans cannot capture the real world. There are always "unforeseen circumstances" and "unintended consequences" in any major engineering project; as discussed in the previous chapter, these are often euphemisms for the "unequal distribution of consequences." Things change as construction progresses and new problems emerge. This is not a problem of the plans per se—the dam was indeed built—but of a real world that exists in time.

The plans don't have to be completely true, in other words, because their purpose isn't to exactly represent the world but to produce a believable imitation of its future. The disjuncture between what the report says will happen and what actually happened is a distinction between writing a fiction and building a reality. The feasibility report is a fiction: what engineers state with great certainty will happen, will not happen—in ways large and small. Instead, just as poets use language to transport a reader to a summer's day or a bleak December, to a songlike river or a raging one, the engineers imitate the natural world in language and diagram, evoking through these methods an imaginative awareness of economy, society, rock, and river.

Calculation, simplification, and abstraction all play their part, but even more important to the EBASCO engineers' story of vast transformation is a complex evocation of believability. So, the "what" of engineers' environmental imaginary is a river to be remade by dam height and cubic meters of impounded water, and a landscape devoted to turning electrical turbines and punctuated by tension wires, while the "how" of the imaginary is a poetics of engineering deploying imagery, metaphor, juxtaposition—all the linguistic tools a poet uses—to make a world for the reader to inhabit, a future for them to envision. This is no small imaginative feat, and the complexity of the act, including the historical and natural detail required, has not been adequately recognized.

Recognizing such complexities is a first step to conceiving how a feasibility report such as this one works as a shaper of rock and an organizer of labor and finance. The feasibility report might have been a fiction in its elaboration of how things would come to be, but its imagination of the dam was enough to set in motion millions of dollars, thousands of laborers, and tons of rock. The question becomes—why is this form of imaginative awareness enough to bring all of that about? We might find our way to an answer by asking an allied question: what makes a poem good? The answer depends very much on time and context—in a word, history—but we might consider that the most adulated poems are transporting to place and time—to the Unreal City or to Adana during the French withdrawal or to the banks of a wild, free-flowing Euphrates—and evocative of experience, sensorial, emotional, or otherwise. We might then consider that engineering documents of this sort convince not merely through the weight of facts but by summoning or relying upon hopes, goals, and understandings—about, say, the vitality of the nation-state, the burdens of debt, the socially transformative quality of electricity, the destructiveness of flood, and the impermeability of rock.

Place, time, experience, hopes, goals, and understandings all matter to a techno-poetics, along with matter. The next section will detail how that material world changed and moved, but it is enough here to note that the believability of the engineers' fiction relied on thinking through a real river and canyon. An abstracted and diagrammed and calculated river, to be sure, but a real river eroding real rock to be plugged by concrete and rockfill transported by machines worked by human bodies. The engineering report's fictions ramified and diffracted—an unreal contingent on the real, an imaginary dependent on the material—and then reconfigured a river valley by entangling a fiction within a changing world.

The river valley to be reconfigured in this way was home to thousands of people, and the goals and understandings animating the engineers' techno-poetics were not shared by all. Thus, the feasibility report also works by unimagining Kurdish claims to nationhood and nonhuman claims to existence, by assuming the transfer of people and property rights, and by forgetting cultural connections to place.[79] So, it's not merely that the river has been simplified and its flows made calculable. To borrow from anthropologist Michel-Rolph Trouillot, the report operates by reconfiguring some features of the world as unthinkable.[80] When the engineers used American standards to categorize social and cultural factors as outside the purview of the report—even as they still considered some of those factors—they used professional and institutional power to set the parameters of believability, of genre, of what the world to be constructed would be allowed to contain.

Excluding the displaced from both poetry and techno-poetics—making their experiences and aspirations into an environmental "unimaginary"—has resulted in a strange historical asymmetry: the world knows more about the people who lived in the river valley thousands of years ago than about those who lived there in the twentieth century. As the dam rose, an international effort to catalog and record archaeological sites commenced in the Upper Euphrates region. A group of students from Ankara's Middle East Technical University (METU), under the guidance of Cevat Erder, conducted the initial survey of the Keban Dam Rescue Project, as it was known. Later surveys included teams from three countries—the United States, Britain, and

Germany—and were funded through government grants, including one from the US National Science Foundation.[81] The Turkish and foreign archaeologists hired local labor to help in manual tasks such as digging and laying out a grid. Yet, the archaeological reports rarely mention local workers by name, reducing them to a numerical quantity.[82]

This marginalization extended beyond the archaeological record. In 1968, the inequities underlying the dam project came to public view when 2,000 workers—1,700 local and 300 foreign—went on strike. The workers protested low wages and a stark discrepancy between the pay of Turkish workers and their French and Italian counterparts. Corruption marred not only the construction contracts but also the compensation program for displaced residents, exacerbating the sense of injustice. In response, government officials formed a "yellow union" to undermine the strike, but the outrage it provoked—combined with the solidarity of French and Italian engineers—led to a settlement.[83]

This strike was not simply a disruption in the project's timeline; it challenged the narrative logic of the feasibility report itself. The report imagined an orderly future of transformation and progress, secured by expertise and capital, yet this vision depended on the erasure of dissent, inequality, and resistance. To borrow again from Michel-Rolph Trouillot, the strike was not just a reaction to injustice—it was an eruption of the unthinkable into the record, a refusal of the genre's exclusions and of a fictional claim to inevitability.

As for the displaced, most of what we know comes from government reports, social scientific studies, or oral histories, with the latter providing insights beyond mere statistics.[84] Joseph D. Lombardo undertook interviews with displaced persons, though the 2016 attempted coup in Turkey made his interlocutors skittish and curtailed his work. Even so, the accounts he collected reveal how the ideas and narratives embedded in the poetry and techno-poetics of the 1960s helped naturalize a national imaginary so persistent that, in the 2010s, the displaced would say the dam "has so many benefits for Turkey." Several of those affected by the dam project noted how the dam was "important for Turkey on the level of national development," even as it had severed their local community connections, deprived many of sustainable livelihoods, and instigated unrest in the city of Elazığ, where the arrival of the newly displaced exacerbated social and political tensions.[85] As for the archaeological discoveries, the material remains of the river valley's ancient civilizations are now on display at the Elazığ Archaeology and Ethnography Museum, which opened in 1982 to preserve the rescued artifacts. The living were displaced; the dead were curated.

The consequences of the Keban Dam's construction are inscribed not only in the physical reconfiguration of the Euphrates River but also in the selective memory of history. While the engineering feats and archaeological efforts garnered international attention and acclaim, they did so at the expense of the lived experiences and histories of the valley's displaced inhabitants. Their stories, overshadowed by the multiple poetics of national progress, demonstrate how the mimesis practiced by poets and engineers can obscure the human costs embedded within such monumental projects. The cultural and environmental engineering of the Euphrates River at Keban worked, yes, by simplifying and abstracting, but also through complex cognitive procedures that evoked cultural notions about the infallibility of human design, a nation's role in development, and the human relationship to rivers and rocks, while simultaneously shaping collective memory and forgetting.

Even as the rocks about Keban gave rise to a cultural response—the artistic and technical narratives of the rockscape—the rock itself acted in ways unimagined and perhaps beyond imagining. In advance of dam construction, as we've seen, poets, storytellers, and engineers all conceived of a huge edifice blocking the river. In 1966, construction commenced. Laborers and all manner of machinery quarried, dynamited, and transported rock around the site to build the dam, which was supposedly completed in 1974. At least this is the date on the website of the Turkish State Hydraulic Works.[86] As we will see, the commencement and completion dates of the dam remain uncertain.

1966 is also the year that Cenan Dökmeci published a story in the local Elazığ newspaper connecting the building of the Keban Dam to the conquering of Anatolia by a Turkic empire. At the Battle of Manzikert in 1071, the Seljuq Turks led by Alp Arslan captured the Byzantine Emperor Romanos IV Diogenes. Turkish historiography over the course of the twentieth century had remade the Battle of Manzikert into the watershed moment of Turkish national ascendance in Anatolia. Dökmeci echoed this point of view in his 1966 story, which equates the capture of the Byzantine emperor to the Keban Dam's "capture" of the waters of the Euphrates River. Five years later, with dam construction fully underway, the Turkish Republic celebrated the 900th anniversary of the Battle of Manzikert. As historian Doğan Gürpınar demonstrates, the historiography of the battle became a way for Turkish scholars to enact and justify different ideas of the Turkish nation-state. By the 1970s, Gürpınar argues, "The refashioning of Manzikert . . . marked a perfect moment for the reconciliation of Kemalism and Islamic discourse at a time when the chief enemy of the republic emerged as the communists."[87] In this telling, the Islamization of Anatolia, traced to Alp Arslan's defeat of the Christian Byzantine Empire, and Anatolia's Turkification, attributed to the martial prowess of the Turkish Republic's founder, Mustafa Kemal Atatürk, were part of a single continuum of historical progress for the Turkish people.

Dökmeci's story of the Keban Dam adds another meaning. In addition to a racial, religious, and national destiny within an Anatolian homeland, "Keban as Manzikert" asserted a material, environmental domination. Manzikert offered not only a historiographical opportunity for defining the Turkish nation but also, through a connection to the Keban Dam, a means of asserting dominion over space and landscape in Turkey's oft-contested eastern provinces. Indeed, the fifty lira coins issued for the 1971 anniversary depict Alp Arslan on one side and a map of Turkey on the other. The map shows a star and crescent at the battle site in eastern Anatolia with thirteen arrows pointing north, south, and west to different Anatolian cities.[88] The map appears strikingly similar to engineering diagrams of the Keban hydroelectric station, which show electricity flowing westward along lines that echo the arrows on the coin. In a cultural framing, then, the dam's completion was more than the culmination of local hopes and international engineering; it became a monument to Islamic culture and Turkish national ascendance in Anatolia. By this measure, the cultural construction of the Keban Dam began not in the 1960s but in 1071; the Battle of Manzikert foreshadowed the events to come. And, according to nearly every consultable source on the internet, the dam—and with it, Turkish environmental dominion in Anatolia—was completed 903 years later.

Except that the dam wasn't completed in 1974. By that year, the dam was certainly huge, one of the largest ever built up to that point. Broad and imposing, the Keban Dam was a wall made half of earth and half of concrete filling the canyon opening. Three great concrete spillways stretched like a forked tongue from the wall to touch the river's newly engineered bed. As poems and stories had foretold, the dam had harnessed the menacing river into a great treasure, and it would no longer burst forth on the plains below. For the engineers, though, the dam had yet to fulfill both its economic and social purposes by transforming the stored water into electricity. For that to happen, the reservoir had to be filled so that water could flow through the penstocks and turn the turbines to produce electrical power. The filling of the reservoir began in November 1973.

Only someone acquainted with the dam's original design would have noted at that time how almost the entire construction had been shifted. During construction of the dam in 1967, excavation of the surrounding rock uncovered karstic conditions—the rock intended to hold the dam and its component structures had over time been eroded or dissolved, producing fissures, sinkholes, and caverns. There was nothing to do but move huge sections of the installation. A 110-meter section at the south end of the dam was moved sixty-three meters downstream, creating a dogleg in the dam's sweeping span. The entire powerhouse also had to be moved, as some 5000 meters of drilling had determined that cavities in the rock extended more than forty meters below the surface. The powerhouse, the primary reason for the dam's existence, was rotated and moved 104 meters closer to the river. The penstocks, the massive five-meter diameter tubes that carry water to the turbines, came to rest on an "egg-crate" structure buttressed at the cavern's edges.[89]

In addition, geologists working on the project had identified several faults traversing the rock at the dam's location. But they had noted the dam's positioning in relation to earthquake zones and found the danger minimal. Despite an earthquake in 1804 that had destroyed Malatya and necessitated the town's reconstruction at an entirely new site, buildings of considerable age around the dam site showed little earthquake damage. Since the dam's construction, several earthquakes have tested the structure, but the most concerning have been the earthquakes caused by the dam itself. Beneath the dam and reservoir, air pockets created by eroded rock explode when filled with water. These have caused moderate intensity earthquakes, and the breakdown of the rock could even threaten the stability of the dam itself.[90]

Earthquakes are not the only threat to the dam's integrity. Too much seepage of water below the dam could undermine its base. During regular construction, a barrier known as a grout curtain was installed below the dam's foundation. Workers injected 90,000 tons of cement and slurry through 462,000 meters of holes to produce the grout curtain. And even so, the dam leaked. As the reservoir filled, water began seeping from several areas around the structure. Some leaks were so large they created whirlpool-like vortexes on the surface of the slowly rising reservoir; at least thirty were observed along its edges. The canyon walls suddenly seemed a lot less stable. Powerful vortexes created waves that further eroded the rocky confines of the reservoir. The nearby Keban Creek swelled to more than three times its typical volume, becoming the Keban River. Springs began to emerge along the hillside downstream of the dam and in another drainage, Zeryan Stream.

The leaks grew worse as the reservoir reached its peak volume. The pressure of a 190-meter-deep reservoir helped the waters of the Euphrates River find permeable layers of rock and several pathways around the dam. In the spring of 1976, the water level reached a high of 844 meters after heavy rains. On May 11, a vortex two meters in diameter revealed a massive cavern, known as Petek, which had been uncovered during construction. Now, however, the leak had grown so large that it threatened the dam. After the spring rains, water levels were lowered and engineers descended into the cavern. They paddled about the flooded grotto in inflatable boats, inspecting the vast hole in the rock. To stem leakage from the reservoir, the Turkish State Hydraulic Works filled Petek cavern with 2 million cubic meters of sand, gravel, and crushed stone, and installed a 17,000 cubic meter concrete cap. This wasn't the only huge hole in the rock, however. For nearly eighteen months, from November 1977 until March 1979, Turkish engineers worked to fill another cavern, known as Yengeç (Crab), with nearly 600 million cubic meters of material.[91]

Dealing with the leaky reservoir nearly doubled the cost of the dam project. In other words, the leaks cost as much to fix as the dam that caused them. In a sense, engineers built two dams, one above ground and another below ground, and despite the Keban Dam's immense size, the unseen dam below ground is larger (and wasn't subjected to a feasibility assessment). The fact that the dam built to block up the rock is larger than the one built to block up the water underscores the problems of ecological, cultural, and historical premises that assert control over nature and presume the possibility of completion and totality in environmental engineering. The underground dam took five more years to build; only then could the installation be considered safe from major leaks. So, while the Keban Dam was ostensibly completed in 1974, the filling of the Yengeç cavern—the underground dam—was not finished until March 1979. At what point, if ever, might we consider the dam "finished"? At what point was it begun? The dam perhaps does more than trap silt or leak water—it also traps and leaks time.[92]

Just as the water of the Euphrates River moved through the rocky world below and around the dam, so too does the rock itself around the Keban Dam continue to move. As huge as it is, the dam may not ever be truly "done," at least if that word means complete control of the landscape. The rocks themselves have their own nonhuman unintention, disrupting the human intentionality embodied in the narrative rockscape in ways both large and small. On a grand scale, there is the movement of continents. At the dam's location in eastern Anatolia, the continental plate containing Africa and Arabia collides with the one containing Eurasia. The continental crust around Keban is eight kilometers thicker than the crust near Ankara in central Anatolia. The crust grows even thinner at the Aegean Sea, where the seafloor subducts, pulling Anatolia to the west. At the surface Anatolia's movement westward registers as earthquakes. Below ground, the shaking can cause liquefaction, when rock turns from a solid state to something akin to a liquid, and result in slippage in the dam's foundation. In the fabled land of the two rivers, rock can turn into water, no miracle necessary. While earthquakes in the twenty-first century have so far done only minor damage to the dams on the Euphrates River, the movement of the earth has been enough to level cities using the dams' electricity and to open gaping fissures in the beds of the dams' reservoirs.[93]

Along with caverns in the rock, leakage, and earthquakes, droughts in the twenty-first century have caused the Keban Dam reservoir to recede, baring parts of the rocky canyons to view once again and returning land to other uses. The newly available land has led nearby farmers to reclaim fields where they grow beans, melons, and chickpeas.[94] A reservoir that at one time displaced people is now itself being repositioned by climate change, driven in part by the burning of coal, a sedimentary rock.

If we include rocks in our thinking about agency and the dam, we must reconsider the words and concepts we use to describe this infrastructure. The hydroelectric power station produces electricity and the wall of earth and concrete impounds water, but the Keban Dam is more aptly described as temporarily operational than as "complete." If change is an inevitable pattern for nature, then so too is it for the dam itself. Engineering tropes announce what will be and envision a mastery of the rocks but from supposed start to supposed finish this is indeed an imaginary, one conjured through both poetry and a poetics of engineering, that will always be confounded by the rocks themselves.

Four

RESERVOIR

The reservoirs created by the dams on the Tigris and Euphrates Rivers are massive, akin to inland seas. While al-Tabqa Dam in Syria is a giant construction—200 feet high and 2.8 miles long—the dam's footprint has had a relatively small impact on the landscape, because behind the massive wall of rock sits an even more massive reservoir covering hundreds of square miles. Like other languages, neither Arabic nor Turkish make a strong linguistic distinction between these huge man-made bodies of water and those formed by natural processes. The reservoir behind the Keban Dam is known as the Keban Baraj Gölü, or the Keban Dam Lake. Meanwhile, in Arabic, the body of water behind al-Tabqa Dam is known as Buhayrat al-Assad, or the Assad Lake. On English-language maps, both are rendered as lakes.

Collecting so much water in one place has myriad effects on local and global ecosystems. Some of these changes are visible in the landscape, as vast swathes of the desert become irrigated lands. Others are less visible. Immense quantities of water in these reservoirs regularly turn into a gas. In arid regions, the combination of heat and dry air produces high evaporation rates. The rates in the basin vary, with those in Syria and Iraq much higher than in Turkey. Still, the vastness of Turkish reservoirs means that on average two cubic kilometers of water (two billion cubic meters) evaporates into the atmosphere each year. In Syria, the reservoir behind al-Tabqa Dam loses over one cubic kilometer of water to evaporation, with estimates ranging as high as 1.3 billion cubic meters.[1] The substantial water loss from Lake Assad is due to its shallow depth. Though the reservoir is three quarters the area of the one behind Turkey's Atatürk Dam, Lake Assad contains less than one-fifth of the water. A shallower reservoir heats up more quickly and reaches a higher temperature than a deeper reservoir, driving a greater evaporation rate. Considering Syria's low annual per capita water consumption of 300 cubic meters, just the water evaporating from Lake Assad could support average consumption for an additional 4.3 million Syrian citizens.

Such high rates of evaporation create climatic changes in the areas around the reservoir. Scientific studies conducted in North America in regions similar to those of the Tigris-Euphrates River Basin show how increased evaporation can lead to higher humidity, temperature changes, and, eventually, to new patterns of rainfall.[2] Though long-term local climate data is difficult to come by, studies suggest that humidity levels have increased around the Keban Dam Lake, producing some of the same effects.[3] Moreover, the size of these reservoirs and the way they drastically alter land use can have wide-reaching ecological effects.[4] Among these is the emission of greenhouse gases. Reservoirs contribute to greenhouse gas emissions in several ways. By inundating vast areas of carbon-rich organic material, they trigger decomposition processes that emit climate-warming gases such as carbon dioxide, methane, and nitrous oxide. While deep reservoirs may act as sinks and keep those gases underwater, those managed according to human needs often experience significant fluctuations in water level. Drawdowns increase methane emissions. Intensive land use, deforestation, and other human processes around the reservoir also increase the delivery of organic material, further escalating decomposition and the release of greenhouse gases. In some cases, a reservoir may release as much or more climate-warming gases as a thermal power plant, negating the supposed climate benefit of hydroelectricity production.[5]

Natural lakes undergo many of the same physical processes as reservoirs. While they do not inundate entirely new landscapes, they still receive inflows of organic material that decomposes over time. Their water levels fluctuate, though typically less dramatically than those of reservoirs. Lake water may also be diverted for irrigation, reshaping the surrounding terrain. Yet the bodies of water behind the Keban and al-Tabqa Dams are not natural formations, despite what their names suggest. They are artificial, reliant on human engineering of the rivers. Over time, more and more of these reservoirs were created along the Tigris and Euphrates Rivers, and as their presence and effects proliferated, the rivers themselves ceased to behave as they once had. So, what we call lakes are engineered bodies; what we call rivers are controlled and diverted flows. Perhaps the rivers too are now misnamed.

If the rivers are no longer rivers, it is not only engineers and ecologists who have noticed. "There is no river . . . the river is gone." So concludes Syrian filmmaker Omar Amiralay's final film, *A Flood in Ba'th Country*.[6] The camera in these last moments focuses on the broad expanse of water that was once the Euphrates River Valley in north central Syria—the opposite bank is too far away to be seen. The speaker is a member of one of the families displaced by the reservoir. They are known as the *maghmurun*, the submerged, a designation referring more to their homes and the culture of place than to their physical bodies. Yet, who better to say whether the river is still there?

The disappearance of the river and the transformation of residents into *maghmurun* began in 1973. That year, engineers in Syria closed the gates at al-Tabqa, a few months before Turkish engineers began filling the reservoir behind the Keban Dam. If the term "river" connotes a free-flowing watercourse, then 1973 marked the beginning of the Euphrates's transformation into something other than a river, dammed as it was by these two huge concrete and earthen plugs (see Figures 14 and 15). Over the succeeding decades, all three states in the river basin built additional facilities along the Euphrates. By the mid-1990s, the river had become, more accurately, a long series of reservoirs. A traveler with exceptionally long legs could step from one dam's reservoir to another, walking down a set of watery steps from the Taurus Mountains to the plains and deserts, progressing from one long, sinuous "lake" to another. As a result, an ecological history that began in the first decade of the twentieth century with flowing water ends in this chapter with stilled water, the reservoir.

The Tigris River is slowly coming to resemble its neighbor and partner. In 2019, the Turkish government inaugurated the huge Ilısu Dam on the Tigris River in southeastern Turkey and began to fill the dam's reservoir. Now, the Tigris for much of its length appears like the Euphrates, more as a series of reservoirs than a river. The high canyon walls leading to the dam have become

FIGURE 14. Landsat imagery of the Euphrates River in eastern Anatolia in 1973 and 1977, before and after the filling of the reservoir behind the Keban Dam. Courtesy of the U.S. Geological Survey.

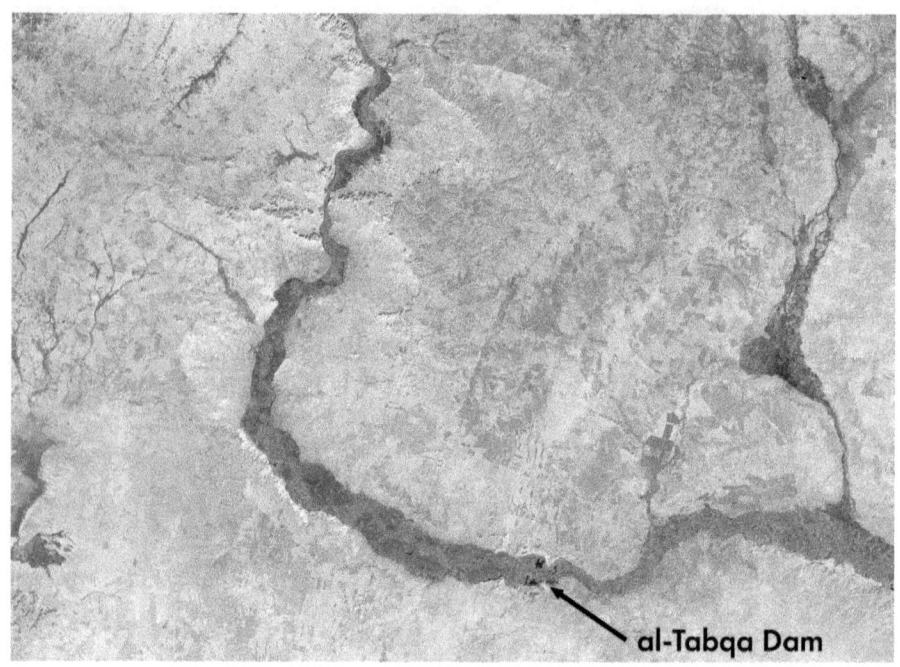
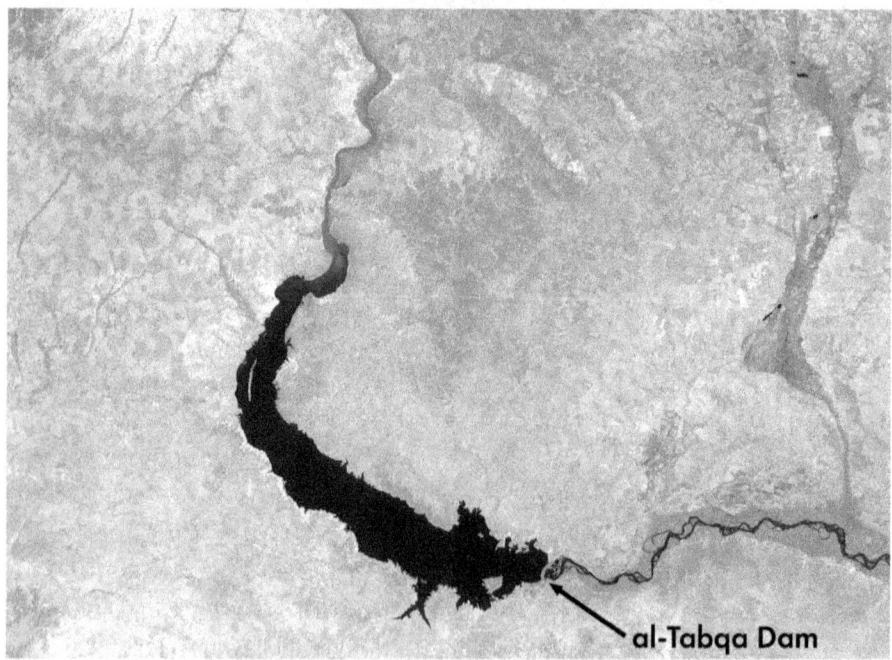

FIGURE 15. Landsat imagery of the Euphrates River in northern Syria in 1973 and 1976, before and after the filling of the reservoir behind al-Tabqa Dam. Courtesy of the U.S. Geological Survey.

a shoreline. The water's narrow rush toward the dry flatlands beyond Mardin eddy instead behind a dam, a good portion of the flow traveling first through turbines before continuing the journey south. The communities along the river's edge, including ancient Hasankeyf with its Abbasid bridge and other historical treasures, have been inundated, lost to residents and terrestrial visitors alike.[7]

How do we tell the history of a river that is no longer a river? It seems an offense to all the Euphrates was in the millennia prior to 1973 to call what it has become since "a river." This massive ecological shift in the rivers' regime heralds a historical change in view, making this chapter somewhat different than the preceding ones. To consider the reservoir's ecological, political, and social effects means contending not with the history of a natural process, such as a seasonal flood or the formation of limestone, and its growing entanglement with human social processes, but with a nature so altered, so entangled, that attempts at differentiation break down. With the construction of so many dams and the diversion of so much of the rivers' water for power, irrigation, and other uses, ready distinctions between human design and the unintentional forces of nature grow murkier, submerged if you will beneath the rising waters of the dammed rivers. The danger, then, is in disentanglement or simplification, in supposing the dam and reservoir as only a human construction and losing sight of the natural processes that continue in, around, and through it.[8]

As the scale of human transformation of the rivers expanded and the Tigris and Euphrates Rivers were turned more and more into reservoirs, it becomes tempting to think of the basin as a compound of nature and technology, as an "organic machine," akin to Richard White's conceptualization of the Columbia River's history.[9] The dams, reservoirs, and related infrastructure appear as a kind of cyborg in Donna Haraway's usage, sitting at that "leaky distinction... between animal-human (organism) and machine."[10] Such ideas of technological transformation are a kind of gestalt, a way to understand the rivers as an organized whole—an "assemblage"—that cannot be reduced merely to the sum of its parts. It may be useful, though, to note the quality or valence of the gestalt: the concepts used to describe the transformation of rivers have tended to emphasize forms of intention, agency, and control that the previous three chapters of this book have worked to question. As historian Sarah Pritchard writes, "nonhuman nature may be profoundly mediated and constructed, both literally and metaphorically, but it is not wholly reducible to culture."[11] If the installations meant to facilitate irrigation, produce electricity, and reduce flooding instead salinize the land, dry up or silt up, or, worse, collapse, then are concepts like assemble and machine the best ones to use? Further, as Pritchard argues, analyzing the reshaping of a river means "open[ing] two black boxes, nature and technology, *simultaneously* without resorting to determinism or realism..."[12] Confronted with such problematic concepts, it has been easier in many cases to narrate histories of progress overcoming natural constraints in order to avoid stories of decline or as a hedge against determinism. However, this has meant situating environmental crises as "unintended consequences" (a product of agency subverted), rather than as the malfunction, muddle, or snarl emerging from specific processes of entanglement. Yet, those latter terms often seem more apt for describing the mix of intention and unintention at work in damming rivers.

Harnessing the waters of the Tigris and Euphrates ultimately made them into something else—flowing and unflowing, trapped and free, directed and undirected—with new physical properties and sociocultural meanings not so easily conceived of as rivers. That "something else" ought not to be displaced by efforts to naturalize—reservoirs called lakes—or by concepts that in some way preserve fictions of human control. Memories of the river as it once existed—its ghost—and rumors about its eventual return have their own histories, of course, but it is also necessary to contend with what a century's worth of entanglement brought about, such that a man who once lived along the Euphrates claims: "There is no river... the river is gone."

Thus, this chapter contends with the ramifications of replacement, with the ways entanglement disappeared the rivers into reservoirs, turbines, irrigation canals, revolutionary myths, and civil conflicts. In this telling, the reservoirs, and particularly the one behind al-Tabqa Dam in Syria, became vapor (the spectral manifestation of the river, perhaps?) and produced power, but they also became places, symbols, geopolitical tools, and critiques. The reservoirs, then, are the outcome of a transformation beyond the organic and mechanical into wider realms: the martial, the revolutionary, the emotional, the imaginative. Further, by the end of the twentieth century, each reservoir could not be thought of in the singular, as merely a lake or a repository of potential energy. The scale of transformation requires a related scale of description and narration. To borrow some collective nouns from nonhuman river dwellers, a bevy, a gaggle, a siege of reservoirs now haunts the Tigris and Euphrates Rivers. And these reservoirs are both the product and producer of a wide array of other material, environmental, political, and social changes.

At the center of this haunting was a dream, an entangled series of illusions or even hallucinations, that fortified and advanced the transformation of the rivers into reservoirs. And at the center of that dream was an understanding of history.

Discussions of engineering the rivers relied on emphasizing one period of history above all others—that of ancient Mesopotamia. The Cradle of Civilization. The Fertile Crescent. The genesis of agriculture. The birthplace of cities. This emphasis on ancient Mesopotamia served several purposes. For one, processes of change, natural or otherwise, could be ignored or elided in favor of a straight line drawn from ancient wonders to modern infrastructures. But even more important was the way a focus on the rise of civilization in Mesopotamia's deeper past underpinned a dream of Mesopotamian civilization in the present and future. What is remarkable is how this dreaming of twentieth-century (CE) civilization consistently happened through the rivers. While civilization may be construed differently in an empire, colony, mandate, monarchy, terroristic utopia, or an authoritarian socialist republic, in nearly all cases the logics of civilizational progress (what civilizations do) have in each sociopolitical manifestation found their way back to dams and reservoirs.

Hence, the process of reservoirization draws together a multitude of histories into an entangled whole—the amalgam of human technology and nonhuman nature, yes, but also economic concepts of progress, cultural definitions of civilization, and the violence of controlling people and determining ways of life. Moreover, the sedentarization[13] of the Tigris and Euphrates Rivers required more than just physical reservoirs; it also depended on reservoirs of diplomatic support, financial and institutional know-how, and military assistance and weaponry. Finally, once constructed, this multitudinous reservoir, brimming with accumulated water and meaning, obtained both a physical and cultural presence. The reservoirs became a contested terrain, supplying not just water vapor and pressure but also propaganda and opprobrium.

What is more, the human actors involved in making the reservoirs were not shy about their intentions. As we will see, they spoke often about civilization before, during, and after the filling of reservoirs. Invoking civilization in relation to dams has sometimes been construed as mere justification for construction. But to treat civilizational claims as rhetorical cover implies that the "real reason" for building massive infrastructure was something more utilitarian, like electricity production, or more insidious, such as the extension of capitalist relations in agriculture, and that these aims needed to be ennobled by lofty, insincere rhetoric. To be sure, the huge installations on the Euphrates and Tigris Rivers achieved ends both banal and troubling. Yet, this way of interpreting civilizational discourse offers a peculiar conceptual luxury—that the solution to socially and ecologically destructive infrastructure is simply a "better" version of the same: an economic or technological "fix" that might achieve similar goals more cleanly. It implies that industrial civilization could be something else, look some other way, if only its "real" components—not its ideas, which are dismissed as window-dressing—could be found or fashioned through different means.

In this chapter, however, I treat concepts of civilization not merely as justifications but as productions—as a kind of world building where meanings and materialities co-emerge. This shifts civilization from the realm of rhetorical afterthought to that of paradigmatic measure and goal. In other words, civilization is not simply a slogan affixed to dams and reservoirs, or a concept hovering above the world, but a method for fitting seemingly disparate things—bodies, landscapes, technologies, ethics—together into particular arrangements or ways of life.

Advocates of environmental engineering, then, dream civilization through rivers. This sort of dreaming elicits a few binaries—the civilized and the uncivilized, most obviously, but also the watered and the desiccated, the connected and the disconnected, the powered and the powerless—that articulate the direction of cultural and material change, together. We may then add to our understanding of the genre of technical documents and engineering plans; they are descriptions of civilizational dreams.

The goal of building massive dams and reservoirs was never merely the transmission of electrons or the shifting of relations of production—these edifices in the Tigris-Euphrates River Basin were meant as actual, physical repositories and manifestations of civilization. In other words, state officials intended to build civilization through these projects, and rhetoric along these lines was more than a political gloss or rationale. After all, it would be absurd to call the Great Pyramid of Cheops or the Taj Mahal, "a couple of big tombs." By the same logic it would be absurd to call al-Tabqa Dam, "a big power plant." Resistance to the massive reengineering of landscapes, which has grown over time, helps to illustrate the greater valence of these projects, as does the response to that resistance. In the last decade of the twentieth century, proponents of dam building revived an earlier claim that the great reservoirs would "regenerate" lost civilizations of the past—as if through industrial magic, past glory could become a contemporary triumph and ecological crisis could become progress (for the earlier history of regeneration, see Chapter 1).

Throughout history there have been various and diverse forms of civilizational dreaming, but they all rely on linkages between human society and nonhuman nature, connections that not only constitute what civilization is (control over both), but also what it looks like (in the twentieth century the exertion of industrial power over the landscape).[14] The reservoir is the purpose of the dam—without the water behind it, the dam is a pile of rocks with some holes for plumbing. The dam is the purpose of civilization—without the dam behind it, civilization is a people living in supposed disorder, distress, disturbance, dishonor, or despotism. In critical ways, then, the question becomes less about the kinds of civilizations that built dams and reservoirs and more about how processes of entangling human society and nonhuman nature made huge dams seem essential to civilization. While broad theories of change—high modernism, empire, or concrete revolutions—offer important insight and imply "universal" themes or even a teleology by which these things come to appear obvious to civilization, dreaming—civilizational or otherwise—is always subject to complexity and contingency, and all the troublesome details of place and time.

To illustrate the complexity of civilizational dreaming through rivers, the following chapter relates three interrelated histories that tell of the dreamer, the dreaming, and the dreamt. First, I explore the life of one of the dreamers: Süleyman Demirel, the Turkish engineer, prime minister, and president. Demirel earned his degree and began practicing as an engineer during World War II. In the 1950s Demirel worked in the Turkish bureaucracy and came to be known as the "King of Dams" (*Barajlar Kralı*). By the mid-1960s, he had ascended to the premiership and was present at the placement of the Keban Dam's foundation stone. Demirel's writings and public statements are filled with civilizational rhetoric. He used the concept throughout his career, in and out of public office, to describe and structure the work of the central state and its bureaucratic organs. The first section on the dreamer traces the evolution of his thoughts, analyzing how the dam and reservoir were meant to function as mechanisms of civilizational transformation.

The second part of this chapter explores the act of dreaming—the process and components involved in financing and constructing the dams. In the early 1960s, Demirel twice traveled to the United States to learn how to make a water reservoir using other kinds of reservoirs. This section considers a few of those repositories, exploring the diplomatic and financial reservoirs that facilitated construction of the first large dams on the Euphrates River. Here, through a comparison of cases and the various processes and elements of the dream, we can observe civilizational dreaming across distinct ideologies and forms of state. In the middle of the Cold War, Western powers and construction companies worked on the Keban Dam in Turkey while Eastern bloc countries and companies financed and constructed al-Tabqa Dam in Syria. Despite the alleged profound differences between the so-called east and west, between a capitalist politico-economic order and a communist one, there is a striking resemblance between the two projects.[15]

The third and final section of the chapter turns to the dreamt: the time after the filling of the reservoirs, when artists began to reckon with what had been lost and what remains. It examines how a Syrian filmmaker and a novelist repurposed the reservoir as a site of critique, using creative media to unsettle the civilizational rhetoric that had produced the dam and its displacement of residents. The Syrian filmmaker Omar Amiralay made a trilogy of documentaries about al-Tabqa Dam spanning his entire career—from his first film in 1970 to his last work in 2003.[16] The trilogy charts both his growth as an artist and his deepening disillusionment with the the Baʿth Party's vision of socialist progress. This section also considers the work of Syrian novelist ʿAbd al-Salam al-ʿUjayli, whose 1979 book *The Submerged (Al-Maghmūrūn)* tells the story of Nada and her driver ʿUthman, two government workers whose brief love affair unravels as the rising waters of Lake Assad swallow their future.[17] In these artistic works, what was dreamt—a socialist civilization invigorated by a dam and reservoir—appears full of promise on the surface, but ultimately hollow and lifeless, as empty of meaning as the reservoir was full of water.

Darkness, deprivation, and drought. A middle-aged Süleyman Demirel dwelt on these factors when he reflected on his young life in and around the town of Isparta in the Lakes region of southwestern Anatolia. For Demirel, these three factors confirmed that civilization was not inherited, but built—through engineering the environment, raising infrastructure, and mastering nature. His early life in rural Anatolia, shaped by scarcity and isolation, instilled in him a conviction that civilization could only be achieved by overcoming the limitations of geography and history. As this narrative will show, Demirel came to define civilization as "what you have," and primarily as the expansion of electricity, water control, and industry—forces that, in his view, could elevate Turkey from rural poverty to urban prosperity. Demirel tempered some of these views later in his life but ultimately defined civilization as technical, and technology and infrastructure as not simply a means to an end but as an end in itself.

Demirel was born in a small village, İslamköy, in the Isparta province on November 1, 1924, about three months after the Treaty of Lausanne, the international instrument that recognized the new Turkish Republic, came into force. His was the first "Republican" generation, and the top-down political and cultural transformations wrought by Mustafa Kemal Atatürk and his Kemalists shaped Demirel's youth. This fact was not lost on the older Demirel, who noted how the era of his upbringing and early education laid the foundation for the Republic's later challenges. To Demirel, "[u]nless you know the 1930s, it is not possible to understand Turkey."[18] During that decade the young man witnessed the arrival of electrification and a new rail line, while also experiencing the privations of post-war economic distress compounded by a drought.

The drought began in the Lakes region in 1934, five years into the economic depression, and underscored the precariousness of what Demirel called the "closed economy and closed life [*kapalı ekonomi ve kapalı hayatı*]" of the rural countryside. According to Demirel, villagers at that time had little interest in urban life and "had little to gain from the city," although the city's cultural and infrastructural connections enchanted him. In some ways, Demirel's notion of the family farm conveyed the autarkist national economy that the central government espoused in the 1930s: self-sufficiency was a key value and often a necessity. The drought, however, revealed the limits of such self-sufficiency. In Demirel's village, young and old prayed for rain, and the prayers did little good. Demirel later recalled how the ears of grain dried before they were full: "It is a sign of despair. It is the suffering of the peasant that winter that pushed me into the struggle for civilization [*medeniyetçilik*]."[19]

Darkness was also a key theme for Demirel as he reflected on his youth. In 1935, he left his small village school to attend a secondary school in the provincial capital of Isparta. He made the twelve-kilometer trek from family farm to the town on the back of a donkey. Unlike the primitive conditions of his primary school, which at first occupied the upper floor of a barn and where students were asked to bring bundles of wood each morning for heat, Isparta benefited from electrification. The young Demirel watched the construction process and was inspired by the lights flickering on in the city each night, while his own village was illuminated only by the stars. Within a year of Demirel's arrival, the city obtained its first rail connection to the rest of the country. The rail line brought Demirel a glimpse of the Turkish Republic's elite. İsmet İnönü, hero of the Independence War and prime minister for most of Mustafa Kemal's presidency, opened the new rail station near Isparta in 1936. The new station included a plaza that became a gathering spot [*mesire*] for urban dwellers.[20]

The contrasts between city and village—traveling by rail or by donkey, gathering in public plazas or in mosques and homes, engaging with money and trade or merely subsisting—shaped Demirel's vision of what modern civilization ought to be. Civilization was the opposite of rural poverty, a condition that ought "to be rebelled against [*karşı isyan etmiş*]." Moreover, in these memories of his education, Demirel clearly distinguished between the educated and the rural peasant, a distinction he placed on a continuum of enlightenment. He recalled fighting in secondary school over the question of whether "we" (the educated) could bring "these people [the rural peasantry] to civilization."[21] Urban life, in contrast, and even impoverished urban life, seemed to Demirel a natural outcome of the development process.

To be sure, such a contrast between urban and rural, a place of civilization and a place of want, had antecedents. The distinction between urban and rural has a long history in the Ottoman Empire. The Kemalists who guided the young Republic espoused an urban-rural binary and were hardly shy about Turkey's adherence to Western models.[22] However, it is important to consider Demirel's particular intellectual and political genealogy. The Kemalists' cultural formation had occurred within the Ottoman Empire; Demirel's occurred squarely within the Republic. When Demirel entered primary school in 1930, his was the first class to be taught Turkish in the Latin alphabet. The students ahead of him had learned the old Arabic script and had to be retaught. When later in life Demirel took up a trademark fedora on the campaign trail, he didn't have to first put down a fez. While the Kemalist break with the Ottoman period was most assuredly not absolute, Demirel's generation contended with a different set of conditions and articulated a vision for Turkey's future in their own way.

The concept of civilization Demirel espouses, then, is not one of reaction to a previous historical model or one freighted with past contestation. The Westernizing impulse of Kemalism was fully underway when Demirel began his education, and it was clear to him, even from his middle school principal, that "Europe" was the goal. By the 1930s, Kemalism had in sometimes brutal fashion enforced the ascendance of "western" over "eastern" concepts and embraced a reductive view of national aspiration. When Demirel looked back to his upbringing with a hagiographic impulse, he could narrate a path of development—his own and his country's—with no sharp counter-revolution. For Demirel, his biography demonstrated that the adoption of Western modes and models was not fraught, not manifestly incompatible with other political or revolutionary goals, and not subject to revision or even to an understandable resistance. Indeed, the civilizational struggle Demirel most reflexively confronted when he later ascended to the political firmament was the one between Soviet communism and liberal capitalism, both ideologies largely of European origin, and, as we will see, the distinction in many ways was often little more than rhetorical.

The young Demirel personally benefited from the new republic's emphasis on education, doing well enough to earn state-sponsored grants to attend high schools in Muğla and then in Afyon. By the time he took his university entrance exams, Demirel had seen well beyond his small village. In 1941 he crossed the Bosporus into a city much closer to the European war than the towns he'd visited while in high school. He'd been accepted into the Higher School of Engineering, which would be renamed the Istanbul Technical University in 1945. The new nomenclature was illuminating: a technical education was a necessity, Demirel was beginning to see, for bringing about a technological civilization. In the coming years the university would train several young men who would later play significant roles in Turkish politics, including Necmettin Erbakan, Turgut Özal, and his brother Korkut. All four men studied engineering, and, when they eventually turned toward politics, each joined or founded right-wing parties.[23]

Turkish sociologist Nilüfer Göle notes Demirel as a key figure in this "political ascendancy of engineers," and traces a synthesis between "technocratic consciousness" and a "revived Islamist movement."[24] Göle shows how, despite efforts to secularize the public sphere, Islamic practices found their way into technical education institutions. Turgut Özal, an ally of Demirel who served as Turkey's prime minister and eighth president, notes how he learned of Islam and connected with other students in breaking the university's strictures around its practice: "The call to prayer was in Turkish; we used to recite the prayer in Arabic, but we recited it secretly. We used to pray in a hut in the Gümüşsuyu building. It was also hidden. It was quite dangerous for such students to perform a prayer."[25] Demirel himself referenced Islamic ideals obliquely, as "spiritual characteristics [*manevî hasletler*]" or "morality [*ahlak*]," and often suggested that these directed economic development.[26] Of course, the implication in Demirel's usage is that those who resisted the state's development efforts also opposed a revival of Islamic ideals.

Demirel's technical training earned him a place in the Turkish state's growing bureaucracy and a way to learn more about American notions of economic development. In 1949, he began working at the Electric Power Resources Survey and Development Administration (*Elektrik İşleri Etüt İdaresi Genel Müdürlüğü* or EİE), the government organ most involved in electrification projects. As part of his work there Demirel escorted delegations of American engineers sent to study and assess Turkey's hydroelectric potential. His ingenuity in ferrying US government personnel across the country attracted the attention of Demirel's superiors. In 1950, he became the first engineer sent by the Turkish government to study in the United States. While in America, Demirel worked for nine months with the Bureau of Reclamation on water and energy projects. This visit had a significant impact on Demirel, who recalled spending three days gazing at the Hoover Dam.[27]

American influence extended well beyond straightforward technical knowledge. Demirel noted how his time in the United States fueled a new political ambition: "Water engineering's peak had been attained in the USA's western states. It was my primary goal to find an opportunity to realize in my country some of the things that I learned and witnessed in those territories."[28]

When Demirel returned to Turkey, he was posted to the Seyhan Dam near Adana, a project planned by Turkish engineers in coordination with the American firm Morrison Knudsen. The firm's cultural capital was at its height in the 1950s with a feature in *Time* magazine on its chief executive, Harry Winford Morrison, declaring him "the driving force behind the Hoover Dam," and "the man who has done more than anyone else to change the face of the earth."[29] Demirel's connections to Morrison Knudsen proved useful later in life. When forced out of government by the 1960 coup, Demirel entered private practice as the American company's representative. When Turkey-US relations soured, Demirel's detractors nicknamed him "Morrison Süleyman."[30]

Shaped both by Turkey's Westernization efforts and American engineering influences, Demirel began at this time to forge a vision of economic development that fused technical expertise, political ambition, and Islamic moral undertones, an ideological concoction that helped him serve as a technocratic "bridge" between Turkey and the United States. Recalled to Ankara in 1951 to work as a project engineer, Demirel soon became the head of the Dams Administration (*Barajlar Dairesi*) in the newly created State Hydraulic Works (*Devlet Su İşleri* or DSİ). Turkish officials deliberately modeled the new DSİ on the US Bureau of Reclamation; Demirel, one of the young engineers who had undergone training at the Bureau's office in Denver, Colorado, was a natural choice for a leadership position. Demirel worked for the next couple of years helping to manage the rapid expansion of Turkish dam-building, overseeing several dams and hydroelectric projects. It was not long, however, before he was sent back to the United States.[31]

In 1953, a group of American businessmen established the Eisenhower Exchange Fellowships and Süleyman Demirel was a member of the inaugural group. For much of 1954, Demirel and his wife toured the United States, investigating how various US organizations operated. When asked about his visit, Demirel replied:

> During my visits, I saw how natural resources were used, how investments that were made in this area were directed, and how money was allocated. Truly, I saw a lot of things out there. My enthusiasm, after all, was civilization [*medeniyet*]. Civilization, well; I sought it, I sought its elements...[32]

One of Demirel's biographers, Hulusi Turgut, notes how Demirel's second visit to the United States was about placing "a photograph of the works of civilization, so to say, into his mind." Put differently, his tour equipped him with several of the imaginative components of a dream. This photographic record would help provide the basis for the "concrete revolution" to sweep Turkey; Turgut credits Demirel's second trip to the United States as the inspiration for Demirel's support of the Southeast Anatolia Project and its more than twenty dams.

The elements of civilization Demirel identified in 1953 came into clearer focus over time as, later in the 1950s, he began to write and speak in more sophisticated terms. In these works, several facets of his biography come into play. His earlier identification of rural living with want or lack continued, as Demirel contemplated civilization through a juxtaposition of urban and rural, with civilization associated with the former. Meanwhile, the problems and solutions he pinpointed reflected Demirel's upbringing in a milieu that clearly aspired to replicate European models. Once equipped with a technical education and bureaucratic leadership, though, a peculiarly American version of how to develop agriculture, tame rivers, and conquer aridity, while suppressing any resistance to those aims, shaped the vision of civilization that Demirel later expounded as a politician.

Such a combination of expertise and political acumen in Demirel's story may lend itself to a Weberian notion of bureaucratic leadership: Demirel as a member of a "managerial elite."[33] This conception of Demirel's sociological position focuses more on social implications than environmental ones. It dwells more on his political shrewdness than the context and specificities of his technical expertise or how these two aspects worked in concert. The histories that combined in Demirel's life and worldview entailed specific forms of entanglement of human society with nonhuman nature. In effect, Demirel's view of village life is one little removed from natural conditions, with its "closed economy and closed life." Rebellion against such conditions was justified to achieve an urban civilization modeled on great European metropolises, many of which, it should be noted, were centers of imperial and national power. Finally, as Donald Worster has pointed out, the conquest of the western United States involved feats of engineering, the amassing of capital, social violence and legal regulation, some of which had hardly been seen thus far in Turkey as an aspirational new Europe.[34] What we see in Demirel's biography, then, is how a layering of different histories produced a dream of technical civilization, meaning one defined more by infrastructural interventions in the environment than by the military, religious, or scientific factors that were ascendant in earlier Turkish and Ottoman history.[35] Rather than merely excelling at technocratic management, Demirel refined and developed a Turkish concept of technical civilization over the course of a career during which he reached the heights of political power.

In 1955, fellow engineer and Minister of Public Works Kemal Zeytinoğlu appointed Süleyman Demirel to the position of General Director of the State Hydraulic Works. Demirel was a mere thirty years old and now in charge of managing the DSİ's entire portfolio of public works.[36] He held the post for five years until the military coup of May 27, 1960. During that period, Demirel applied many of the lessons he had learned during his visits to the United States. He also had the opportunity to travel throughout Turkey and "get to know Turkey's geography, natural resources and people." Unsurprisingly, thirst dominates Demirel's recollections of those travels, particularly in relation to southeastern Anatolia. Of a visit to Mardin, he noted: "People in these areas were thirsty. The birds were thirsty. All the animals were thirsty. The land was thirsty."[37] Like darkness, deprivation, and drought in the 1930s, need, want, and lack were the concepts Demirel used to define the Turkey of the 1950s. In so doing, he promoted a theory of Turkey's social and economic problems and represented his concept of technical civilization as the way to resolve those problems.

Demirel's leadership at DSİ marked the moment when his vision of technical civilization became a tool of governance. By merging infrastructure with political strategy, he transformed water management into both a development agenda and a means of consolidating power. Along with supervising several dam building and planning projects, including the Keban Dam, Demirel's five-year tenure as head of DSİ helped him make vital political connections and hone his message. Demirel made positive impressions on several high-level officials in the ruling Democrat Party, including the prime minister at the time, Adnan Menderes.[38] He also pursued his own political profile. In his public writings and appearances as General Director, Demirel entwined technical, political, and religious language to the extent that Murat Arslan, in his biography, notes how difficult it became "to understand whether the author was a bureaucrat or a politician."[39]

The coup d'état at the end of May 1960 delayed Demirel's ascent to leadership by only a few years. From late 1960 to early 1962, he completed his military service requirement within the Turkish government's new State Planning Organization [*Devlet Planlama Teşkilatı*], an experience that provided both an understanding of, and a particular philosophy toward, state-driven central planning.[40] Demirel reentered politics in 1962 as vice chairman of the new Justice Party (*Adalet Partisi*, or AP), seen as heir to the pre-coup Democrat Party. Tensions with the military drove him back to private consulting with Morrison Knudsen for a time, but Demirel returned to take the chairmanship of the AP in late 1964. The Justice Party's win in the October 1965 election propelled Demirel to his first term as prime minister, the youngest elected in Turkish history, at forty-one years old. Demirel would serve as prime minister for the rest of the 1960s and through the 1970s.

Over the course of his career in the Turkish bureaucracy and in its parliament, Demirel further elaborated and discussed his dream of technical civilization. In these writings, Demirel championed industrialization as the foundation of civilization, but his vision was more complex than a simple embrace of modernization theory. His civilizational ideals blended material progress with political ideology, balancing enthusiasm for economic development with anxieties about history, spirituality, and the limits of technological transformation. Since Demirel loomed so large over Turkish politics for much of the latter half of the twentieth century, his writings and speeches also represent an approach to nature and society with significant popular support.

In his 1975 book *Great Turkey* [*Büyük Türkiye*], Demirel aligns his views with modernization theory in a manner that reflected his earlier career, yet *Great Turkey* also reveals emerging tensions in his thinking about the consequences of material progress. In a nod to Rostow's stages of economic growth, Demirel argues that civilization occurs at different levels.[41] To ascertain the status of civilization at each level, Demirel writes, "people ask what you have." One cannot make a claim to be "a civilized country" without having certain technical and material accoutrements. For example, he notes "[a] civilization without a motor is unthinkable," and urges Turkish industrialists to pursue the construction of a Turkish engine.[42] Thus, only certain forms of economic development animate Demirel's definition of, and enthusiasm for, civilization. In *Great Turkey* economic development focuses on by now familiar tropes—"making the village a city" and "liberating the peasant from the peasantry."

However, Demirel also notes that "it is a wrong idea to explain history only in terms of technology and material."⁴³ As *Great Turkey* unfolds his vision, the prime minister concedes a dark side to material progress. He reflects on how "the effects of technology on production and economic life . . . did not fail to shape non-economic social and cultural life at the same time."⁴⁴ He evinces concern about human societies making themselves subject to their own technological progress:

> We should not allow the development of science and technology—which by freeing people from the bondage of nature makes them dominate nature with great power and possibilities—to take humanity under its own domination and bondage, as if in fact to avenge nature.⁴⁵

In this passage, we can see the concerns of the Cold War nuclear age. After all, the destruction of all life on earth would be fitting vengeance for "the bondage of nature." In considering the limits of progress, Demirel appears to recognize how the pursuit of technology can render the dreamer a vassal to the dreamt, to the engineered. Only in other, later writings does Demirel admit more about his reservations regarding technical progress, conceding that merely producing light, water, and food cannot prevent harm or promote solidarity. In a collection of interviews published in 1987 under the title *The Spiritual Side of Development* [*Kalkınmanın Manevî Yönü*], Demirel notes how "one of the conditions of achieving happiness is material wealth. But it has been seen how material wealth alone is not enough to provide happiness." Instead, "spiritual elevation is of paramount importance." Why it's important, though, is telling. "Because if people harm each other," Demirel observes, "life loses its meaning. There will be no peace, no brotherhood, no unity, no togetherness. We cannot achieve unity and solidarity with financial means."⁴⁶

Yet, in learning from his American tours how to draw together reservoirs of technocratic expertise, foreign capital, and repressive governance, and all the while espousing the social balancing effects of infrastructure, Demirel and the governments under his leadership very much thought that unity and solidarity could be acquired "with financial means." The dams, according to Demirel, were the remedies to darkness, drought, and deprivation. In her book *Hotels and Highways*, Begüm Adalet argues that while modernizing reformers (and, at times, later analyses) often touted a triumphalist narrative, a careful detailing of the politics suggests that "fragilities and anxieties . . . mark expert thinking and practice."[47] In Demirel's case, we might consider such fragilities and anxieties—the contradictions in his thought—not merely as deviations from a modernizing dream but as essential to it. His vision of technical civilization held both danger and hope. Yet, as in his earlier claims that dams would cure darkness and thirst, Demirel framed these tensions not as opposites but as interlocking problem and solution. By pairing the harms of materialism with the remedy of spirituality, he positioned himself as the arbiter of both.

So, one can conclude that Demirel did not see the contradictions in his dream as undermining but as intrinsic. Technical civilization was a force that could empower society, literally and figuratively, for material wealth was connected to "spiritual elevation." Meanwhile, an aim to do no harm ought to guide a society bound to its own technological momentum—that such technological momentum was in itself harmful was the cost of material progress.

This formulation of technical civilization proved lasting. Demirel and Turkey's central government retained throughout the troubled 1980s, 1990s, and beyond, a fascination with the expensive water engineering project, whether in the form of a giant dam or a huge canal.[48] At the same time, Turkey's politics shifted away from the secular models of Kemalism.[49] Moreover, in admitting that financial means could not bring peace, brotherhood, unity, and togetherness, Demirel voices the true quality of his dream: the reproduction of urban life and the transformation of the human relationship from a material connection with nature to a technical connection with an engineered nature. This is how Demirel, King of Dams, dreamt civilization through rivers. His Tigris and Euphrates were not just conduits of water and electricity; they were the manifestations of a civilization defined by the relentless logic of engineered transformation.

Süleyman Demirel's two trips to the United States convinced him that engineering expertise alone was not enough to bring about technical civilization. The American example—postwar superpower and self-described economic miracle—revealed what the dreaming would require: technical know-how had to be coupled with flows of capital and political power. The United States offered not only a model of scientific prowess but also one of socio-material formation. It was not enough to measure water or generate electricity; one also had to direct investment and control the institutions and territories through which modernity could be materially constructed.⁵⁰

Demirel's biography traces not only the elaboration of a dream, then, but its entanglement with material systems and political hierarchies. Realizing his dream required international support, which at that time meant navigating the politics of the Cold War. While those politics were defined by ideological division, building a technical civilization—the dreaming—produced similar material patterns on both sides of the Iron Curtain. Both the United States and the Soviet Union promoted a fusion of politico-military alliance and industrial-scale environmental engineering to reshape landscapes, societies, and political life.

Thus, the construction of Keban and al-Tabqa dams linked water, development, and military power in ways that transcended ideology, even as competition characterized the relations between the two blocs.⁵¹ For example, the case for building the Keban Dam was strengthened when Syria announced that it would also build a mega dam on the same river but with Soviet assistance.⁵² Like the space race, dreaming of technical civilization resulted in similar material forms; convergences that, from spatial and material perspectives, largely undermine the idea that a capitalist "west" and communist "east" represented two distinct civilizational models. Instead, what emerged was not a shared ideal per se but a common ontology: a world shaped by the premise that power could—and should—be embedded in infrastructure.⁵³

Using the lens of Cold War politics, we could depict the Keban Dam as an illustration of Turkey's orientation toward the West. The dam makes sense as a symbol of Turkey's progress toward democracy and capitalist development, underpinning its status as a NATO member. Meanwhile, to the south, this same Cold War lens would render al-Tabqa Dam as an example of Syrian-Soviet prowess and the progress of Arab socialism. The dam heralds the rise of Ba'thist Syria as an industrial power and as a bulwark against the Israel-US beachhead in the Arab World. In these ways, the dams may be seen, and indeed have been subsequently commemorated as, paragons of industrial progress achieved by opposing ideological systems.

Historian Kate Brown, in her studies of nuclear technology development in the US and USSR, has analyzed such interpretations and instead argues that the two Cold War foes had a lot in common. Working from

Henri Lefebvre's observation that "in the history of space, communism and capitalism have produced no qualities that distinguish one from the other," Brown details the similarities in the settlement landscapes of Kazakhstan and Montana, comparing the US railroad town of Billings to a USSR counterpart, Karaganda.[54] Brown wonders how supposedly opposed ideological and socioeconomic systems could have produced such similar places, contending that historians have largely "ignored the parallels produced by the industrial-capital expansions of the twentieth century."[55]

The key historical processes in Brown's account are, first, industrial centralization, and second, a distrust of the nomadic peoples American and Soviet governments sought to colonize (Native American and Kazakh, respectively) and of organized labor in general. The American and Soviet frontier towns are thus connected by the "utopian wish for gridded order and discipline," a wish shared by elite decision makers in the economy commanded from Moscow and the one structured around large corporations regulated from the District of Columbia.[56] Brown avoids the pitfalls of ideological narratives by focusing on the Foucauldian nature of the gridded city, which served as an "apparatus for conquest, as a way to dominate space," following and expanding on Lefebvre and James C. Scott. Spatial analysis allows Brown to sidestep vying ideological rhetoric and to foreground instead material realities.

Could the large dams on the Euphrates River likewise be less the product of contrast and competition—the building of a Soviet sphere in the Middle East versus an American one—and more about common, underlying desires for the advantages of technical civilization? Could Keban and al-Tabqa, large dams built at the same time on the same river, be like Karaganda and Billings, very nearly the same place? Very nearly the same dream?

Resituating the dams' history in this way changes the questions we might ask of their origins. Instead of focusing on how the dams reflect Cold War alignments, we might consider how they function, through environmental transformation, as instruments for extending industrial power and spatial control. The Keban and al-Tabqa reservoirs, like Brown's gridded towns, reshaped geography, transportation and other networks of connection, and the lifeways of local communities in ways that reinforced (but never quite accomplished) state control.[57] While grids, radial or wheel designs, and other geometric forms have long signified forms of discipline, the reservoirs worked in a different way, sending water oozing like a slow infiltration into the nooks and crannies of a topographically complex region. By filling in the terrain's low points, the engineered water represented a "utopian wish for *hydraulic* order and discipline," which, as we learned in the previous chapter, entailed a hydrographic fiction of control—a carefully constructed image of stillness and containment imposed on a geography that remained mobile, unpredictable, and never fully mastered.

This vision of nature reimagined as compliant infrastructure was never enough on its own, which is why the dreaming of technical civilization underpinned more overt efforts to surveil, control, and displace local communities—some quite distant from the reservoirs themselves. The Euphrates dams were built in the less-developed and less-populated east of both countries, areas that both governments saw as frontier regions. Those areas are home to sizable Kurdish minorities that were equally viewed as potential threats to national integrity.[58] As a result, and much like Brown's frontier towns, the dams were used to enact state policies of expropriation and exclusion.

In Turkey, Kurdish human rights groups in the country's southeast have noted how the dams and reservoirs appear not as "economic or social development but . . . [as a] project [that] will weaken Kurdish identity in Turkey and will potentially allow for a military victory in the ongoing armed conflict in the Kurdish regions."[59] The displacement of Kurds in Turkey has removed people from their lands, while resettlement in amalgamated villages or in cities—many in the western part of the country—has dispersed solidarity groups and diluted Kurdish culture.[60] Just as gridded towns determined how the people of Billings and Karaganda might move and congregate, so too did this dam-driven process of displacement and resettlement.

In Syria, construction of al-Tabqa dam was essential to the government's efforts to displace, disenfranchise, and denigrate Kurds. A few months after the March 1963 military coup that brought the Baʿth Party to power, a Syrian army officer, Lieutenant Muhammad Talab al-Hilal, published a report that denied the existence of a Kurdish people and called for the creation of an "Arab belt" along 280 kilometers of the Turkish border. According to the report, in this region 140,000 Kurds would be deported to the interior. If they stayed, they would be denied citizenship, education and employment. Supposedly loyal Arabs would then be settled on newly constructed farms. As historian

Jordi Tejel notes, the plan was not implemented until 1973, for "the 'colonization' of Jazira [the land between the rivers in northern Syria] depended upon certain favorable conditions, such as the construction of the Tabqa dam on the Euphrates basin." With the filling of the reservoir, "around 4,000 Arab families of the Walda tribe, whose own lands had been submerged, were settled (and armed) in forty-one of the 'model' farms in Jazira, in the very heart of the Kurdish region, as well as in fifteen 'model' farms north of Raqqa."[61] In the regions around Keban and al-Tabqa, Kate Brown's gridded towns are thus replaced by dams and reservoirs, which enact similar patterns of eviction and deportation with irrigation works and "model" settlements allocated to groups perceived as loyal to the state. Further, in both cases, military action and arms worked in concert with dam building to enforce the will of the central government.

Still, even as the Syrian and Turkish dams functioned as tools of internal control, their construction helped implement a second layer of imperial influence—one shaped by the financial and military backing of Cold War superpowers. The financial resources to build the dams on the Euphrates River came from the Soviet Union and the United States. Thus, a dual imperialism characterizes the Euphrates reservoirs: one in the US-Turkey and USSR-Syria relationships and one in Turkish and Syrian central government views of their own frontiers. In this double layer of empire, there is a dual conception of spatial control, one localized to the areas around the reservoirs—the extension of an industrializing logic into the hinterlands by a centralizing state—and the other ascribed to the security of the bloc, whether construed as "containment" or the export of a "revolutionary" model. The convergence of imperial ambitions—both foreign and domestic—at multiple sites along the Euphrates River suggests that the reservoirs must be understood not only as tools of socialist or capitalist "development" but as mediums for fantasies of power across multiple scales, and for entangled (and quite similar) dreams of technical civilization, always vulnerable to the shifting currents of nature, history, and resistance.

The dreaming, of course, was not just an imagined set of connections. These were dreams funded, built, armed, and poured in concrete. These were dreams institutionalized through funding mechanisms that paired military assistance with development finance—two categories treated as distinct in policy but deeply entwined on the ground. National governments and international organizations often categorized and managed these two forms of aid—military support and development assistance—separately, and analysts have gone on to study them so, writing separate treatises on military spending and economic aid.⁶² But just as industry and capital restructured landscapes in the US and USSR, so too did the joint logics of military assistance and development finance converge in the dams and reservoirs reshaping the Euphrates River. Concrete, currency, and command met to materialize a dual layer of imperial ontology: technical civilization as a fusion of empire, engineering, and enforced order.

Moreover, the desire to harness the shaping power of dams and reservoirs proved surprisingly durable, further evidence that these infrastructures transcended ideology or economic system. Even as both Syria and Turkey convulsed politically during the planning and construction of the Euphrates dams, the flow of capital and the dam projects themselves remained constant, all but unaffected. From the time the dams at Keban and al-Tabqa were first conceived to the time of initial construction, Turkey transitioned from a single-party democracy to a multi-party one and then experienced a military coup d'état in 1960, followed by a new constitution and the restoration of democratic rule in 1961. During roughly the same period, Syria experienced four coups and was at one point ruled from Cairo. Through it all, the efforts to build these dams continued, even as the politics and the alliances around them shifted. Just as the organization of space served as an inescapable, common component in Brown's study of capitalist and communist systems, so too did these connected reservoirs of international capital, military assistance, and Euphrates River water.

The entanglement of military strategy and environmental engineering in the Middle East reveals another crucial similarity between Cold War adversaries and between their regional allies: a shared logic of domination that linked weapons and dams. Scholars have noted the connection between military weaponry and environmental engineering in other places. A growing literature has observed the substantial correlation between humans waging war on one another and on the earth itself. This literature has examined direct connections—the use of nonhuman nature to wreak havoc on the enemy—and indirect connections—the development of chemicals such as DDT that helped fight a war.⁶³ In the stories of dam financing in Cold War Turkey and Syria, the connection is more subtle but nonetheless reinforces the relationships these scholars have demonstrated—military and environmental modes of domination were mutually reinforcing. As we learned from the dreamer Demirel, technical civilization was not about solidarity and brotherhood. Through the entanglement of arms and dams, the landscapes around Keban and al-Tabqa came to be defined not only through reservoirs of water but also reservoirs of the instruments of war.

In Turkey, financing for dam building after World War II quickly became tied to loans for weaponry and the obligations of a military alliance—the North Atlantic Treaty Organization (NATO). From a position of armed neutrality during the war, the Turkish Republic afterward sought greater ties with Western Europe and the United States to counterbalance Soviet claims to the all-important straits, the Bosporus and Dardanelles [*Çanakkale Boğazı*], connecting the Black Sea to the Aegean and Mediterranean Seas beyond. US policy was vital in Turkey's move from armed neutrality to NATO ally. The Truman administration included Turkey among the "Truman Doctrine" states in 1947, and the country won its first allotment of American economic aid. A year later, the US government included Turkey in the Marshall Aid program. By 1952, Turkey had joined a trio of institutions headquartered on the western side of the Cold War political divide: the Organization for European Economic Cooperation, the Council of Europe, and the North Atlantic Treaty Organization.

A deliberate move toward democratization helped cement Turkey's western alliances. In 1950, Adnan Menderes and the Democrat Party (DP) won Turkey's first multi-party elections. Menderes amplified relations of economic dependence on the United States by using Cold War geopolitics to secure ever-increasing aid packages. In return, Ankara liberalized economic policies and embarked on public works in the rural areas that had brought the Democrat Party to power. Before too long, the DP discovered that Western influence could in fact cut both ways—ideological affinity and the opening of Turkey's economy did not prevent the use of aid as leverage.[64]

The first such encounter involved the Seyhan Dam, the project to which Süleyman Demirel was assigned after his first trip to the United States. Diplomatic wrangling over aid for the Seyhan Dam shows how dams and arms could become linked within the context of alliances. In early 1952, British officials met with their Turkish counterparts to discuss £32 million in outstanding armament credits provided to Turkey in 1938 and 1939. The loans had paid for military equipment and experts to assist in Turkish war preparedness. The Turkish government stopped payment on the loans in July 1951 after disbursing nearly £9 million in capital and £8 million in interest.[65] The Turkish delegation asked for cancellation of the remaining payments as an act of generosity between two allies, or at least that the British accept a token amount instead of full repayment. The Turks also requested the same of the French, who had extended similar credits early in World War II.[66]

Burdened by postwar reconstruction and the costs of Cold War rearmament, the British were reluctant to agree. They were left with two options: accept Turkey's offer and set a bad precedent for their other debtors, or recall the delegation and wait for a better moment to press their claims. That moment came sooner than expected.[67] In February 1952, UK Treasury officials contacted British representatives in Washington to ask when the International Bank for Reconstruction and Development (now the World Bank) would consider a $20 million loan for Turkey's Seyhan project. They noted that "the only way we might bring them [the Turkish government] to reason at the present time is through the IBRD." With Turkey in the process of joining NATO, direct pressure was politically risky; British officials instead sought leverage through development aid.

American officials responded with apprehension, suggesting that the Bank should not be used as a "collecting agency" and that holding up a development loan because of wartime debts would be unethical.[68] The IBRD president, Eugene Black, noted that a veto or abstention by the United Kingdom would make the Seyhan Dam project the first development loan to be voted down by the Bank's Board of Directors. To give the two sides more time, the Bank agreed to hold up the loan on some "technical hitches" in the hope that a delay would encourage the resumption of negotiations.[69]

The Turkish government responded that connecting the armament credits dispute to the development loan was unacceptable because "armament credits were in a separate category."[70] But most countries disagreed; only the Pakistani and American representatives on the Board supported Turkey's definition. American control of the Bank proved decisive, and the United States forced a vote on the Seyhan Dam aid package. Britain and France, reluctant to set the precedent of the Bank enforcing disputed military loans, voted with the other representatives to approve the loan.[71]

Though Turkey won aid from the IBRD, the dispute pointed to how postwar development assistance could be used as political leverage in other contexts, particularly security related ones. Moreover, it showed that newly established postwar international organizations, ostensibly operating according to some high-minded principles, could be manipulated to coerce developing states. Further, the politics of the IBRD held overtones of earlier European imperial arrangements against the Ottoman Empire, and the Turkish government responded by severing ties with the IBRD in 1954.[72]

Officials in Ankara instead sought aid through bilateral negotiations rather than through international forums, an approach that made the connection between dams and arms even more apparent. Throughout much of the 1950s, the Menderes government obtained both kinds of assistance, particularly from the United States, and aid agreements padded the government's budget. The United States, in turn, received benefits, one of which was cementing a military relationship with Turkey through the construction of a strategic air base near Adana that came to be known as İncirlik. The base is located less than ten miles from the Seyhan Dam and uses the hydroelectricity the dam produces. The İncirlik base was critical to the deployment of US troops to Lebanon in 1958 and to Gulf War operations in 1990–1991.[73]

The Turkish experience with aid from the 1940s to the 1960s highlights the creation of a military-development complex, an outgrowth one could argue of a burgeoning military-industrial complex in the United States and other nations.[74] Turkey exemplified the way military and economic aid was becoming intertwined during this period and the political tensions that emerged as a result. In 1963, Robert Komer, later to become the American ambassador to Turkey, wrote the following memorandum for assistant secretary of state Phillips Talbot:

> We have never really decided in our own minds whether to treat Turkey primarily as a NATO partner (whose main need was military aid for the defense of Europe) or as an underdeveloped country whose primary need was to become a going concern. As a result we have pursued both aims—and fully succeeded at neither.[75]

The British reached a similar conclusion about the relationship of military aid and economic development assistance. A report for the British Ministry of Overseas Development described the problematic relationship of Turkey toward the foreign aid regime:

> It is sometimes said by people concerned with Turkey that the Turks want to have it both ways. As a member of the western alliance, Turkey wants to be treated as an equal partner. As an underdeveloped country, Turkey claims the right to fulfillment of all its needs on concessionary terms. There is some truth in this, but what is said less often is that the Turkish attitude is the mirror image of the attitude of the western powers, who also want to have it both ways. As a member of the western alliance, Turkey is required to show all the sophistication and responsiveness to change that was expected, for instance, of Britain during the sterling crisis of 1964–65. As an underdeveloped country, Turkey is required to accept all the paraphernalia of investigation and review, the unending stream of admonition and advice, that is heaped upon the "real underdeveloped countries" of Asia and Africa. In the field of economic aid, what this means is that the western powers have set up all the machinery for attacking a long-term problem, while protesting all the time that from their point of view no long-term problem exists, and that it is therefore unnecessary for them to enter into any open-ended commitment.[76]

Thus, the Turkish plank of NATO deterrence relied upon a convenient fiction that Turkey was capable of exerting military-industrial power like Britain and France, even as the latter countries provided economic and military aid to Turkey with the expected oversight and control. Put another way, the larger powers in the NATO alliance sought to exert forms of imperial control while pretending to a coequal partnership that required no such forms. Although officials on both sides of the Atlantic ostensibly saw military and economic development aid in separate categories, in practice the two were connected and codependent.[77]

Recognizing that foreign aid often came with political and security strings attached, Turkish officials sought to dilute external influence by assembling diverse funding sources for the Keban Dam, creating a syndicate that both secured resources and preserved Ankara's negotiating power.[78] The syndicate included several banks, including the European Industrial Bank and the IBRD (which Turkey rejoined after the 1960 coup) and the American, West German,

French, and Italian governments. The four governments were also members of the NATO alliance.[79] Moreover, much of the aid money spent on the Keban Dam paid for the services of American and European organizations and companies, either in the form of expertise or equipment. United States Agency for International Development (USAID) funds were funneled to EBASCO Services, Inc., the New York–based engineering firm that supervised construction after producing the feasibility report.[80] The US government also paid its own Army Corps of Engineers to complete an assessment of the Keban Dam project—another connection between development aid and the military. French and Italian monies went to the construction company that undertook the work, SCI-Impregilo.[81]

This method of granting foreign aid points to how the dam projects produced a broad range of benefits not only to the receiving country but also to the donors. Foreign governments could use the dams as leverage against Ankara in bilateral disputes, could use the projects to influence Turkish leaders in the development of the country's economy, and could use the dams to support a certain military and economic order at home while creating it elsewhere.

Syria's path to dam financing, though similar in some respects to Turkey's, was far more precarious, shaped by decolonization, military threats, and Cold War competition. Lacking at first the security of a long-term alliance, Syrian leaders courted foreign aid for infrastructure projects while attempting to avoid political subjugation. Negotiations with Western European states, the United States, and eventually the Soviet Union reveal how economic development and military security became inseparable concerns in Syria, with the Euphrates Dam project over time becoming both a means of securing external support and a statement of national independence. The ever-increasing connection between these two goals—building the dam and securing the country's sovereignty—led Syria to play both sides of the Iron Curtain, both sides of a split Germany, and to ally with Egypt and then break up.

During these changes the ambition to build the dam remained constant. One historian noted the project was "isolated from national opinion," tied as it was to an aspiration for greater and more decisive political autonomy.[82] Likewise, the story of al-Tabqa Dam demonstrates again how these infrastructure projects had a value based less on the competitive ideologies of the day than on the pursuit of the power and practice of technical civilization. So, even as the Soviets played the decisive role in financing and building al-Tabqa Dam, that is not where Syria's search for dam financing began.

After gaining independence in 1946, Syrian leaders initially sought the support of the British Middle East Office (BMEO) Development Division, based at the time in Cairo. The BMEO facilitated a connection to a British engineering firm, Sir Alexander Gibb and Partners, and in 1947 the firm analyzed a Euphrates River dam site near the village of Yusuf Pasha, approximately fifty kilometers south of the Turkish border. The company's report built on earlier French investigations and offered hydroelectric and irrigation possibilities for the site.[83] Syrian officials asked the BMEO for additional technical experts, and there was even discussion of establishing a development commission like the one being debated simultaneously in Iraq.[84]

The British, however, were not able to translate their studies into reality, which irked officials at the BMEO. After the Arab defeat in the 1948 war with Israel, Syria looked more toward the modernization of its military than its water infrastructure. The head of the BMEO Development Division, W. F. Crawford, thought the country's push for weapons was misguided:

> It is heartbreaking to think that £50 million would develop the country so that it could be independant [sic] and stable. In the meantime who knows what is going in arms? (Some at least seems to be not much better than scrap). The vanity of human wishes on the part of a colonel.[85]

While Crawford seemed to believe Syria's political independence could be brought about economically, his disparaging remarks suggest that Syrian officials saw weapons as a more viable strategy. Crawford also notes the dismal failure of the British to translate development work into political influence in Syria, a situation that might change if the British added arms to the aid package:

> So far as we, the Development Division, is concerned Syria is in a mess. British Council is practically closed down.... If we can sell them military equipment... everything will change. The Foreign Office expect too much of the Syrians. They won't sell them military equipment which they want, we have no economic assistance to give and yet we expect miracles of friendship and obedience from them.[86]

Crawford's bluntness demonstrates how Syria was not that different from Turkey in the expectation that aid—military or economic—should facilitate an imperial-style relationship. Britain remained an important trading partner for Syria during much of the Cold War, but the opportunities for British engineering in Syria after 1948 were limited.

As the 1950s progressed, Cold War contests and military threats on Syria's borders pushed the Euphrates dam-building project more firmly into the security context: arms and dams became the central concerns in foreign aid. In 1955, the IBRD conducted another survey of the Yusuf Pasha site.[87] However, the political landscape shifted when Khalid al-'Azm's left-leaning government, advocating a policy of neutrality, refused exclusive Western partnerships including with the IBRD. Instead, in March 1955 the new Syrian government signed a military and economic agreement with a similarly situated country, Gamal Abdel Nasser's Egypt, an action that brought the threat of Iraqi and Turkish military intervention. This saber rattling, along with aggressive actions by Israel in Gaza, propelled Syrian foreign policy toward deeper integration with Egypt and cooperation with the Soviet Union—the latter for the first time articulating an interest in protecting Syria from outside interference. Moscow's statement of support declared that the threat to Syria

> comes not from the Soviet Union but from those powers which, on the pretext of "guaranteeing security," are creating aggressive blocs in the Near and Middle East in an attempt to convert the countries of this area into their military and strategic springboards and to reduce them, economically, to the status of colonies or dependent territories.[88]

The unfolding of the 1956 Tripartite Aggression, also known as the Suez Crisis, which Nasser provoked by seeking economic aid to build the Aswan High Dam, seemed in many ways to confirm the Soviet statement. The invasion of Egypt by Britain, France, and Israel demonstrated the lengths to which Western powers would go to protect lucrative interests within the prevailing political and economic order, confirming Syrian reluctance to accede to economic assistance without accompanying military aid.[89]

After an arms shipment from Czechoslovakia via Egypt arrived in late 1955, Syria's connection with the Soviet Union grew to the point that in 1957, Soviet engineers were sent to study the management of the Euphrates River. The Soviet survey was quite extensive, occupying twelve volumes, and recommended that the Syrians abandon the Yusuf Pasha site in favor of a dam emplacement farther downstream at al-Tabqa.[90] However, regional politics drove a wedge between the two parties. In March 1960 the Soviets bowed out of the project because the three major basin states—Syria, Turkey, and Iraq—could not reach an agreement on sharing the water of the Euphrates River. While Soviet leaders courted Syria with the Euphrates survey, they also extended to Iraq a $137.5 million credit in 1959. Rather than aggravate a budding relationship with Iraq by supporting Syria's ambitions on the river, the Soviets pulled out. The US Central Intelligence Agency noted several years later that this was the "only major aid commitment to a less developed nation on which the USSR has ever reneged."[91]

The loss of Soviet sponsorship left Syria's leaders scrambling for alternative funding, and the path back to Soviet aid was long and circuitous. In 1958, Egypt and Syria merged into the United Arab Republic (UAR), and in 1961 the new state settled on an economic aid agreement with the Federal Republic of Germany that included 500 million marks ($120 million) for the dam at al-Tabqa. The UAR's Damascus bureaucracy at this stage took concrete action, creating a new agency, the General Organization for the Euphrates Dam. For a few months in 1961, the United Arab Republic was pursuing two of the largest dams ever constructed: the Aswan High Dam on the Nile River (which became operational in 1970) and the Euphrates Dam at al-Tabqa. Then, in September, a revolt by Syrian army officers sundered the Egyptian-Syrian union.[92]

The animosity between Damascus and Cairo ultimately voided the dam financing agreement with West Germany.[93] The Bonn government refused to fund the dam at the same level since Syria without Egypt was a smaller economy.[94] Then, on May 13, 1965, the United States succeeded in pressuring West Germany to recognize Israel. According to American diplomats, it wasn't so much the building of embassies in respective capitals that irked officials in Damascus, but West Germany's financing of an Israeli purchase of American arms.[95]

In a bid to regain leverage, Syria attempted to play East and West Germany against one another. Damascus sent the Minister of Agrarian Reform, 'Abd al-Karim al-Jundi, to East Berlin for meetings with the leader of the German Democratic Republic, Walter Ulbricht, in mid-July 1965.[96] Then, in October 1965, Syria's Minister of Foreign Affairs Ibrahim Makhus appeared on West German television to argue that the Syrian government was not using the dam as "an opportunity for political bargaining" and that Syria was "trying to generate for the Euphrates dam project the interest of all states without exception . . ."[97] But an economic agreement with one of the German states was not to be, as an arms buildup and aggressive actions in the Arab-Israeli conflict altered yet again the calculus of dam building.

Since the dissolution of the UAR, Syria had attempted to divert water from one of the Jordan River tributaries, the Baniyas, in response to Israel's use of the Sea of Galilee. Israel attacked the diversion sites, and in retaliation, Syria supported Palestinian groups in cross-border guerilla attacks on Israeli targets. Meanwhile, the United States sold arms to Syria's neighbors to the south, Israel and Jordan.[98] In the midst of these events, the Director General of the Agency for the Euphrates Dam, Ibrahim Farhoud, traveled to the Soviet Union on December 17, 1965, to visit "Soviet dams that offer some analogy to the work projected on the Euphrates." After three weeks, Farhoud returned to Syria without fanfare or announcement. Sources at the Syrian Ministry of Foreign Affairs refused to say anything, but it seemed that the sticking point in rapprochement was still the agreement for equitable sharing of Euphrates water. Months passed, and it appeared as if nothing had come of Farhoud's visit. However, in February 1966, a Ba'thist military cadre launched a coup that paved the way for closer relations with the Soviet Union.[99] In April, a Syrian delegation, including Hafez al-Assad as Defense Minister, traveled to Moscow. There, they secured both military and economic aid, including a $150 million loan for the construction of the Euphrates Dam.[100]

Then, just when it seemed that Syria's saga of finding weapons and a dam together had concluded, a German offer underscored how environmental engineering could transcend ideological constancy. West Germany proposed financing the dam alongside the Soviet Union, with the German commercial counselor commenting that "some simultaneous participation of the USSR and West Germany for the construction of the Aswan Dam had been considered for a moment." The French ambassador to Syria speculated:

> Should we conclude that certain officials in Bonn do not exclude the possibility of a German-Russian collaboration on the Euphrates? One can ask moreover whether Syrian leaders themselves, while condemning a posteriori the German offer, would not basically want to use it at the same time as the Russian offer, or at least maintain the opportunity to again appeal to the Germans if the Soviets were to impose difficult to accept conditions.[101]

The ambassador's observations point to the strategic balancing act Syria was performing—leveraging Cold War rivalries to maximize its autonomy rather than simply falling into the Soviet orbit. By securing Soviet backing for both weapons and infrastructure, Damascus shored up its defenses against regional rivals while anticipating that the dam's economic benefits would reinforce Syria's political and economic independence. However, Syria's willingness to entertain a West German offer, even after accepting Soviet aid, reveals that its commitment to Moscow was still conditional. Rather than ideological alignment, Syria's calculus was about the advantages of technical civilization, advantages that both sides might offer. Syria played the Cold War game of dams and guns not to join a bloc but to build an independent future, transforming infrastructure into strategy and technical civilization into a means for asserting sovereignty.

The Soviet deal would stick this time, though, as the connection between weaponry and the Euphrates Dam only became stronger after the military disaster of the 1967 Arab-Israeli war. Syria's military and territorial losses necessitated stronger ties with the Soviets. The war revealed the slippery slope in connecting military and economic aid, as Syria could no longer afford a relationship with a state that couldn't provide copious quantities of both. In 1971, the Syrian-Soviet relationship became even stronger, and Syria's independent future dimmed somewhat when the USSR established a naval base at Tartus on the Mediterranean coast. Moreover, building the dam became a counterpoint to the humiliation of losing the Golan Heights to Israel and a symbol of the Ba'th Party's commitment to invigorating the economy and revolutionizing Syrian society.

These two tales of financing the first megadams on the Euphrates River reveal a set of common and ideologically transcendent features—shared not only between the United States and Soviet Union, but between Turkey and Syria as well. Despite their different geopolitical positions, both Syria and Turkey treated dam-building as a tool of national power and autonomy, even as intertwining development aid with military strategy compromised the very sovereignty it was meant to secure. Turkey, as a NATO member, juggled multiple Western donors to maximize control, while Syria, navigating a less stable international position, exploited Cold War rifts to secure promises of aid from shifting patrons. The financing of the Keban and al-Tabqa dams underscores a fundamental reality of Cold War development: the separation between economic aid and military assistance was largely rhetorical. Large dams became vehicles for securing external support while reinforcing internal state power—material expressions of a geopolitical order that conflated infrastructure with influence.

Syria and Turkey became entangled not in separate systems or civilizations, but in a shared Cold War system, one in which the industrial control of territory was achieved through a blend of environmental engineering and military force. The answer to questions of sovereignty in Syria and Turkey was the same: buy weapons and build the dam. However, the consequences of trying to engineer sovereignty through a river emerged immediately, as once al-Tabqa Dam was built, Iraq threatened Syria with war over the diminished supply of water.[102] Fear of intervention fueled years of diplomatic maneuvering for financing, but it also linked arms and infrastructure in the engineering of rivers and regimes. These were not separate domains—they were mutually reinforcing tools of technical civilization.

Ultimately, the weapons procured through these projects were turned inward. In the late 1970s and early 1980s, the Damascus government used Soviet-supplied arms to suppress a rebellion by the Syrian Muslim Brotherhood centered around the country's northern cities. In 1979, thousands of government troops occupied Aleppo and other cities in the north prior to the brutal 1982 suppression of the revolt in Hama.[103] More recently, the Syrian state used chemical weapons—developed with the help of Egypt and the Soviet Union—against rebel forces in Aleppo.[104] In Turkey, a rebellion led by the Kurdistan Workers Party (*Partiya Karkari Kurdistan* or PKK) beginning in the 1980s was met by brutal repression, much of it carried out using American-made aircraft, helicopters, and munitions.[105] According to Human Rights Watch, the "most egregious examples" included "the punitive destruction of villages" and indiscriminate targeting of civilians.[106] In the twenty-first century, Turkey has continued to use American weapons against Kurdish groups in Syria and to arm regional proxies.[107]

By considering the financing of large dams, it becomes clear that these acts of state violence were not incidental to Cold War development—they were part of its long afterlife. The very technologies acquired to protect sovereignty were ultimately deployed to silence those displaced or marginalized by the promises of engineered rivers. The greatest threats to the governments based in Damascus and Ankara did not come from their Cold War enemies, but from within—from the people whose lives had been reshaped by the very projects meant to secure the state.

Dreaming of civilization through rivers, then, meant seeing military and environmental interventions not as separate realms, but as variations of the same act—not as a product of one ideology or another, but as a shared grammar of Cold War modernity. The reservoir and the rifle belonged to the same dream of engineered landscapes: a dream of civilization as containment, of security as control, of rivers dammed and diverted not for peace, but for power.

While the construction of Cold War-era dams on the Euphrates River gave physical form to the dream of technical civilization, their operation laid bare the tensions that dream had tried to conceal. Once the reservoirs filled, the dams ceased to be abstract symbols of national progress and became lived landscapes that altered daily life for those who had long relied on the river's seasonal rhythms. These disruptions could radiate well beyond the reservoir's edge, as displaced communities were often relocated to unfamiliar terrain. For planners and politicians, then, the dams and reservoirs stood as triumphs of engineering; for those whose villages disappeared beneath the rising waters, they marked a landscape of rupture and loss. In response, Syrian artists, writers, and filmmakers gave form to this dissonance, reckoning with the chasm between dreamt-of transformation and lived dispossession.

Few places embodied the Baʻthist vision for revolutionary transformation in Syria more fully than the Euphrates Dam at al-Tabqa. In fact, the dam is known by another name: Sadd ath-Thawra, the Dam of the Revolution. The name references the March 8, 1963, military coup d'état that brought the Arab Socialist Baʻth Party to power. The regime that emerged from the coup pursued what it believed were leftist, revolutionary policies, policies that grew more radical after the party congress in October 1963. A political struggle then ensued over what a Baʻthist Syria should become. Interparty strife brought about another coup in 1966 and further turmoil in 1970, through which Hafez al-Assad consolidated power.[108] When the dam became operational in 1973, al-Assad was there to inaugurate it, and he would rule Syria until his death in 2000.

The dam project survived the political perturbations of the 1960s to become an essential revolutionary effort as its provision of water and electricity supported the regime's promises of modernizing, socialist projects, such as collective farms and land reform. When Hafez al-Assad came to power many of the socialist components of the so-called 1963 "revolution" had been scaled back, but as Lisa Wedeen has shown, the Assad regime found appearances tactically useful to maintaining power and social equilibrium.[109] As the Ba'thist state consolidated power, the dam's significance extended beyond its economic and infrastructural role. It became a visual and ideological centerpiece of the regime's revolutionary narrative, a carefully curated symbol of socialist progress and national unity.

This revolutionary rhetoric was inscribed not only in political speeches but also in the very geography of the project, with both the dam and the town built to support it renamed ath-Thawra (revolution). Together, and more than anything else, they signaled the kind of Syria the Ba'th Party set out to construct. The regime actively shaped the dam's public meaning, presenting it not only as a technical feat but as a political achievement, and turning its construction into a cultural performance. The government invited artists to participate and disseminate its preferred symbolism, culminating in an exhibition at the dam's opening in July 1973, entitled *Visual Artists Live Through the Stages of the Dam*. *Al-Ba'th* newspaper, mouthpiece for the party, noted how the exhibition "represents the happiness of the artists about the dam, reflecting on the most important achievements of the 8th of March." The newspaper used patriotic, hyperbolic terms to describe the art displayed, "committed art (*al-fan al-multazim*) is the art that comes from the artist himself and his freedom entirely, committed with all his convictions to the work he paints. The most wonderful kind of commitment (*iltizam*) is the commitment to freedom..."[110] The visual works on display evoked this freedom by focusing on the dam's various components, with paintings of the electric station, the turbine gates, and different views of the main structure. The dam embodied—symbolically and materially—liberation from division and oppression.

Among the visual artists of the time was the filmmaker Omar Amiralay, who was inspired to make films while studying in Paris, where he witnessed the 1968 student uprisings. His return to Syria in 1970 coincided with the dam's construction, which he captured that year in his first film, *A Film Essay on the Euphrates Dam*. During its thirteen-minute run time, the film illustrates the hope and optimism of social transformation through infrastructure. The camera focuses on the movement of machinery and contrasts the modern, rectilinear forms of the dam to a mud wall being fashioned by a farmer in a nearby village. Like the symbolism of a "new Euphrates" in Turkish-language poetry, Amiralay's point of view is clear: the dam will not only transform the Euphrates River but also improve the lives of those touched by the project.[111]

Amiralay's optimism for the dam project did not last long. In the next two films of his "Euphrates Trilogy," Amiralay dismantles the Baʿthist regime's grand narrative of revolution, showing how the Euphrates Dam, once a symbol of national transformation, became an emblem of economic disparity, displacement, and political disillusionment. In *Everyday Life in a Syrian Village* in 1974 and *A Flood in Baʿth Country* in 2003, the filmmaker expresses his disappointment at the bankruptcy of Baʿthist rule by placing the films' protagonists—the people of the river valley—in a web of nature, society, and government. The 1974 documentary follows the rhythms of agriculture, interspersing scenes of tilling, weeding, and seeding with commentary on social and political conditions. Two scenes recur throughout the film: a group of children in the desert sand playing around a camel carcass and an older man tearing at his clothing to reveal an emaciated torso (the meaning of these two recurring scenes becomes apparent at the end of the film).

Between these recurring scenes, the film works like Amiralay's first—using juxtaposition and contrast but, this time, to reveal the great disparity between the realities of peasant life and the central government's rhetoric of revolution and reform. In one such comparison, a teacher in a classroom slowly recites a story about a farm cooperative with a combine. The next scene presents a view of young people not in the classroom but working on the land. Instead of new machinery, they struggle to direct a half-dozen donkeys. In another contrast, a Sufi religious ceremony with music and recitation of the Qur'an is followed by a Ministry of Culture film screening. The Ministry representative asserts that cinema provides a "true idea about the progress that advanced nations have achieved." The film-within-a-film then presents that great progress: families at leisure on a beach, the men bare-chested and the women in bikinis. To drive home the contrast, Amiralay includes a view of children at the beach playing in the sand—there are no camel carcasses. The reality of cultural life in cities along the Mediterranean could not be more divergent from that of the farms and villages of the Euphrates River Valley.

At the end of the film, the children have dug up the camel's bones. They drag pieces of the carcass across the sand to disappear behind a dune. In the other recurring scene, after tearing his clothing the older man shouts about how landowners are kicking people off the land they had farmed since the Ottoman period. "The only roof we have is our blankets," he cries, "we are hungry and dying!"[112] Despair, death, and disappearance are the essential characteristics of "everyday life" in the lands shaped by the dam, Amiralay argues, especially with a government so involved in its ideology and exercise of power that it has no concern about the entangled economic, environmental, and social crisis affecting its rural citizens and created by its own policies.

Between the last two films of Amiralay's "Euphrates Trilogy," Syrian writer 'Abd al-Salam al-'Ujayli also responded to the newly engineered Euphrates River in his book *The Submerged (Al-Maghmūrūn)*, published in 1979. Born in 1918 in Raqqa, Syria, a town on the Euphrates River downstream of al-Tabqa Dam, al-'Ujayli studied medicine at the University of Damascus and became one of only two doctors in his hometown. In 1948, after a stint in the Syrian parliament, al-'Ujayli fought in Palestine against the newly-declared state of Israel. He returned home from the war disenchanted, much like his contemporary Gamal Abdel Nasser, also born in 1918. Unlike Nasser, who would become president of Egypt, al-'Ujayli declined to return to politics until the early 1960s when he briefly served in several government positions. From that time until his death in 2006, al-'Ujayli focused on his medical practice and his writing.[113]

His novel, *The Submerged*, is a love story set in the Euphrates valley during the filling of the reservoir behind al-Tabqa Dam. A young Syrian fellah, a peasant named 'Uthman, meets a member of the well-educated urban elite, Nada, while working on the "great project." Nada's job focuses on newly built model villages where the state intends to move dislocated people to escape the rising waters of the reservoir.[114] 'Uthman drives Nada around to work sites in a Land Rover and represents the project's cooperatives management. They fall in love and once secretly betrothed, Nada tells 'Uthman that she wants to move with him into a house in one of the model villages. At first, the pair appear to represent the socialist revolution's hope of equality through their transcendence of class division.

Their hopes are dashed, however, when it later becomes clear that the government has changed its plans, telling 'Uthman that his family and others in the Euphrates Valley will not be moving to the model villages. They will be relocated away from the river entirely, to a place in the country's far east. Al-'Ujayli does not directly discuss the politics around the relocation, referring instead to a "higher government interest." The context, though, clearly references how the Syrian state at that time resettled displaced Arabs in places of Kurdish majority along the Turkish border in what came to be known as "the Arab belt."[115] The government strategy intended to reduce Kurdish influence and geographically divide Kurds in Syria and Turkey by implanting a supposedly loyal population.[116] Hence, the river engineering project provided an excuse and a material basis for social engineering.

As the river's water submerges the homes of the villagers, al-'Ujayli voices the state's attitudes toward the people of the river valley. In a confrontation between dam bureaucrats and local people a state functionary announces:

> You speak as if the land we want to move you to is in China or on the surface of the moon. Listen, uncle.... The state is the daughter of the dam, and the water is gathering behind it, and you know that you no longer have a place in your old villages, which day after day are flooded by the lake. Like it or not, you will move.... What does it matter to you if your move is ten kilometers or a thousand kilometers?[117]

The dam in this view gave birth to the state, not the other way around. The dam is the revolution, and the liquid inexorably rising behind the dam is not just water. It is the power of the state itself, which can erase all the villagers' ties to their homes.

The irony of the relocation is not lost on 'Uthman, who acts throughout the novel as an intermediary between the state and the villagers. It is an uncomfortable position, for the Damascus government makes 'Uthman into a hypocrite and a liar, undermining his earlier efforts to encourage acceptance of the model villages. Nada accuses 'Uthman of fatalism, the death knell of any revolutionism, but he retorts:

> I told you that we are happy with the great project because it will deliver us from a long injustice that we have endured in this valley and the surrounding valleys ... the injustice of nature, and the injustice of customs and traditions imposed on us by our tribal society.... Now I opened my eyes and saw the reality. We are doomed to remain oppressed. When I saw the condition of our people [already in the east], I realized that all that happened to them was that they fled from one oppression to another. They sought refuge with the state

to obtain justice from nature and society, and the state lifted their oppression and brought them under its own oppression . . .[118]

'Uthman here sounds surprisingly like Süleyman Demirel in articulating the "injustice of nature," but unlike Demirel, al-'Ujayli positions the state, not technology, as the ultimate oppressor. Further, the submerged in al-'Ujayli's novel are like the peasant farmers in Amiralay's *Everyday Life in a Syrian Village*. They have no escape; they will either be immersed in a "higher government interest" or deluged by the rising waters of the reservoir.

The villagers make a final stand at the project's administrative center, but the highest official there is merely "a technical employee who implements the stages of a plan . . ."[119] At the demonstration, 'Uthman is again forced to represent the state. This time the villagers reject him, exposing the betrothed couple's expectation of special treatment, for Nada has arranged for them to occupy one of the model houses barred to the villagers. 'Uthman, torn by loyalty to his people and his promises to Nada, breaks down and flees to the east before his parents move. He adheres to his principles but loses his love. Nada, upset and longing, leaves for Aleppo, but she already has another suitor—a corrupt Lebanese businessman siphoning money from the dam project.

The final scene of the novel places the separated lovers and the displaced together as refugees from the rising waters of the reservoir. In al-'Ujayli's telling, what is submerged is not just those who lose their homes but also the very concept of revolution—each drawn into the web of relations created by the dam:

> The only one who did not travel and did not move away, neither in that week nor in the following weeks, months, and years, is the flood. Those submerged by the water and those immersed in the events that the water made, they traveled, but the flood remained in place, with all its elements and landscapes, and increased in size and intensified. Day after day the carpet of water woven by the deluge expanded and deepened. The low ground disappeared underneath, then the higher ground, then higher and higher. The houses with their thresholds washed by water at first girded themselves after water rose to their middle, then their walls collapsed, and their bricks turned into lumps, and the sludge was deposited at the bottom of the lake of the flood, while ceiling beams, window timbers and light utensils that the submerged people did not carry with them before their departure floated on the surface of the lake.[120]

As 'Uthman and Nada's betrothal dissolves so too does revolutionary possibility. The slow collapse of a house—the couple's object of desire—mirrors the revolution itself, submerged and rendered invisible under the encroaching waters. The expanding reservoir represents two powers imbricated, both indifferent to human needs and meanings: a river halted in its flow to the sea, and an unresponsive state. In al-'Ujayli's account, both are creeping, inviolable forces that act on the landscape, leaving behind individual lives as detritus.

Al-'Ujayli deepens his critique by dismantling the illusion that the revolution can remake the world by severing society from nature. *The Submerged* shows how environmental forces—whether scorching thirst or overwhelming flood—are inextricable from human fate, collapsing the boundary between forms of human violence and environmental destruction. Al-'Ujayli accomplishes this through the careful use of narrative structure and a key bit of syntax. The novel begins with Nada's first encounter with 'Uthman's family. During the visit, she learns a family secret—'Uthman's father is a murderer, having killed his cousin years before at a well. After confessing to Nada that he did the murder "with my dagger and this hand," 'Uthman's father recants, seeking exculpation in a confused jumble of natural imagery:

> Then I knew I wasn't the killer. . . . The heat of that infernal day killed him, and the flocks of sheep thirsty in the desolate desert, and the water of the far valley, the deep, the bitter-tasting, and the sun killed him. . . . The sun that was blazing in the middle of the sky, melting the brains of men in their heads. . . . The sun, the well, the thirst, and the desert are the ones that killed my cousin, and killed me with him . . .[121]

Al-'Ujayli opens the novel with this story of a man driven to violence by the heat and thirst and desperate animals and bitter-tasting water, but he complicates the murderous action itself. The hand holding the dagger cannot, it seems, be decoupled from the context of arduous natural conditions.

The Submerged then closes with the opposite of thirst and unrelenting sun. There is plenty of water—a cool, refreshing Euphrates River backing up behind al-Tabqa Dam. But the imagery al-'Ujayli deploys is no more comforting. Instead of the heat that once tormented 'Uthman's father, the reader witnesses rising water and collapsing homes. Human action is never separate from nature, al-'Ujayli's narrative asserts, and in *The Submerged*, the entanglement of human and natural powers at al-Tabqa does not redeem in revolution but obliterates. The novel's final line blurs the boundary between physical and existential submersion: what the displaced knew of their lives "becomes submerged, and they become submerged ones." Until this final line, al-ʿUjayli uses the term "the submerged" to describe the displaced, but in the novel's closing, a distinction dissolves—identity *is* trapped under water and identity *as* trapped under water. Knowing and being are entangled with the waters of the reservoir, inseparable and inescapable.

The submerged lives in al-ʿUjayli's novel find their cinematic counterpart in Amiralay's later work, which similarly exposes how the state's promises of progress masked physical displacement and existential loss. A quarter-century after *The Submerged*, Amiralay's final film, *A Flood in Baʿth Country*, returns to the Euphrates Valley to examine the long shadow cast by the dam and the Baʿth Party's revolutionary rhetoric. Released in 2003, the film suggests the stubborn persistence of triumphalist concepts of environmental engineering. Amiralay's documentary begins with scenes from his 1970 film on the dam's construction and a *mea culpa*: "Thirty-three years ago, I was a staunch defender of the modernization of my homeland Syria. . . . Today, I regret this mistake I made as a youth."

The filmmaker notes that it was the shock of the collapse of the Zeyzoun Dam and concerns about other dams built by the Baʿth that drove him to make another film. The Zeyzoun Dam on the Orontes River in northeastern Syria burst in early June 2002, flooding more than 20 square miles and killing at least ten people. In the days leading up to the disaster, local inhabitants had called the provincial government about cracks in the dam.[122] Amiralay returns in *A Flood in Baʿth Country* to consider the implications of the combined failure of infrastructure and revolution.

Upon returning to the Euphrates River, the filmmaker finds an ongoing indoctrination program in the Baʿthist schools. Students read from an approved text about how an anthropomorphized Euphrates River went to a "new school" on the day the dam was inaugurated (July 5, 1973) to learn "in a modern way." At the river's new school, president Hafez al-Assad grooms the muddy Euphrates and gives it some tools, so "he writes his diary as a civilized river." The schoolchildren's text turns the river into a revolutionary: "On July 5, a human was born in Syria: the human of the Euphrates, the Victorious Human." As in the Turkish poet Arif Nihat Asya's writing, the river is associated with new life, with the arrival of a new human. However, this human being is not a blank slate—they have a purpose. With the students neatly ensconced behind rows of desks, the instructor asks: "Was the Euphrates Dam an engineering project exclusive to Syria or a nationalistic liberation project?" Thirty years after the dam's opening, the students chant the correct answer like a mantra—a nationalistic project for liberation. In the regime's propaganda, the Euphrates River is a life-giver, while the dam itself is a liberator, not just for Syria but for the wider Arab nation. By practicing good hygiene and going to school the river is liberated and civilized; it must be the same for the people of the valley.

Amiralay counters the regime's story of the river, grounding his account in the memories and experiences of those displaced. Unlike the teacher who uses a map to show students the revolutionized river, Amiralay takes his audience to the river itself, to the spot where he made his first film. There, one of the displaced points out where his house was once located and tells the documentarian what the displaced villagers teach their children about the river:

> Of course, our children are not aware of what we have experienced. They have known this lake since they were born; they think it is the Euphrates. We teach our children that the Euphrates was different, that it used to run down a valley; a river we used to swim in and walk along. These kids don't know how to swim; you know why? Because they're afraid they might drown in this lake. This is the first time in history a river is changed completely and is no longer a river! As you can see this is a sea. There is no river... the river is gone.[123]

The waters of the lake stretch to cover the entire frame during this account, a counterpoint to the simple state parable of living alongside a free-flowing river. In this account, there is no triumphant overthrow of history through revolutionary fervor and of nature through technical prowess. Denying the extensive changes to the river means denying the man's memory and the presence of a new sea, perceived now as more dangerous than the river it replaced, stretching nearly to the horizon.

Amiralay and al-'Ujayli dismantle the state's narrative by turning its triumphalist symbolism inside out. The regime portrays the dam as a liberator, a force for revolutionary modernization that brings prosperity to the people. Yet, the filmmaker and writer reverse this perspective, making the supposed beneficiaries the true subjects of the story and exposing the dam's role as an instrument of dispossession. Amiralay and al-'Ujayli show how the dam's waters do not uplift but instead engulf—submerging homes, histories, and the identities of the people it claimed to serve. In their telling, the dam does not empower; it constrains, determining and shaping education, cultural expression, and everyday life. In the process, the author and filmmaker show the co-imbrication of the state with nature; their separation is a conceptual tool for enacting domination. The drowned valleys, displaced communities, and broken promises of prosperity are thus not mere byproducts of technical civilization; they are its foundation as the dam brings about a deep, material entanglement of political power, environmental forces, and the fabrics of social existence. What was once imagined as the pinnacle of modern engineering and revolutionary change becomes in Amiralay's trilogy and al-'Ujayli's novel an enduring testament to the costs of dreaming civilization through rivers.

The dreamer, the dreaming, and the dreamt—each phase of the Euphrates dams' history reveals a complex interplay of civilizational ambition, geopolitics, and the consequences for lived experience. The dreamers, whether Turkish engineers like Demirel or Ba'thist leaders, envisioned a technical civilization that would transform rivers into engines of national vitality and sovereignty. The dreaming—the process of funding and constructing these projects—was shaped as much by political alliances, arms deals, and violent repression as by river surveys, concrete, and engineering expertise. Dreaming civilization through rivers was neither wholly capitalist nor communist but reflected a shared faith in industry's power to master rock and water—and, through them, to reshape social and communal life. The dreamt—the people and places reshaped by rising reservoirs—confounded the dream's coherence, exposing the ideological illusions and contradictions that made submerged homes, suppressed identities, and displaced communities seem necessary to national progress.

Rather than a linear tale of success or failure, then, the story of Keban and al-Tabqa is one of entanglement, of how deeply human social and political life became embedded in the natural world through these projects. The dams were built to engineer the environment and transform landscapes, to as 'Uthman puts it, "obtain justice from nature and society," yet in merging state, technology, and nature, the dams and reservoirs became much more than that: vehicles for dreams and revolution, encapsulations of 'development' and geopolitical orientation, terrains of contestation and war. Through this entanglement, the dams and reservoirs themselves became forces in history, their effects ramifying outward into state and society. They brought new forms of governance, culture and community into being—just as these had first made their construction possible—influencing lives, landscapes, and ecologies long after their construction.

And, as the reservoirs kept rising, the entanglements only grew tighter. In the decades following the construction of Keban and al-Tabqa, the march of dam-building continued, each new project—Karakaya Dam, the Mosul Dam, the vast Atatürk Dam—another chapter in this ongoing entanglement of politics, culture, and environment spanning multiple levels and scales. By the 1990s, the discourse around water management had shifted, with statesmen and engineers alike invoking sustainability and another round of "regeneration." But the history of these dams and reservoirs suggests that most of the familiar framings—geopolitical, cultural or technical—cannot fully contain the realities set into motion. In the twenty-first century, as climate change alters the hydrology of the region, the entangled historical powers of these reservoirs will evolve, forcing new reckonings with the dreams of technical civilization.

Conclusion

REGENERATION AND A NEW CENTURY

Throughout the 1980s and 1990s, dam building and reservoir making continued in Iraq, Syria, and Turkey.[1] But the reservoirization of the two rivers reached its apex in the upper basin, where dams, irrigation works, and hydroelectric power plants in Turkey underpinned a much larger multi-sector development project, known as the "Southeast Anatolia Project" [*Güneydoğu Anadolu Projesi*]. The project is commonly referred to as the GAP according to its Turkish acronym. Conclusions typically eschew new material, but in this case, a discussion of the Southeast Anatolia Project provides a way to review this book's primary arguments—demonstrating how the GAP reflects earlier twentieth-century histories, while also showing the continued relevance of this book's frameworks and arguments for the twenty-first century.

The Keban Dam may be excluded from the official boundaries of the Southeast Anatolia Project, but it was the prototype—technically, institutionally, and symbolically—for the massive transformation that followed. As in Iraq in the late 1940s and 1950s, a piece of water infrastructure grew into something much larger: dam building not only as infrastructure but as a blueprint for economic, social, and political change. Initially, various parts of the Turkish bureaucracy viewed the Tigris and Euphrates Rivers separately, but the development plans written from 1961 onward considered schemes for broader regional development.[2] In 1977, the Turkish government connected its plans for the two rivers into a single project, and in 1988 commenced plans to integrate the rivers into broader investments in the economy, public services, and cultural initiatives.³ The GAP envisioned the construction of twenty-two dams and nineteen hydroelectric power plants in the rivers' basin, as well as the irrigation of thousands of square kilometers.⁴ The massive Karababa Dam, renamed for the Turkish Republic's founder Mustafa Kemal Atatürk, is the GAP's keystone project. In 1992, the dam became operational, producing electricity and diverting nearly one-third of the Euphrates River flow toward irrigation systems on the plains surrounding Urfa and Mardin. Though the GAP began in the last decades of the twentieth century, it continues as of this writing with additional dams under construction.⁵

The "balanced social order" that Turkish planners dreamt for eastern Anatolia through the Keban Dam in the 1960s has taken material and social form in the GAP's transformations of the southeast. The development project extends far beyond hydraulic structures, contributing to the construction of airports, hospitals, schools, and universities. Cultural interventions accompany infrastructural ones. The GAP is self-consciously attempting to remake economic and social life in southeast Anatolia, with programs that target different communities and classes. Geographer Leila M. Harris has done significant ethnographic work in southeastern Anatolia and argues, "... in the GAP case understandings of gender, particularly related to the status and situation of women, are part and parcel of what fuels the alteration of the agrarian and socioeconomic landscape of southeast Turkey from the outset." Some of these interventions are overt, as in the construction of women's centers, while others are a product of infrastructural transformation, as Harris notes: "As GAP water-related development proceeds, men and women of the region are forced to literally renegotiate their locations, positions, identities, livelihoods, and knowledges in the face of rapid and extensive waterscape changes."⁶ The wholesale remaking of land and water compels a redefinition of everything rooted in those places—identity, labor, belonging, and being itself.

Gender is not the only sociocultural factor undergoing change through the GAP. Memory and history are also implicated and, like gender, shift in large and small ways. As in the case of the Keban Dam, flooding from the GAP's reservoirs has imperiled important archaeological sites, most notably the ancient Roman mosaics at Zeugma, which were threatened with inundation by the Birecik Dam. Millions of dollars were spent to excavate and preserve the mosaics, which, as of 2011, are housed in a 320,000 square foot museum in Antep.[7]

The effort and expenditure to preserve ancient culture stands in contrast to the efforts to erase living cultures, also linked to the GAP. As the two rivers became connected in planning documents the Turkish military instigated another coup in 1980 and set out to destroy a nascent Kurdish nationalist movement, and, indeed, any form of "Kurdish ethnic awareness."[8] During the 1980s, tens of thousands of Kurdish citizens were arrested; many faced torture. Mass trials sentenced hundreds to death. The relative opening for Kurdish self-expression that had existed in the 1970s slammed shut, and bans on Kurdish language, clothing, and even the word "Kurd," were strictly enforced. Kurdish groups resisted these measures, leading to a militarization of southeastern Anatolia at the same time as Turkey prepared to dam the rivers.

Cultural and material violence mixed in southeastern Anatolia as the dams rose. Wolfgang Wohlwend tells of the town of Halfeti, inundated by the Birecik Dam in 2000. Prior to the dam's construction, Halfeti residents had been involved in the political turbulence and violence of the 1980s. Wohlwend states: "[O]ral reports suggest that a member of almost every family was carried off to Diyarbakır prison, which was notorious for unspeakably horrific conditions and systematic torture."[9] With the filling of the reservoir behind the Birecik Dam, Halfeti residents lost prized orchards, which served not only as an economic resource but also as a gathering place and a locus of solidarity. Those displaced from Halfeti obtained new homes, but the newly constructed town absorbed a Kurdish village and consolidated several others, while also being cut off from its previous transportation links to the outside world. Thus, the GAP broke the community and material ties of the old Halfeti, and then what remained was used to disrupt the community and material ties of other villages. This process was replicated again and again across southeastern Anatolia with the construction of each new dam and the filling of each new reservoir.

Meanwhile, in place of these cultural and material ties, the GAP sought to produce a new cultural conception of southeastern Anatolia, just as Turkish poets had sought to engineer new cultural understandings of the Euphrates River at Keban. In other words, the GAP sought to overwrite the Kurdish script of the region, while producing entirely new physical landscapes. The project published a guidebook entitled *The Ascending Place of Light* in multiple languages to encourage tourism from both internal and international visitors. Accompanying the guidebook is a small pamphlet of legends from different parts of the region. Atilla Koç, the Minister of Culture and Tourism at the time of publication, notes in the preface: "Thus, the book in your hands can be viewed as the materialized intention of bringing old and new, museum and bazaar, history and this day together."[10] Koç's comments present a rhetoric of continuity between past and present—a key idea for the GAP's proponents—even as the project has been deeply disruptive to physical and cultural landscapes. Moreover, by conjuring the museum, Koç suggests that the history to be explored is one curated, a controlled vision of the past, perhaps in the form of a museum full of Roman mosaics. The GAP even has a cookbook, the product of a "GAP Gourmet Tour and Disappearing Local Tastes Recipe Contest," held in mid-December 2011. The publication does not specify why or whose local tastes are disappearing, instead extolling the GAP for preserving the recipes (if not the chefs), since "many of the countries that are successful in tourism introduce themselves by highlighting their cuisines."[11] The GAP is thus repackaging culture as a commodity to go along with the other commodities produced through the extension of irrigation. Further, just as Iraq's poets used metaphors conjured by a flood to deride the government's empty promises, so too may we look upon the word "disappear" in the cookbook's title as a verb, the active creation of absences in cultural memory that the book's glossy pages cannot possibly remedy.

And, speaking of the bazaar—a place of exchange—while the GAP has injected significant capital into the economy of southeastern Anatolia, its benefits have not been spread evenly, replicating the inequalities that existed before the project. Political scientist Arda Bilgen has noted that "just 6.2% of large agricultural enterprises cultivated almost half of the land," a ratio similar to the disparities in land ownership and cultivation prior to the GAP.[12] Meanwhile, in Harris's ethnography, Kurdish communities disadvantaged by state policy since the advent of the Turkish Republic have been left out of irrigation provisions. Under earlier economic conditions, some landless Kurds had enjoyed grazing rights on fallow land. The advent of irrigation eliminated access to those lands, cutting off opportunities to earn a living through animal husbandry.[13] As in Iraq's political ecology of salt, the so-called "balanced social order" has through violence, water infrastructure, cultural erasure, and displacement mostly extended and entrenched the previous status quo and its original inequities.

Both the lack of progress in socioeconomics and the "disappearing local tastes" may be tied to the profound transformations in land use that have occurred under the GAP. To bring Euphrates River water from the mountains to the steppe, the GAP employs two massive tunnels, more than 7.5 meters in diameter and 25 kilometers in length, which transfer irrigation water from the Atatürk Dam Lake (read: reservoir) onto the plains. The system provides irrigation water to 476,000 hectares of land and drinking water to the city of Urfa.[14] The GAP Master Plan envisioned a total of 1.8 million hectares under irrigation by 2013; as of 2021, about a third was operational.[15] This dismal result has partly to do with the top-down nature of the GAP's implementation. While the large-scale works are impressive, the smaller-scale ones—one might say the "human-scale ones"—where small canals deliver water to the fields, have sometimes been inappropriately designed.[16] At these fields, in what had been a rainfed agricultural zone primarily dedicated to the growing of food staples such as wheat, barley, and lentils, large enterprises now produce export-ready and water-hungry crops such as cotton. South of Urfa on the Harran plain, where Adam and Eve supposedly fell after being expelled from paradise, cotton represents 85% of summer irrigated crops. This shift in land use has meant a threefold increase in water use, from 370 million cubic meters in 1993 to over one billion cubic meters in 2002.[17] Recalling the discussion in the previous chapter, these figures mean that cotton cultivation uses nearly the same amount of water annually as evaporates from the GAP's reservoirs.

Extracting that much water from the basin has myriad consequences downstream, as we will soon discuss. But, here, let us consider where that water goes. Evaporate is subject to the vicissitudes of atmospheric currents, while the water used to grow the cotton is exported as "virtual water." A good part of Turkey's raw cotton production goes to China, Vietnam, and Bangladesh, countries with significant textile production. The GAP was supposed to fulfill Turkey's domestic cotton needs, but the modernization of the textile industry now requires Turkey to import half of those needs. So, in a sense, Turkey is exporting the water of the Tigris and Euphrates Rivers in order to import water from elsewhere.[18] Meanwhile, as in Iraq, expanding irrigation has contributed to salinization, particularly on the Harran plain, where more than ten percent of fields have experienced lower yields as a result. Reclaiming salinized lands by washing out the salts results in higher salinity in the rivers' water. Further, decreasing soil productivity has led to an increase in fertilizer use; producing fertilizer requires an enormous amount of energy and overall contributes five percent of global greenhouse gas emissions.[19]

Climate change, weather, land use, infrastructure, and economic policies have all come together to produce a decline in cotton production from the GAP. In the 2020–2021 growing season, rain on the plain, now more an irony than a benefit, damaged much of the irrigated cotton crop, while the devaluation of the Turkish lira increased fertilizer and pesticide prices. Reduced fertilizer application caused a decline in yields on the depleted soil. Pests, a great danger to a monoculture, then destroyed some production, so farmers turned to counterfeit products that caused even more damage to the crop.[20] Finally, all of the salt and chemicals must end up somewhere, and the result has been a serious deterioration in water quality, affecting downstream flows into Iraq and Syria, and contributing to further agricultural decline in the rest of the basin. As a result of the water engineering built in all three states over the course of the past several decades, the Fertile Crescent is becoming increasingly infertile.

The Southeast Anatolia Project, like earlier efforts across the Tigris-Euphrates River Basin, was built not only out of concrete and planning documents but also out of narratives—stories of neglect, loss, and regeneration. As in Syria, where the Baʿth framed al-Tabqa Dam as a form of national liberation and revolutionary commitment, these narratives were not mere rhetorical flourishes. They provided the moral, historical, and civilizational logic through which massive infrastructure projects could be imagined, justified, and materially realized. The GAP has drawn on two such narratives in particular. The first is a story of national neglect—the idea that southeastern Turkey has long suffered from economic and political abandonment. The second is a revivalist story that has been articulated since (at least) the beginning of the twentieth century: the call to restore the splendor of ancient Mesopotamian civilization. These two connected narratives frame the southeast as a space out of time, either fallen behind or suspended in ruin, awaiting rescue through state-led technical intervention.

From the early years of the Turkish Republic, state officials and development proponents invoked the narrative of neglect, positioning the east as a disordered space in need of rescue. Mustafa Kemal, as president of the Turkish Republic, referred to Ottoman rule in eastern Anatolia as "apathetic" or "indifferent" [*nemelazımcı*]. In a 1931 visit to eastern provinces, he expressed a wish to see "children whose faces are not pale, whose spleens are not swollen," insisting that "whatever civilization exists in Istanbul, whatever modernity we are trying to bring to Ankara, I want our entire homeland to experience the same level of progress."[21]

Süleyman Demirel, Turkey's "King of Dams" and civilizational dreamer, extended these logics in the language of post-war development. In 1975, he wrote:

> The dire need for the economic, social, and cultural development of Eastern and Southeastern Anatolia is a vital national issue that will unite us all as a nation. The fact that our eastern and southeastern provinces are less developed than other regions is not an interregional issue that pits the interests of the people of one region against the interests of our citizens living in other regions. This is an issue arising from centuries of neglect and the convulsions of wars, invasions, and migration movements throughout history.[22]

For Demirel, centuries of neglect could be rectified by dams and reservoirs *qua* technical civilization. He imagined these infrastructural interventions in the environment as the cure for historical trauma, as a way to calm the convulsions of the past and bring order to a supposedly unruly geography and restive populace.

These framings echo imperial rhetoric, for British engineers and bureaucrats in Iraq in 1919 argued about the rivers' supposed degradation. Moreover, as Diana K. Davis has shown in the North African context, narratives of neglect have abetted practices of political and economic domination.[23] Would-be empire builders along the two rivers have scarcely been different, juxtaposing stories of local disorder—tribesmen murdering governors and "animated by the true Semitic desire for the immediately profitable"—with ones evoking the supposed splendor of earlier civilizations. The British in Iraq and the French in Syria both sought to remake the basin in the image of a lost Mesopotamian civilization. The rise of nationalism did not end these visions; it extended them.

Turkish planners, too, placed the GAP within such civilizational terms. Arda Bilgen has shown how Turkish politicians and bureaucrats linked the project to a "glorious past" and "imagined the GAP region as the continuation of ancient Mesopotamian civilization." References to Mesopotamia, Bilgen argues, served to emphasize regional disparity and to reinforce "the negative image of poverty and a sense of dissimilarity in the eyes of Turkish society."[24] The contrast between past splendor and present poverty supported the legitimacy of state-led transformation.

Yet narratives of neglect and restoration should not be seen merely as political rationales, external to the engineering process. These narratives do not frame the technical—they are folded into it, materialized in forms, diagrams, flows, and, ultimately, dams. Too often, development is treated as having two distinct dimensions: the stories we tell about it, and the technical work that realizes it. Each form of knowledge is widely assumed to exist in its own epistemological sphere. This division—between discourse and practice, meaning and materiality—is a fallacy.

Demirel himself is proof. He was not just a politician nor just an engineer. He was both: a maker and a dreamer, a builder of hydroelectric systems and a theorist of civilization. To separate his rhetoric from his technical work is to overlook how his vision emerged through their entanglement—spoken, drawn, and set in concrete. His biography shows not conception and then implementation, but a vision forged through the act of building itself—in fellowships, blueprints, budgets, and the reshaping of land and water. Much of the development literature functions according to a separation of discourses—narratives as justification in one box and data and measurements as technique in another—as do the institutions designed to facilitate its analysis.[25] But as a new materialist lens insists, worldmaking is always simultaneously conceptual and material.

As Karen Barad puts it, "we know because we are *of* the world."[26] And because of that, technical documents can equally embed dreams, operate materially, and act as instruments of imagination. They are both tools of measurement and artifacts of dreaming. The same narratives that framed the southeast as neglected or as the heir to Mesopotamian grandeur do not merely precede or justify technical plans—they inhabit them. Engineering reports, design specifications, and feasibility studies encode civilizational ambitions and assumptions about history, geography, and identity. These documents do not occur outside of political rhetoric; they formalize it into timelines and technical benchmarks. Thus, narratives and technical plans belong alongside the rocks and water, the shovels and backhoes—not just enabling or justifying dam construction but making the dam thinkable in the first place and real in the second. Put another way, civilization is both a plan and a material goal or project, thought and built in concert, realized into being together.

The fact that technical documents help materialize civilizational visions means that those visions are mobile—they may be transported across time and territory through infrastructure itself. This is perhaps most evident in the reappearance of the term "regeneration," a concept revived in official GAP discourse decades after the British left Iraq. The continuity of the concept across space and time suggests that, even as the landscapes around the rivers have changed in staggering ways, modern ways of dreaming and building have not. British engineers in 1919 Iraq used the term regeneration to advocate "saving" the rivers from the people living along them. One cannot help but wonder if its introduction in Turkey seventy years later was likewise an effort to "save" the rivers from the Kurds, for the concept returned due to political resistance to the GAP both at home and abroad.

Kurdish nationalism and separatism intensified after 1984, when many of the dams were under construction, belying Demirel's claims that development is an "issue that will unite [the] nation." Kurdish resistance originated in ethnic solidarity and in response to long-standing grievances against the Turkish state and so was not solely aimed at the GAP. However, Kurdish groups have sabotaged and attacked dam construction sites, and the project specifically has provoked ire among the people of southeastern Anatolia.[27] Local resistance and concerns about the transboundary effects of the GAP's dams and irrigation systems also resulted in major international donors refusing to fund dam building, pulling out of projects underway and providing monies only for other aspects of the GAP.[28] The Turkish government therefore has financed a good deal of the water infrastructure on its own. The increased international scrutiny gave rise in the late 1980s and 1990s to additional claims that the GAP would be environmentally sustainable and a net positive for national security; it is in these contexts that the concept of regeneration found new usage.[29]

In 1989, as Japanese and Turkish engineers devised the multi-volume GAP Master Plan, Dr. Ali İhsan Bağış published a book entitled *GAP Southeastern Anatolia Project: The Cradle of Civilization Regenerated*. Bağış, one of the consultants to the Turkish State Planning Organization, notes the book as "the first attempt to make GAP known to the rest of the world."[30] Far from an individual effort, the text includes several prefatory notes from Turgut Özal, president of Turkey; Kamuran İnan, the State Minister for the GAP; and Melih E. Araz, the CEO of Interbank, which funded the title's production. Each author evokes a shared vision: that the great and wondrous past of Mesopotamia could be reborn through modern engineering, and that the GAP was the vehicle for that regeneration. İnan predicts that completing the GAP in the 2000s would mean "the rebirth of the prosperity which Mesopotamia enjoyed thousands of years ago, accompanied by modern technology." Araz goes further:

> GAP aims to reanimate the Tigris-Euphrates basin by providing irrigation and electricity to the arid, sparsely populated area. In the 17th century according to the traveler Evliya Çelebi, the area was a huge oasis; "I passed through such a basin, that I did not see the sun all the way to Baghdad for the citrus trees which covered the region."[31]

The reference to the Ottoman traveler and his journey appears as a statement of fact, not as an exaggeration or wondrous exclamation on the part of a traveler.[32]

The book's historical narrative works along the same lines, presenting a selectively curated civilizational genealogy. Assyrians, Seleucids, the Parthians, a series of Christian kingdoms, and then Arab and Turkish Muslim states are arranged into a seamless procession culminating in a Turkicized Anatolia. According to the text, the "cultural intermingling" that happened in Mesopotamia and Anatolia continued "from the 8th millenium [sic] to the present, [and] has resulted in a colorful mosaic of cultures," leading naturally toward Turkish rule. Religion also became mixed, with Urfa as an important historical center where Abraham "fought paganism" and where Nestorian Christianity was propagated. Under Muslim rule, the plains of Harran were irrigated and a great center of learning established. Finally, though rich in culture, the "entire Region entered into a process of Turkicization" from the mid-1100s to the present.

More recent history, meaning in this case the past 500 years, is merely glossed. The Ottoman period is treated in a single paragraph, omitting the "wars, invasions, and migrations" of the immediate past. No mention is made of the Armenian genocide or the suppression of Kurdish aspirations, events of rather more profound import to twentieth-century Anatolia than the ruins of Harran University. The omissions are striking, but they are not accidental. As with the GAP itself, this history is not merely discursive: it is a material project of replacement. In regenerating Mesopotamia, the GAP displaces villages, redirects rivers, disappears "local tastes," and transforms cultural suppression into tourist offerings. Regeneration in this context is both a symbolic and physical process.

Such a selective invocation of the past is not unique to Turkey. The Ba'th regime in Syria imagined al-Tabqa Dam as a revolutionary break—a literal and ideological rupture that would remake the Arab world. Its construction was framed as an act of national liberation, binding infrastructural development to political sovereignty. By contrast, the Turkish framing of the GAP—especially in Bağış's formulation—portrays it as a return rather than a break. Where the Ba'th positioned their dam as a new beginning, Turkish officials used "regeneration" to invoke the GAP as a return to a latent civilizational destiny. In doing so, they aligned their project more closely with British efforts in Iraq during the mandate period, where colonial engineers proposed "regenerating" the rivers by rescuing them from local use and restoring them to the imagined heights of an ancient imperial order. In both Turkey and Iraq, regeneration served as a conservative gesture—a recasting of modern infrastructural transformation as the fulfillment of a lost and mythologized past.

Understanding regeneration in this century-spanning comparative context illuminates the broader stakes of the concept. As Barad argues, material practices are never purely technical operations or symbolic gestures—they are entangled enactments of knowledge, matter, and ethics.[33] Thus, the discourse of regeneration in Turkey materializes in the built landscape, in the bureaucratic systems that manage the rivers, and in the historical narratives that render displacement and dispossession into civilizational necessity. Bağış's claim that the GAP is the culmination of nearly a thousand years of Anatolian cultural formation reframes modern dam building as the visible expression of a latent Turkish civilization, the culmination of intermingling and mosaic-making with a Turkish inflection. Just as the Keban Dam was retroactively tied to the events of 1071, the officials involved in the GAP imagine its own deep past, asking us to locate the roots of its future in a mythic continuity of national becoming.

In this dream of regeneration, the dams are not symbols of modernity imposed on a resistant geography—they are instruments for revealing a civilizational destiny long concealed by supposed underdevelopment. The GAP is thus framed not as a generative act but as a regenerative one, a recovery of what was always meant to be. A similar logic animated American-authored plans for the Keban Dam decades earlier, where technical documents insisted on the inevitability of a prosperous future, and appeared again in Syria, where Ba'thist dreams of dam-driven revolution encoded visions of Arab greatness within concrete and turbines. In each case, infrastructure functions not only as a physical intervention but also as an ontological claim about what civilization is, where it resides, and who has the right to shape it.

Seen in this light, the concept of regeneration does not simply reappear in GAP discourse—it endures as a transnational, transimperial technique of worldmaking. Regeneration draws together political design, material transformation, and historical narrative into a single apparatus. To build a dam is not only to redirect water but to enact a vision of the world, to write history in rock and steel, and to remake human and ecological relations under the sign of progress. The GAP is not merely Turkey's project; it is the latest expression of a broader and older dream of civilizing rivers—of taming flow in the name of order, unity, and the resurrection of imagined pasts.

To tell entangled histories is not merely to write differently—it is to dream differently. For if the future is made from the shards of memory, then the stories we tell about rivers, civilizations, and the ruins of development must include not only what was thought and built but what was broken, flooded, absorbed, shifted, desiccated, and reconstituted. This book has traced such entanglements across the Tigris-Euphrates River Basin, showing how dreams and data, dams and documents, coproduced landscapes and states now imperiled by the very visions that shaped them.

The Marxist theorist Raymond Williams argued in a 1980 essay that our concepts of nature are shaped by the social world—and in turn shape it:

> Out of the ways... we have interacted with the physical world we have made not only human nature and an altered natural order; we have also made societies. It is very significant that most of the terms we have used in this relationship—the conquest of nature, the domination of nature, the exploitation of nature—are derived from the real human practices: relations between

men and men.... We need different ideas [of nature] because we need different relationships.³⁴

But Williams's "old materialist" formulation, for all its clarity, still relies on a neat division between idea and relation, mind and world. Our very conceptual categories can defeat us before we begin—the gendered language of Williams's essay offers further evidence. Instead of social relations on one side and ideas on the other, new materialism insists that no such clean boundary exists: knowing, building, dreaming, and living happen in the same moment. Entanglement challenges us to recognize the inextricable qualities of thought and being happening in concert with the world we inhabit. As a result, the stories told in technical documents are no less world-making than the stories told by novelists, artists, or activists. The difference lies in what kind of world they call into being.

A good example of such entangled stories is Khalid Suleiman's *Guardians of the Water: Drought and Climate Change in Iraq*, a work that moves between history, science, and literature to explore how Iraqis have understood and navigated the natural world. Unlike studies driven by "numbers, government data, and regional and international agreements," Suleiman focuses on the lived experiences of communities, connecting these in creative ways to scientific data and government policy.³⁵ In 2019, he traveled with environmentalist Jassim al-Asadi to al-Dalmaj Marsh to learn about the condition of Iraq's southern wetlands. Suleiman writes of his journey through Iraq's watery areas, comparing them to the parts of a living organism:

> If the land of Iraq forms its body, then these wetlands—rich with reeds—are its lungs, generating oxygen, mitigating heat, and supporting diverse forms of life: birds, mammals, reptiles, and insects.
>
> ... [T]he southern marshes ... are also Iraq's kidneys, as the reeds found in the marshes and basins ... filter pollutants from aquatic ecosystems.³⁶

It is tempting to see these quotations as metaphors, but Suleiman's descriptions resist such an easy categorization. His vision of Iraq's wetlands is not merely symbolic; the actual breath and health of humans and nonhumans alike relies on these ecosystems. Just as engineers' writing must be read as materialized dreams, Suleiman's account is an embodied way of knowing, entangled in landscape, ecosystem, infrastructure, and social struggle. Like lungs and kidneys, these connections are not ornaments to the body politic—they are constitutive of life and its viability.

Such stories matter not only for what they depict but for what they *do*. Al-Asadi, Suleiman's guide in the marshes, is not just an observer but an activist, engineer, and co-founder of Nature Iraq, the country's most important environmental organization.[37] His work is part of a long struggle to defend Iraq's fragile ecosystems—one that has put him in real danger. In 2023, al-Asadi was kidnapped by armed men and held for more than two weeks. He was tortured, electrocuted, and beaten during his captivity. He is not alone. Iraqi government agencies have also harassed and imprisoned environmental activists as they draw attention to polluting activities, water theft, and state-backed degradation and contamination.[38] That such violence and persecution targets those who seek to protect water, soil, and air underscores the potency of entangled narratives. These stories reconfigure the terrain of environmental politics by using the very tools of state expertise—hydrology, engineering, planning, and ecology—to critique state practice.[39]

As this book has demonstrated, entangled stories and technical documents share more than a rhetorical structure: both function as technologies of the future. Before meeting al-Asadi, Suleiman visited the writer Muhammad Khudayyir, whose fiction depicts his home "water city" of Basra and surrounding wetlands.[40] But in a recent story, "Graffiti 2042," Khudayyir imagines a dystopian future for Basra, one where the city's "inhabitants moved underground after temperatures soared beyond the limits of human endurance and surface water became scarce."[41] The protagonist, a postman, tries to deliver the mail in a city suffocating under the weight of climate disaster. Suleiman argues this short narrative is not a piece of science fiction [*al-khayāl al-'ilmī*] but "a futuristic story [*qiṣṣa mustaqbaliyya*] that foretells climate change—a phenomenon confirmed by scientific studies . . ."[42]

In other words, Khudayyir's story is not so much speculative as grounded, less fantasy than forecasting. Climate models suggest that Persian Gulf cities could experience heat indexes of over 165 degrees Fahrenheit (wet-bulb temperatures of over 35 degrees Celsius) for several hours multiple times a year—long enough to kill a human being, not to mention other animal life.[43] Meanwhile, overuse and diversion of the Tigris and Euphrates threatens to stop the rivers from reaching the sea. Under such conditions, living underground in the summer may soon become not a metaphor in a prominent writer's story but a reality.

Against the backdrop of such precarity, what does an entangled history offer? Across its several chapters, this book has insisted on a different kind of environmental historicism. Unlike Karen Barad's agential realism, which remains largely in the realm of philosophy and theoretical physics, or Dipesh Chakrabarty's planetary history, which risks dissolving the local within the global, *Two Rivers Entangled* offers a historically grounded, ethically engaged materialism.[44] It does not abstract rivers into metaphors or planetary signs. It follows their course through salinity and sediment, blueprint and battlefield, bureaucracy and belief. It offers entanglement not as a metaphor, but as a method.[45]

It is not enough, though, for this method to apply only to history. The engineers at Keban wrote as if the dynamism of human pasts and natural processes did not matter—the dam will be thus, they wrote, even though it would not. Their reports constructed a future immune to contingency, built from a vision of nature as controllable and of history as progressive. But entangled history resists such certainty. It insists that the past, like the future, is as much the product of unintention as intention—shaped by salt as much as by statecraft, by floodwaters as much as by ideologies.

Perhaps a step toward a different kind of environmental engineering—and with it, a different dream of civilization—would be to insist that every study of dam building also include plans for dismantling. An un-engineering, then—not just how to build, but how to undo, and what might be gained in the process. What if more engineers became like Jassim al-Asadi, not the builders of civilization but the guardians of mutual adaptation? What if more people learned to see marshes as breath, to read riverbanks as archives, and to question the stability we have embedded in rock? In such practices, new relations might emerge—relations founded not on mastery or control or progress, but on humility, reciprocity, and the difficulties of standing within the flow of a world in motion.

Notes

Introduction

1. Lena Hommes, Rutgerd Boelens, and Harro Maat, "Contested Hydrosocial Territories and Disputed Water Governance: Struggles and Competing Claims over the Ilisu Dam Development in Southeastern Turkey," *Geoforum* 71 (May 1, 2016): 9–20, https://doi.org/10.1016/j.geoforum.2016.02.015.

2. For more on the history of al-Jazira, see Samuel Dolbee, *Locusts of Power: Borders, Empire, and Environment in the Modern Middle East* (Cambridge: Cambridge University Press, 2023).

3. Originally published in 1964, the classic English-language introduction is Wilfred Thesiger, *The Marsh Arabs* (London: Penguin Classics, 2008); later works have revisited the subject and discuss the transformation of the marshes and the people who relied on them. See Gavin Young, *Return to the Marshes: Life with the Marsh Arabs of Iraq* (London: Faber & Faber, 2011); Sam Kubba, *The Iraqi Marshlands and the Marsh Arabs: The Ma'dan, Their Culture and the Environment* (Reading, UK: Ithaca Press, 2011).

4. Historical statistics for the two rivers vary widely depending on source, date range, gauging station, and other factors, a fact that most hydrological studies of the rivers recognize. Figures comparing the Nile, Tigris, and Euphrates Rivers come from Nurit Kliot, *Water Resources and Conflict in the Middle East* (London: Routledge, 1993), 22–27 and 108–11. See also Nadhir Al-Ansari, Nasrat Adamo, and Varoujan K Sissakian, "Hydrological Characteristics of the Tigris and Euphrates Rivers," n.d., 12–20; John F. Kolars and William Mitchell, *The Euphrates River and the Southeast Anatolia Development Project* (Carbondale: Southern Illinois University Press, 1991), 3–8.

5. Daniel Hillel, *Rivers of Eden: The Struggle for Water and the Quest for Peace in the Middle East* (New York: Oxford University Press, 1994), 95–97.

6. Marc Van de Mieroop, *A History of the Ancient Near East, ca. 3000–323 B.C*, 2nd ed. (Malden, MA: Blackwell Pub, 2007), 7–16.

7. There are several such accounts during the early years of the mandate. See, for example, Kenneth Mason, "Notes on the Canal System and Ancient Sites of Babylonia in the Time of Xenophon," *The Geographical Journal* 56, no. 6 (1920): 468–81, https://doi.org/10.2307/1780469.

8. Heavy rains have caused flash flooding in Baghdad and other areas of southern Iraq. See, for example, "Floods Cause Severe Damage in Southern Iraq," *Al-Monitor*, May 9, 2013.

9. Geologists face the same problem in narrating prehistoric change, and their models offer opportunities for historians. See Dale J. Stahl, "The Dam as Catastrophe: Connecting Geological Models to Modern History," *Water History* 13, June 21, 2021, https://doi.org/10.1007/s12685-021-00278-4.

10. Richard White, "Environmental History, Ecology, and Meaning," *The Journal of American History* 76, no. 4 (1990): 1115, https://doi.org/10.2307/2936588.

11. Azzam Alwash of Nature Iraq has noted, "For a more permanent solution, two construction companies have suggested supporting the foundation of the dam on the lower bedrock formations. This means an unprecedented 200-meter (600 feet) cutoff slurry wall under the dam. This solution, estimated to cost approximately $3 billion, may not even solve the problem as such a deep cutoff wall has never been attempted before and special equipment would need to be designed and built to accomplish such an engineering feat." Azzam Alwash, "The Mosul Dam: Turning a Potential Disaster into a Win-Win Solution," *Viewpoints* 98 (2016).

12. Nadhir Al-Ansari et al., "Mystery of Mosul Dam the Most Dangerous Dam in the World: The Project," *Journal of Earth Sciences and Geotechnical Engineering* 5, no. 3 (2015): 16–17.

13. Dave Philipps, Azmat Khan, and Eric Schmitt, "A Dam in Syria Was on a 'No-Strike' List. The U.S. Bombed It Anyway," *International New York Times*, January 22, 2022, Gale Academic OneFile (accessed May 27, 2025).

14. The group's 2014 declaration of a caliphate reads: "Here the flag of the Islamic State, the flag of tawḥīd (monotheism), rises and flutters. Its shade covers land from Aleppo to Diyala. Beneath it, the walls of the tawāghīt (rulers claiming the rights of Allah) have been demolished, their flags have fallen, and their borders have been destroyed." SITE Intelligence Group, "ISIS Spokesman Declares Caliphate, Rebrands Group as 'Islamic State,'" June 29, 2014, https://news.siteintelgroup.com/Jihadist-News/isis-spokesman-declares-caliphate-rebrands-group-as-islamic-state.html, accessed Nov 1, 2016.

15. Darryl Li writes, "Jihadi groups may invoke an authority above this formal legal system (and they are hardly alone in doing so), but such universalist messages must always contend with and often work through actual institutions such as states." Darryl Li, "A Jihadism Anti-Primer," *MERIP*, no. 276 (Fall 2015), https://merip.org/2015/12/a-jihadism-anti-primer/.

16. The Center for International Security and Cooperation at Stanford University noted in 2014 that the group was "the richest terrorist organization in the world." "The Islamic State," Mapping Militant Organizations, accessed April 21, 2023, https://cisac.fsi.stanford.edu/mappingmilitants/profiles/islamic-state.

17. Jeremy Ashkenas et al., "A Rogue State Along Two Rivers," *The New York Times*, July 3, 2014. https://www.nytimes.com/interactive/2014/07/03/world/middleeast/syria-iraq-isis-rogue-state-along-two-rivers.html.

18. The reference here is to James C. Scott, *Seeing like a State: How Certain Schemes to Improve the Human Condition Have Failed*, Yale Agrarian Studies (New Haven, CT: Yale University Press, 2008).

19. Li, "A Jihadism Anti-Primer."

20. A previous state built that infrastructure, but only a very few states in human history can claim a pristine provenance, with little to no reliance on the knowledge and works of previous societies.

21. Benedict R. Anderson, *Imagined Communities: Reflections on the Origin and Spread of Nationalism* (New York: Verso, 1983).

22. Faisal Husain, *Rivers of the Sultan: The Tigris and Euphrates in the Ottoman Empire* (New York: Oxford University Press, 2021).

23. See, for example, Frederick M. Lorenz and Edward J. Erickson, *The Euphrates Triangle: Security Implications of the Southeastern Anatolia Project* (Washington, DC: National Defense University Press, 1999); Ayşegül Kibaroglu and Sezin Iba Gürsoy, "Water–Energy–Food Nexus in a Transboundary Context: The Euphrates–Tigris River Basin as a Case Study," *Water International* 40, no. 5–6 (September 19, 2015): 824–38, https://doi.org/10.1080/02508060.2015.1078577; Greg Shapland, *Rivers of Discord: International Water Disputes in the Middle East* (London: Hurst, 1997); Hillel, *Rivers of Eden*; Zekâi Şen, *Sınır aşan sularımız* (İstanbul: Su Vakfı Yayınları, 2002); John Bulloch and Ādil Darwīsh, *Water Wars: Coming Conflicts in the Middle East* (London: Gollancz, 1996); Asit K Biswas, *International Waters of the Middle East: From Euphrates-Tigris to Nile* (Bombay: Oxford University Press, 1994), 44–116.

24. Examples of this research include Ahmed Sousa, *Wādī al-Furāt wa-mashrūʿ saddat al-Hindīyah*, al-Ṭabʿah 1 (Baghdad: Maṭbaʿat al-Maʿārif, 1944); Maḥmūd Shawqī Ḥamdānī, *Lamaḥāt min taṭawwur al-rayy fī al-ʿIrāq* (Baghdad: al-Maṭbaʿa al-Saʿdūn, 1984); Walīd Riḍwān, *Muškilat al-miyāh baina Sūriyā wa-Turkiyā: (asbāb al-muškila–al-mašāriʿ al-māʾiya as-sūriya–āfāq al-ḥall)*, Ṭabʿa 1 (Bairūt: Šarikat al-Maṭbūʿāt li-t-Tauzīʿ wa-ʾn-Našr, 2006); *Orta-Doğu'da Su Sorunu*, EIUK (Ankara: Dişişleri Bakanlığı Bölgesel ve Sınıraşan Sular Dairesi, 1994); Refik Akarun, *Zor ve Sorunlu Temel Üzerinde Yapılan Bir Büyük Baraj Keban Barajı / A Large Dam on Difficult Foundation: Keban Dam* (Yapı Teknik Engineering and Consultancy Co., 1999).

25. The promising side of this development is a deeper engagement with the Ottoman archive in Istanbul, which has offered scholars a rich source of information on environmental themes and engendered several important recent studies. See, for example, Michael Christopher Low, *Imperial Mecca: Ottoman Arabia and the Indian Ocean*

Hajj (New York: Columbia University Press, 2020); Elizabeth R. Williams, *States of Cultivation: Imperial Transition and Scientific Agriculture in the Eastern Mediterranean* (Stanford, CA: Stanford University Press, 2023).

26. Studies of state formation foregrounding modernization theory or developmentalist discourses fall into this category. See, for example, Edith Tilton Penrose and E. F. Penrose, *Iraq: International Relations and National Development* (Boulder, CO: Westview Press, 1978); Bernard Lewis, *The Emergence of Modern Turkey* (London: Oxford University Press, 1961).

27. Diana K. Davis, *Resurrecting the Granary of Rome: Environmental History and French Colonial Expansion in North Africa* (Athens: Ohio University Press, 2007); Diana K. Davis, *The Arid Lands: History, Power, Knowledge* (Cambridge, MA: MIT Press, 2016).

28. The approach is akin to Faisal Husain's book on the rivers in which he notes, "Writing a history of the Tigris and Euphrates is an attempt to piece back together a jigsaw that time has torn apart." Husain, *Rivers of the Sultan*, 5.

29. The diversity of these projects suggests that it may be useful to think along the lines of "new materialisms." Thomas Lemke, "Varieties of Materialism," *BioSocieties* 10 (December 1, 2015): 492, https://doi.org/10.1057/biosoc.2015.41.

30. Diana Coole, "Agentic Capacities and Capacious Historical Materialism: Thinking with New Materialisms in the Political Sciences," *Millennium: Journal of International Studies* 41, no. 3 (June 2013): 452, https://doi.org/10.1177/0305829813481006.

31. Donald Worster, *Rivers of Empire: Water, Aridity, and the Growth of the American West* (New York: Pantheon Books, 1985); Karl August Wittfogel, *Oriental Despotism: A Study of Total Power* (New Haven, CT: Yale University Press, 1957).

32. Jane Bennett, *Vibrant Matter: A Political Ecology of Things* (Durham, NC: Duke University Press, 2010), 23.

33. Bruno Latour, *Reassembling the Social: An Introduction to Actor-Network-Theory*, Clarendon Lectures in Management Studies (Oxford: Oxford University Press, 2007), 72. Italics in the original.

34. Hans Schouwenburg, "Back to the Future?" March 28, 2015, 60, https://doi.org/10.18352/22130624-00301003.

35. Many of the scholars within new materialism emerged from feminist studies and had scientific training, and so adopted some of the techniques of the cultural turn to articulate critiques of the history and sociology of science and technology. Karen Barad, *Meeting the Universe Halfway: Quantum Physics and the Entanglement of Matter and Meaning* (Durham, NC: Duke University Press, 2007), 30.

36. White's approach has also influenced scholars not writing about rivers. Samuel Dolbee adopts White's perspective with his study of locusts in the Jazira region of the river basin; "[T]he insects were in many ways creatures of humans, just as humans were creatures of locusts." Richard White, *The Organic Machine* (New York: Hill and Wang, 1995), xi; Dolbee, *Locusts of Power*, 16.

37. Sara B. Pritchard, *Confluence: The Nature of Technology and the Remaking of the Rhône*, Harvard Historical Studies 172 (Cambridge, MA: Harvard University Press, 2011), 17.

38. Pritchard, 18 and 20; Gabrielle Hecht refers to techno-politics as "the strategic practice of designing or using technology to enact political goals." For more, see Gabrielle Hecht, *The Radiance of France: Nuclear Power and National Identity after World War II* (Cambridge, MA: MIT Press, 1998); Gabrielle Hecht, ed., *Entangled Geographies: Empire and Technopolitics in the Global Cold War* (Cambridge, MA: MIT Press, 2011), 3; The use of techno-politics in relation to rivers in Middle Eastern history may be found in Timothy Mitchell, *Rule of Experts: Egypt, Techno-Politics, Modernity* (Berkeley: University of California Press, 2002).

39. Latour, *Reassembling the Social*, 16 and 54–55.

40. Coole, "Agentic Capacities and Capacious Historical Materialism," 457.

41. Barad, *Meeting the Universe Halfway*, 91.

42. Mitchell, *Rule of Experts*, 27 and 53.

43. Bennett, *Vibrant Matter*, 119.

44. Mitchell offers up the question but sidesteps some of its implications by focusing on expertise, calculation, and simplification. This research instead spends effort considering the complexity of producing believable visions of the future. Mitchell, *Rule of Experts*, 53.

45. Coole notes three theories of agency, each of which corresponds fairly well to historical fields: a liberal humanism that emphasizes individual action, a Marxist view that centers on social class, and a realist approach focused on the state. Coole, "Agentic Capacities and Capacious Historical Materialism," 458.

46. By hydraulic determinism, I refer to the theories of Karl Wittfogel about Mesopotamian civilization. Wittfogel, *Oriental Despotism*; quotations are from Mitchell, *Rule of Experts*, 30. For critiques of Wittfogel's approach, see Alan Mikhail, *Nature and Empire in Ottoman Egypt: An Environmental History* (Cambridge: Cambridge University Press, 2011); David Gilmartin, *Blood and Water: The Indus River Basin in Modern History* (Oakland: University of California Press, 2015).

47. Barad, *Meeting the Universe Halfway*, 140.

48. Sara Pursley, *Familiar Futures: Time, Selfhood, and Sovereignty in Iraq* (Stanford, CA: Stanford University Press, 2019); Pursley draws on the work of Reinhart Koselleck, *The Practice of Conceptual History: Timing History, Spacing Concepts*, trans. Todd Samuel Presner (Stanford, CA: Stanford University Press, 2002); Reinhart Koselleck, *Futures Past: On the Semantics of Historical Time*, trans. Keith Tribe (New York: Columbia University Press, 2004).

49. Some environmental histories have done this kind of work, as in Mark Fiege's analysis of the plantations of the American South. In Fiege's account aspects of enslaved resistance worked with and through the life cycle of the cotton plant. Mark Fiege, *The Republic of Nature: An Environmental History of the United States* (Seattle: University of Washington Press, 2012), 100–155.

50. Anna Lowenhaupt Tsing, *The Mushroom at the End of the World: On the Possibility of Life in Capitalist Ruins*, First paperback printing (Princeton, NJ: Princeton University Press, 2017), 19–25.

51. During the revision of this book I was pleased to discover a similar approach in Chris Gratien's history of the Çukurova plain. While Gratien emphasizes "percussive repetitions... that can be conceptualized as 'refrains,'" I have tended to conceptualize the twentieth century using other qualities of music—atonality, uneven meter, and, like Tsing, polyphony—a small difference and one that perhaps owes more to a distinction in subject than approach. See Chris Gratien, *The Unsettled Plain: An Environmental History of the Late Ottoman Frontier* (Stanford, CA: Stanford University Press, 2022), 5.

52. The terms are so ubiquitous I will provide but a few examples from Middle Eastern history. Fawaz Gerges, *Making the Arab World: Nasser, Qutb, and the Clash That Shaped the Middle East* (Princeton, NJ: Princeton University Press, 2018); Uğur Ümit Üngör, *The Making of Modern Turkey: Nation and State in Eastern Anatolia, 1913–1950* (Oxford: Oxford University Press, 2011); Toby Dodge, *Inventing Iraq: The Failure of Nation-Building and a History Denied* (New York: Columbia University Press, 2003); Matthew F. Jacobs, *Imagining the Middle East: The Building of an American Foreign Policy, 1918–1967* (Chapel Hill: University of North Carolina Press, 2011). For more on reclaiming the terms, see Eduardo Kohn, *How Forests Think: Toward an Anthropology beyond the Human* (Berkeley: University of California Press, 2013); Tsing, *The Mushroom at the End of the World*.

53. Thomas draws in part from observations made in Arjun Appadurai, ed., *The Social Life of Things: Commodities in Cultural Perspective* (Cambridge: Cambridge University Press, 1986).

54. Thus, in Thomas's account, a gun given in trade as a commodity can become over time "an historicized artifact," a representation of cultural relations and individual works. In this way, Thomas sought to collapse the easy dualism of colonizer versus colonized to show how the history of colonization was a shared one in which both peoples were entangled. Nicholas Thomas, *Entangled Objects: Exchange, Material Culture, and Colonialism in the Pacific* (Cambridge, MA: Harvard University Press, 1991), 4, 98–100.

55. Thomas, 208. Since the time of Thomas's work, entanglement has been deployed by other scholars, particularly in literary theory. See Sarah Nuttall, *Entanglement: Literary and Cultural Reflections on Post Apartheid* (Johannesburg: Wits University Press, 2009). Other uses of entanglement may be found in Bill Brown, "Thing Theory," *Critical Inquiry* 28, no. 1 (2001): 1–22; Ian Hodder, *Entangled: An Archaeology of the Relationships Between Humans and Things* (Hoboken, NJ: John Wiley & Sons, 2012).

56. William Cronon, *Nature's Metropolis: Chicago and the Great West* (New York: Norton, 1992).

57. This method is, in part, what new materialists refer to as "transversal" readings. Tiago Saraiva, *Fascist Pigs: Technoscientific Organisms and the History of Fascism* (Cambridge, MA: MIT Press, 2018); Edmund Russell, *Evolutionary History: Uniting History and Biology to Understand Life on Earth* (Cambridge: Cambridge University Press, 2011).

58. Kate Brown, "Gridded Lives: Why Kazakhstan and Montana Are Nearly the Same Place," *The American Historical Review* 106, no. 1 (2001): 17–48.

59. Diego Gambetta and Steffen Hertog, *Engineers of Jihad: The Curious Connection between Violent Extremism and Education* (Princeton, NJ: Princeton University Press, 2016). Nilüfer Göle, *Mühendisler ve İdeoloji: Öncü Devrimcilerden Yenilikçi Seçkinlere*, 2nd ed. (İstanbul: Metis Yayınları, 1998); Rudolf Mrázek, *Engineers of Happy Land: Technology and Nationalism in a Colony* (Princeton, NJ: Princeton University Press, 2002).

60. Christopher Sneddon, *Concrete Revolution: Large Dams, Cold War Geopolitics, and the US Bureau of Reclamation* (Chicago: University of Chicago Press, 2015).

61. Joseph Morgan Hodge, *Triumph of the Expert: Agrarian Doctrines of Development and the Legacies of British Colonialism* (Athens: Ohio University Press, 2007); see also Helen Tilley, *Africa as a Living Laboratory: Empire, Development, and the Problem of Scientific Knowledge, 1870–1950* (Chicago: University of Chicago Press, 2013).

62. Some examples include Erez Manela, *The Wilsonian Moment: Self-Determination and the International Origins of Anticolonial Nationalism* (Oxford: Oxford University Press, 2009); David Ekbladh, *The Great American Mission: Modernization and the Construction of an American World Order* (Princeton, NJ: Princeton University Press, 2010); Arturo Escobar, *Encountering Development: The Making and Unmaking of the Third World* (Princeton, NJ: Princeton University Press, 2012).

Chapter One

1. A. A. M. Aqrawi and G. Evans, "Sedimentation in the Lakes and Marshes (Ahwar) of the Tigris-Euphrates Delta, Southern Mesopotamia," *Sedimentology* 41, no. 4 (August 1994): 757, https://doi.org/10.1111/j.1365-3091.1994.tb01422.x.

2. Faisal Husain, *Rivers of the Sultan: The Tigris and Euphrates in the Ottoman Empire* (New York: Oxford University Press, 2021), 25.

3. Varoujan K Sissakian et al., "Sea Level Changes in the Mesopotamian Plain and Limits of the Arabian Gulf: A Critical Review," *Journal of Earth Sciences and Geotechnical Engineering* 10, no. 4 (2020): 87–110; Galina S. Morozova, "A Review of Holocene Avulsions of the Tigris and Euphrates Rivers and Possible Effects on the Evolution of Civilizations in Lower Mesopotamia," *Geoarchaeology* 20, no. 4 (2005): 401–23, https://doi.org/10.1002/gea.20057.

4. Qur'an 37:76.

5. William Willcocks, "The Garden of Eden and Its Restoration," *The Geographical Journal* 40, no. 2 (August 1912): 137, https://doi.org/10.2307/1778459.

6. A 1923 law upheld the government's right to force landholders to do maintenance, while a 1925 amendment confirmed that the government could do so without paying wages "unless local custom" required it. "Qānūn Al-Rayy Wa-al-Sidād Sanat 1923" (Baghdad: Maṭbaʻat al-Ḥukūmah, n.d.), 2–3; Hanna Batatu, *The Old Social Classes and the Revolutionary Movements of Iraq: A Study of Iraq's Old Landed and Commercial Classes and of Its Communists, Baʻthists and Free Officers* (Princeton, NJ: Princeton University Press, 1989), 145.

7. Abbas Kadhim notes how several disputes between Iraqis and the British may be framed by this difference in terminology. See Abbas Kadhim, *Reclaiming Iraq: The 1920 Revolution and the Founding of the Modern State* (Austin: University of Texas Press, 2012), 7–8.

8. The National Archives, London [hereafter TNA], War Office [WO] 95/4993, H. Walton, "Tigris Floods, 1919," in Irrigation Directorate, "Administration Report for the period from the constitution of the Directorate (6th February, 1918) to 31st March, 1919," 7.

9. Consider, for instance, the differences between Cyrus Schayegh's narration of the "socio-spatial making" of the modern Middle East and Eugene Rogan's more traditional history of "World War I and the postwar settlement." The emphasis on space in Schayegh's history against Rogan's imperial divide-and-rule tactics alters the role of the Versailles treaty in understanding the events of the 1920s. See Eugene Rogan, *The Arabs: A History* (New York: Basic Books, 2009), 156–74; Cyrus Schayegh, *The Middle East and the Making of the Modern World* (Cambridge, MA: Harvard University Press, 2017), 132–91.

10. See for example Charles Tripp, *A History of Iraq*, 3rd ed. (Cambridge: Cambridge University Press, 2010), 30–74; Toby Dodge, *Inventing Iraq: The Failure of Nation-Building and a History Denied* (New York: Columbia University Press, 2003); Peter Sluglett, *Britain in Iraq: Contriving King and Country, 1914-1932* (New York: Columbia University Press, 2007); Ghassan 'Atiyah, *Iraq, 1908-1921: A Socio-Political Study* (Beirut: Arab Institute for Research and Pub, 1973).

11. Batatu, *The Old Social Classes*, 33, 69–71, 132, 145–47, 151.

12. This trope has been used by many, from historians to political commentators. See, for example, Winston S. Churchill, "'My Grandfather Invented Iraq,'" *Wall Street Journal*, March 10, 2003; Clive Irving, "Gertrude of Arabia, the Woman Who Invented Iraq," *The Daily Beast*, June 17, 2014; Gary Sick, "Foreword," in *The Creation of Iraq, 1914-1921*, ed. Reeva S. Simon and Eleanor Harvey Tejirian (New York: Columbia University Press, 2004).

13. Priya Satia, "'A Rebellion of Technology': The British Arabian Imaginary," in *Environmental Imaginaries of the Middle East and North Africa*, ed. Diana K. Davis and Edmund Burke III (Athens: Ohio University Press, 2011), 43; Priya Satia, "Developing Iraq: Britain, India and the Redemption of Empire and Technology in the First World War," *Past & Present* 197, no. 1 (November 1, 2007): 211–55, https://doi.org/10.1093/pastj/gtm008.

14. Dodge, *Inventing Iraq*, 101–29.

15. Sara Pursley, "'Lines Drawn on an Empty Map': Iraq's Borders and the Legend of the Artificial State," *Jadaliyya*, June 2, 2015, https://www.jadaliyya.com/Details/32140.

16. Indeed, Toby Dodge's conclusion in *Inventing Iraq* includes this comment: "It remains to be seen if the UK and U.S. forces have the local knowledge, resources, and staying power to sustain this immense transformative task." Dodge, *Inventing Iraq*, 169.

17. National Archives of India [NAI], Baghdad Residency, Agricultural Adviser to High Commissioner for Iraq, 19 May 1930; Ali Ghalib, *Malaria and Malaria in Iraq* (Jerusalem: The New Publishers Iraq, 1944), 41.

18. For more information on early Mesopotamian irrigation, see Peter Christensen, *The Decline of Iranshahr: Irrigation and Environments in the History of the Middle East, 500 B.C. to A.D. 1500* (Copenhagen: Museum Tusculanum Press: University of Copenhagen, 1993). For more on the Ottoman period, see Husain, *Rivers of the Sultan*.

19. For more on their efforts, see Gökhan Çetinsaya, *The Ottoman Administration of Iraq, 1890–1908* (Routledge, 2006), https://doi.org/10.4324/9780203332467; Roger Owen, *The Middle East in the World Economy, 1800–1914* (London: Methuen, 1981), 180–88; Ebubekir Ceylan, *The Ottoman Origins of Modern Iraq: Political Reform, Modernization and Development in the Nineteenth-Century Middle East* (London: I.B. Tauris, 2011). Several charts and maps of Ottoman plans for the rivers may be found in Cevat Ekici, ed., *Osmanlı Döneminde Irak: Plan, Fotoğraf ve Belgelerle* (İstanbul: Osmanlı Arşivi Daire Başkanlığı, 2006).

20. Prior to Willcocks's detailed surveys, the Ottoman government mainly sought to build a dam on the Euphrates south of the Museyib. The dam was meant to correct problems with the Hindiyya canal, completed in 1803, which by the mid-1800s had become the main channel of the Euphrates River. See Yitzhak Nakash, *The Shi'is of Iraq*, 4 (Princeton, NJ: Princeton University Press, 1996), 30–31.

21. The Ottoman government first sought help from French engineers for a single dam project on the Euphrates River. Çetinsaya, *Ottoman Administration*, 38–39. As for Willcocks, he entered the Egyptian service in 1883 and ten years later he designed and supervised work on the Aswan Dam as Egypt's Director-General of Reservoirs. Frank Unlandherm, "Sir William Willcocks: A Victorian in the Middle East" (senior thesis, Princeton University, 1959), 7–8. For more on the Aswan Dam and Egyptian water control efforts, see Jennifer Derr, "Drafting a Map of Colonial Egypt: The 1902 Aswan Dam, Historical Imagination, and the Production of Agricultural Geography," in *Environmental Imaginaries of the Middle East and North Africa*, ed. Diana K. Davis and Edmund Burke III (Athens: Ohio University Press, 2011), 136–57; Terje Tvedt, *The River Nile in the Age of the British: Political Ecology and the Quest for Economic Power* (London: I.B. Tauris, 2004), 23–26.

22. Sluglett, *Britain in Iraq*, 3; see also Stuart Cohen, *British Policy in Mesopotamia, 1903–1914* (London: Ithaca Press, 1976).

23. Cohen, *British Policy in Mesopotamia, 1903–1914*, 44. The Hindiyya barrage had been a critical component of Willcocks's ignored 1905 recommendations.

24. O'Conor to Grey, Foreign Office [FO] 371/356/41058, as quoted in Cohen, *Mesopotamia*, 45.

25. William Willcocks, *Sixty Years in the East* (Edinburgh: W. Blackwood, 1935), 232. Mesopotamia was not the only object for irrigation development under the Young Turks. Works by the Deutsche Bank also went forward for an expansion of irrigation

around Adana and Konya. Feroz Ahmad, "The Agrarian Policy of the Young Turks, 1908–1918," in *From Empire to Republic: Essays on the Late Ottoman Empire and Modern Turkey*, vol. 1 (Istanbul: İstanbul Bilgi University Press, 2008), 77.

26. Unlandherm, "Sir William Willcocks: A Victorian in the Middle East," 21.

27. William Willcocks, *Irrigation in Mesopotamia* (London: E. & F. N. Spon 1911), 12–20.

28. The CUP's emphasis on public works emanated from its thinking about the role of the state in society, namely that the state should function as a "scientific machine of social intervention." Nader Sohrabi, *Revolution and Constitutionalism in the Ottoman Empire and Iran* (New York: Cambridge University Press, 2011), 61.

29. John Jackson, "Engineering Problems of Mesopotamia and the Euphrates Valley," *Empire Review* 29 (1915): 193–99; for more on the barrage's significance and its construction, see Deniz Akpınar, *Osmanlı'da su projeleri Hindiye barajı* (Istanbul: Arı Sanat, 2017); Ahmed Sousa, *Wādī al-Furāt wa-mashrūʿ saddat al-Hindīyah*, al-Ṭabʿah 1 (Baghdad: Maṭbaʿat al-Maʿārif, 1944).

30. William Willcocks, "The Garden of Eden and Its Restoration," *The Geographical Journal* 40, no. 2 (August 1912): 129–45, https://doi.org/10.2307/1778459; see also Archibald H. Sayce et al., "The Garden of Eden and Its Restoration: Discussion," *The Geographical Journal* 40, no. 2 (August 1912): 145, https://doi.org/10.2307/1778460; William Willcocks, "Two and a Half Years in Mesopotamia," *Blackwood's Edinburgh Magazine*, March 1916.

31. Charles Townshend, *Desert Hell: The British Invasion of Mesopotamia* (Cambridge, MA: Belknap Press of Harvard University Press, 2011); Wilfrid Nunn, *Tigris Gunboats: The Forgotten War in Iraq 1914–1917* (London: Chatham Publishing, 2007).

32. In this scenario, the port city of Basra in southern Iraq was to become a dependency of India, while Baghdad was destined to be the center of an Arab state under the protection of Great Britain. Dodge, *Inventing Iraq*, 10–11.

33. Sluglett, *Britain in Iraq*, 8–41; Arnold Talbot Wilson, *Mesopotamia, 1917–1920; a Clash of Loyalties: A Personal and Historical Record* (London; Oxford University Press, 1931).

34. Batatu, *The Old Social Classes*; Reeva S. Simon, *Iraq between the Two World Wars: The Militarist Origins of Tyranny* (New York: Columbia University Press, 2004); Sluglett, *Britain in Iraq*; Mohammad A. Tarbush, *The Role of the Military in Politics: A Case Study of Iraq to 1941* (London: KPI Limited, 1982); Tripp, *A History of Iraq*.

35. Eugene Rogan, *The Arabs: A History*, 171. This was not T.R.J. Ward's first trip to Mesopotamia. He visited the country before the war and witnessed the construction of the Hindiyya Barrage on the Euphrates in 1913. TNA, WO 95/4993, Irrigation Directorate, "Administration Report for the Period from the Constitution of the Directorate (6th February, 1918) to 31st March," 1919, 1.

36. Irrigation Directorate (1919), 2–3, 15–18.

37. David Gilmartin, *Blood and Water: The Indus River Basin in Modern History* (Oakland: University of California Press, 2015), 40–44.

38. Ahmed Sousa, *Taṭawwur al-rayy fī al-ʿIrāq*, (Baghdad: Maṭbaʿat al-Maʿārif, 1946), 78–79.

39. Irrigation Directorate (1919), 3.

40. "Annual Administration Report, Shamiyyah Division, from 1st January to 31st December 1918," *Reports of Administration for 1918* vol. 1, no.78, as quoted in Batatu, *Old Social Classes*, 174.

41. To put this in perspective, the Hindiyya Barrage is five hundred kilometers (approximately three hundred miles) from the Persian Gulf. Irrigation Directorate (1919), 10.

42. Willcocks, *Irrigation in Mesopotamia*, xii.

43. H. Walton, "Tigris Floods 1919," 7.

44. Ibid.

45. TNA, FO 141/668/6, R.G. Garrow to A.T. Wilson, "Irrigation Policy in Mesopotamia in the immediate future," 4 February 1919, 2.

46. The problems of river navigation during the invasion of Iraq are well documented in Nunn, *Tigris Gunboats*.

47. For more on the Irrigation Department in British-controlled Egypt, see Jennifer L. Derr, *The Lived Nile: Environment, Disease, and Material Colonial Economy in Egypt* (Stanford, CA: Stanford University Press, 2019), 23–43.

48. Buckley noted, "In all deltaic rivers the flood waters command the country and are frequently well above it for the reason that the land is a creation of the river and the silt which it deposits." A. Burton Buckley, "Note on Irrigation in Mesopotamia," Baghdad Government Press, 1919.

49. Howell, "Note on Irrigation," 5.

50. Ibid., 1.

51. Ibid., 7.

52. A similar dynamic occurred in Algeria under French rule. See Diana K. Davis, *Resurrecting the Granary of Rome: Environmental History and French Colonial Expansion in North Africa* (Athens: Ohio University Press, 2007).

53. In a proclamation to Baghdad residents in 1917, British Lieutenant General Sir Stanley Maude declared: "Since the days of Halaka [Hulagu Khan, the Mongol leader who sacked Baghdad in 1258] your city and your lands have been subject to the tyranny of strangers, your palaces have fallen into ruins, your gardens have sunk in desolation, and your forefathers and yourselves have groaned in bondage. Your sons have been carried off to wars not of your seeking, your wealth has been stripped from you by unjust men and squandered in distant places. Since the days of Midhat, the Turks have talked of reforms, yet do not the ruins and wastes of today testify the vanity of those promises?" IOR/L/PS/18/B253, "Baghdad," 19 March 1917.

54. Satia quotes British officials as asserting, "an aerial raid with bombs and machine guns has an overwhelming and sometimes instantaneous effect in inducing submission." See Satia, "'A Rebellion of Technology': The British Arabian Imaginary," 38.

55. Kadhim, *Reclaiming Iraq: The 1920 Revolution and the Founding of the Modern State*, 69–96; Ian Rutledge, *Enemy on the Euphrates: The British Occupation of Iraq and the Great Arab Revolt, 1914–1921* (London: Saqi Books, 2014), 237–379.

56. Sluglett, *Britain in Iraq*, 39.

57. Sluglett notes how many of these constructions had no commercial value. See Sluglett, 62 and 87–91.

58. Such an idea in many ways mirrored British practice along other waterways, particularly those of South Asia. David Gilmartin, writing of the central Punjab, quotes Alfred Deakin's view of imperial irrigation in the region, "What the soldier begins the irrigation engineer continues." See Gilmartin, *Blood and Water*, 69.

59. The Euphrates at Ramadi crested at 49.67 meters and remained above 49 meters from 29 April to 31 May. Ahmed Sousa, *Fayaḍānāt Baghdād fī al-tārīkh: baḥth fī tārīkh fayaḍānāt anhur al-'Irāq wa-ta'thīruhā bi-al-nisbah li-madīnat Baghdād* (Baghdad: Al-Adib Press, 1963), 531–33.

60. "Baghdad in Danger from Floods: Maude Bridge Carried Away," *The Times*, March 24, 1923, 10.

61. Faisal Husain writes of how during Ottoman times, "the bridge floated on some twenty to fifty boats . . . [fixed to] heavy iron chains extended from two large anchors buried in the sand on each bank." Husain, *Rivers of the Sultan*, 31–32.

62. Bell, *Letters*, letter of 11 September 1923.

63. Bury, "Tigris Flood Report," 10–12.

64. "Qānūn Al-Rayy Wa-al-Sidād Sanat 1923."

65. The punishment was not carried out because of the culvert owner's connections to the royal family.

66. TNA, CO 730/94, C. J. Colvin, "Report on the Flood of the Tigris 1926."

67. Bell, *Letters*, 14 April 1926.

68. Report to League of Nations, 1926.

69. While the publisher notes that the poem was written in 1927, Ahmad Sousa states that the events depicted happened during the 1926 flood. Muḥammad Riḍā Al-Shabībī, *Dīwān Al-Shabībī* (Cairo: Maṭbaʿat Lajnat al-Taʾlīf wa-al-Tarjamah wa-al-Nashr, 1940), 165; see also Sousa, *Fayaḍānāt Baghdād fī al-tārīkh*, 536n1.

70. Naji Shawkat, *Sīrah wa-dhikrayāt: thamānīn ʿāman 1894–1974* (Baghdad: Manshūrāt Maktabat al-Yaqẓah al-ʿArabīyah, 1974), 79.

71. Shawkat, 80–81.

72. Bury, "Tigris Flood Report," 31.

73. Timothy Mitchell, *Carbon Democracy: Political Power in the Age of Oil* (London: Verso, 2011), 86–103.

74. The project would require the construction of a barrage and lengthy channel.

75. Baghdad's original flood protections involved huge earthen walls on the right bank above the city to protect cultivation there, while eastern Baghdad was protected by complete fortifications that doubled as flood embankments. Excess flood water was funneled around the city to the east and then through a wide depression between

these fortifications and higher land. The water then rejoined the river farther downstream, leaving the inhabited parts of the city, now effectively an island in the middle of the Tigris, unscathed. See William Willcocks, *The Restoration of the Ancient Irrigation Works on the Tigris, or, The Re-Creation of Chaldea* (Cairo: National Printing Dept., 1903).

76. Bury, "Tigris Flood Report," 38.

77. TNA, CO 730/94, Dobbs to Amery, 4–11.

78. Sousa, *Taṭawwur al-rayy fī al-ʿIrāq*, 80.

79. The shift toward private support of water infrastructure intensified and accelerated because of the 1926 disaster but had been present in some ways because of the budget cuts of 1919. These cuts had forced the Irrigation Directorate to stop all measurements at canals for revenue purposes. No longer did landowners pay for the water they received through government-built canals. This measure reduced revenue once used to pay for water projects. As state investments declined, the government encouraged private parties to pay for large-scale water projects. As an example, a prominent merchant in Arabistan (present-day Khuzestan Province, Iran) donated 300,000 rupees in November 1923 to implement a canal project that was meant to bring irrigation water to the Shiʿi holy city of Najaf. See Muḥammad ʿAbd al-Majīd Ḥassūn. Zubaydī, *al-Amn al-māʾī al-ʿIrāqī: dirāsah ʿan sayr al-mufāwaḍāt qassamat al-miyāh al-dawlīyah*, al-Ṭabʿah 1, Silsilat rasāʾil jāmiʿīyah (Baghdad: Dār al-Shuʾūn al-Thaqāfīyah al-ʿĀmmah, 2008), 42–43.

80. Batatu, *The Old Social Classes*, 224–318.

81. As Samira Haj has noted, "As the main contributor to pump irrigation, mercantile capital was able to complete its subjugation of agriculture by claiming full rights of ownership to both land and labor." Samira Haj, *The Making of Iraq, 1900–1963: Capital, Power, and Ideology* (Albany: State University of New York Press, 1997), 48.

82. Toby Dodge suggests that fully two-thirds of cultivable land came under this designation. Dodge, *Inventing Iraq*, 106–8.

83. The extent to which the Ottoman Land Code was implemented in southern Iraq prior to British control remains a matter of debate. Like Toby Dodge, Doreen Warriner argues that much of the land that ended up under the control of tribal leaders and absentee landowners had been transferred during the time of British rule (not because of Ottoman governance). Samira Haj, on the other hand, asserts that by 1881, "a large proportion of the land had already been legally transferred into the hands of those who had power and wealth." More recent research questions Haj's assertion. Camille Cole found little evidence that the land code had accomplished such a wholesale shift in ownership in ʿAmara. Meanwhile, Nathan Citino suggests debates about the Ottoman Land Code served another purpose entirely, such that "a sharply negative interpretation of the Ottoman legacy in land tenure [helped] to incorporate the Middle East into the Cold War strategy of agricultural development." Haj, *The Making of Iraq, 1900–1963*, 24–27; Doreen Warriner, *Land Reform and Development in the Middle East: A Study of Egypt, Syria, and Iraq*, 2nd ed. (Westport, CT: Greenwood

Press, 1975), 157–58; Camille Lyans Cole, "Empire on Edge: Land, Law, and Capital in Gilded Age Basra" (PhD diss., Yale University, 2020), 12; Nathan J. Citino, *Envisioning the Arab Future: Modernization in U.S.-Arab Relations, 1945-1967* (Cambridge: Cambridge University Press, 2017), 146–56.

84. This was in addition to tax subsidies granted to pump owners, which had cost the Iraqi government nearly 8 million rupees or approximately £600,000. The Iraq Irrigation Directorate also noted that pump owners had to purchase most of their materials from foreign companies, resulting in "an export of about £1,000,000 of the national wealth." NAI, Baghdad Residency, Irrigation Directorate Memorandum by W. Allard, 22 May 1930, 4.

85. W. Allard wrote in 1930, "Nothing that I have learnt so far does anything but create a preference for departmentally-controlled gravity-flow-canals as compared with privately-owned pumps..." Allard went on to show how a ton of barley produced in Kut Liwa through pump irrigation cost nearly three times more than that grown using flow irrigation from a private canal. NAI, Baghdad Residency, Irrigation Directorate Memorandum by W. Allard, 22 May 1930, 3.

86. The 1926 flood cost the government 400,000 rupees. The 1929 flood cost 700,000 and cut the Baghdad-Basra rail line for two months, involving a loss of 100,000 rupees per week in lost revenue. NAI, Baghdad Residency, Irrigation Directorate Memorandum by W. Allard, 22 May 1930, 2. See also Warriner, *Land Reform and Development*, 144–145.

Chapter Two

1. Saline soils typically contain several chemical salts. While common table salt, sodium chloride or NaCl, makes up a good portion of the salts in the soils of the lower Tigris-Euphrates basin, sulfates and chlorides of calcium and magnesium comprise the salts in different areas. For instance, a study of the soil at an experimental farm in the Dujayla area of southern Iraq revealed calcium, magnesium, and sulfate ions. See P.J. Dieleman, *Reclamation of Salt Affected Soils in Iraq: Soil Hydrological and Agricultural Studies* (Wageningen, Netherlands: International Institute for Land Reclamation and Improvement, 1963), 27–35, https://edepot.wur.nl/59923.

2. I. P. Abrol, et al., *Salt-Affected Soils and their Management* (Rome: Food and Agriculture Organization of the United Nations, 1988).

3. *FAO Mediterranean Development Project, Iraq: Country Report* (Rome: Food and Agriculture Organization of the United Nations, 1959), III-2.

4. Vengosh, A. "Salinization and saline environments." *Treatise on geochemistry* 9 (2003): 612.

5. Thorkild Jacobsen and Robert M. Adams, "Salt and Silt in Ancient Mesopotamian Agriculture," *Science* 128, no. 1251–1258 (November 21, 1958): 1252.

6. William S. Gaud, "The Green Revolution: Accomplishments and Apprehensions" (Society for International Development, Washington, DC, March 8, 1968), http://www.agbioworld.org/biotech-info/topics/borlaug/borlaug-green.html.

Notes to Chapter Two

7. Vandana Shiva, *The Violence of the Green Revolution: Third World Agriculture, Ecology and Politics* (London: Zed Books, 1991), 15. For more on the United States's role in agricultural development, see Michael E. Latham, *The Right Kind of Revolution: Modernization, Development, and U.S. Foreign Policy from the Cold War to the Present* (Ithaca, NY: Cornell University Press, 2010).

8. Jack Ralph Kloppenburg, *First the Seed: The Political Economy of Plant Biotechnology* (Madison: University of Wisconsin Press, 2005), 31–32.

9. Hanna Batatu, *The Old Social Classes and the Revolutionary Movements of Iraq: A Study of Iraq's Old Landed and Commercial Classes and of Its Communists, Ba'thists and Free Officers* (Princeton, NJ: Princeton University Press, 1989), 34–35.

10. Other arid zones including the Colorado River Basin in the United States and the Aral Basin in Central Asia face similar problems of saline soils. In each of these places, scholars have noted the power of salinized soils to alter human social life, international relations, and political ecology. See, for instance, Erika Weinthal, *State Making and Environmental Cooperation: Linking Domestic and International Politics in Central Asia* (Cambridge, MA: MIT Press, 2002); April R. Summitt, *Contested Waters: An Environmental History of the Colorado River* (Boulder: University Press of Colorado, 2019).

11. Willcocks, "Two and a Half Years in Mesopotamia," 321.

12. Saleh Haider, "Land Problems of Iraq" (PhD thesis, London School of Economics, 1942), 427n.

13. David Gilmartin, *Blood and Water: The Indus River Basin in Modern History* (Oakland: University of California Press, 2015), 236.

14. J. F. Webster and B. Viswanath, "Report on the Soil Survey of the Diahlah Area—Right Bank," (Bombay: The Times Press, 1921).

15. J. F. Webster, "Alkali Lands in Iraq: A Preliminary Investigation," (Bombay: The Times Press, 1921) IOR MSS EUR F 235 10, 7.

16. J. F. Webster, "Alkali Lands in Iraq: A Preliminary Investigation," (Bombay: The Times Press, 1921) IOR MSS EUR F 235 10, 12.

17. Webster, "Alkali Lands," 18.

18. Constructed prior to World War I, the Hindiyya Barrage rectified a shift in the Euphrates River—the river had taken over a canal and left its main channel—while also providing control for canals in the Hilla District (see Chapter 1).

19. Webster, "Alkali Lands," 12. For more on the Hindiyya Barrage, see Camille Cole, "Controversial Investments: Trade and Infrastructure in Ottoman–British Relations in Iraq, 1861–1918," *Middle Eastern Studies* 54, no. 5 (September 3, 2018): 744–68, https://doi.org/10.1080/00263206.2018.1462164; Deniz Akpınar, *Osmanlı'da su projeleri Hindiye barajı* (Istanbul: Arı Sanat, 2017).

20. The survey was later taken up by Dutch consultant Pieter Buringh after the Second World War. See Pieter Buringh, *Soils and Soil Conditions in Iraq* (Wageningen, Netherlands: H. Veenman & Zonen N.V., 1960).

21. Mahdi, *State and Agriculture in Iraq*, 117.

22. A refinery at Daura in southern Baghdad opened on November 28, 1954. Noam Raydan, "If The Daura Refinery Could Speak (Part I)," Substack newsletter, *The Chokepoint* (blog), August 15, 2022, https://chokepoint.substack.com/p/if-the-daura-refinery-could-speak.

23. Important examples of views at the time may be found in a 1952 International Bank for Reconstruction and Development report and the publications and letters of Michael G. Ionides, an irrigation officer in Iraq from 1926–1937 who later served on Iraq's Development Board from 1955–1958. Both the IBRD analysts and Ionides recognized the importance of drainage. However, the IBRD emphasized the use of fertilizers to "make possible more intensive use of the land." Meanwhile, Ionides in multiple publications asserted the need to expand summer irrigation in Iraq. In one article, he wrote, ". . . there is no doubt whatever that the simplest and quickest way of raising the productivity of Iraq's land and the prosperity of the country-people, farmers and fellahin alike, is by intensifying irrigation . . ." Only the United Nations Food and Agriculture Organization appeared to understand the larger dynamics at work in salinization and drainage in Iraq. See "The Economic Development of Iraq" (Washington, DC: International Bank for Reconstruction and Development, 1952), 234–35 and 256; *FAO Mediterranean Development Project, Iraq*, 2–3; and MECA, GB165-0207, Elizabeth Monroe Collection, "Summer Water," n.d. [early 1958 most likely], 63.

24. Peter Sluglett, *Britain in Iraq: Contriving King and Country, 1914–1932* (New York: Columbia University Press, 2007), 71–86.

25. For more on the delays and prevarications involved in the exploration and exploitation of oil in Iraq, see Timothy Mitchell, *Carbon Democracy: Political Power in the Age of Oil* (London: Verso, 2011), 86–108.

26. Arbella Bet-Shlimon, *City of Black Gold: Oil, Ethnicity, and the Making of Modern Kirkuk* (Stanford, CA: Stanford University Press, 2019), 82–84.

27. Michael Quentin Morton, "River of Oil: Early Oil Exploration in Iraq," *GEO ExPro* 12, no. 1 (February 2015): 60.

28. Bet-Shlimon, *City of Black Gold*, 80.

29. Morton, "River of Oil: Early Oil Exploration in Iraq," 61–62. For more on the political history of the oil companies involved, see Daniel Yergin, *The Prize: The Epic Quest for Oil, Money, and Power* (New York: Simon & Schuster, 1991), 168–89.

30. Joe Stork, "Oil and the Penetration of Capitalism in Iraq," in *Oil and Class Struggle*, ed. Petter Nore and Terisa Turner (London: Zed Books, 1980), 175.

31. Sluglett, *Britain in Iraq*, 137–40. Timothy Mitchell quotes a US State Department official who called the 1931 revision, "one of the worst oil deals that has ever been signed." Mitchell, *Carbon Democracy*, 102.

32. *The Third River* (British Petroleum, 1955), https://www.bpvideolibrary.com/record/385.

33. Ahmed Sousa, *Irrigation in Iraq: Its History and Development* (Jerusalem: New Publishers Iraq, 1945), 41–42.

34. The revenues were expected to also cover a flood escape project at Habbaniyya. NAI, British Residency, Iraq Ministry of Finance to High Commissioner for Iraq, no. 4505/16, 24 May 1931.

35. Joseph Sassoon notes that thirty governments were formed in Iraq between 3 November 1932 and 5 February 1950. Majid Khadduri argues that a coup d'état in 1936 brought about the military's entry into politics, slowing but not ending the revolving door of prime ministers. See Majid Khadduri, *Independent Iraq: A Study in Iraqi Politics since 1932* (London: Oxford University Press, 1951), 45–47; Joseph Sassoon, *Economic Policy in Iraq, 1932–1950* (London: F. Cass, 1987), 42.

36. Susan Pedersen, "Getting Out of Iraq—in 1932: The League of Nations and the Road to Normative Statehood," *The American Historical Review* 115, no. 4 (October 1, 2010): 975–1000, https://doi.org/10.1086/ahr.115.4.975.

37. Charles Tripp argues that Iraqi leaders viewed the state "as an apparatus of power [rather than as] a sense of Iraq as a community." Phebe Marr notes other factors such as the rise of dictatorships in Europe and the influence of Soviet-style social reform in undermining the government's stability and encouraging the entry of the army into politics. Reeva Spector Simon instead focuses on German influence. Eric Davis notes several additional factors including the monarch's ineffectual leadership, tribal revolts, the Assyrian crisis, and the politicization of the army. Charles Tripp, *A History of Iraq*, 3rd ed. (Cambridge: Cambridge University Press, 2010), 104; Phebe Ann Marr, *The Modern History of Iraq*, 2nd ed. (Boulder, CO: Westview Press, 2004), 44–46; Reeva S. Simon, *Iraq between the Two World Wars: The Militarist Origins of Tyranny* (New York: Columbia University Press, 2004), 7–40; Eric Davis, *Memories of State: Politics, History, and Collective Identity in Modern Iraq* (Berkeley: University of California Press, 2008), 59.

38. Rustum Haydar and Najda Fathi Safwa, *Mudhakkirāt Rustum Haydar* (Beirut: al-Dār al-ʿArabīyah lil-Mawsūʿāt, 1988), 62.

39. Naji Shawkat, *Sīrah wa-dhikrayāt: thamānīn ʿāman 1894–1974* (Baghdad: Manshūrāt Maktabat al-Yaqẓah al-ʿArabīyah, 1974), 250.

40. Shawkat, 250.

41. Nuri al-Said attacked Rustum Haydar in the pages of the newspaper *al-ʿUqāb*, which led the Shiʿi president of the Senate, Muhammad al-Sadr, to warn King Ghazi of rising sectarian hostility. Haydar and Safwa, *Mudhakkirāt Rustum Haydar*, 64–65.

42. Habbaniyya would not protect Baghdad, which was threatened more by Tigris River floods, nor was it a perfect solution for the Euphrates River, but adding the project at least gave the appearance that the government appreciated flood risks. Najda Fathi Safwa, *Ṣāliḥ Jabr: sīrah siyāsīyah* (Bayrūt, Lubnān: Dār al-Sāqī, 2016), 46–47.

43. Norman Burns, "Development Projects in Iraq: 1. The Dujaylah Land Settlement," *Middle East Journal* 5, no. 3 (Summer 1951): 363.

44. Plans for these additional works took time to materialize and were not ready when the barrage opened. In addition, the project ended in a financial dispute between the construction firm of Balfour, Beatty and Co., Ltd., and the Iraqi government over

nearly £250,000 in additional costs, including losses from the Tigris River flood of 1937, and £28,000 in "equipment and materials stolen by the police." TNA, FO 371/23217, E. Houstoun-Boswall, Baghdad, to Secretary of State for Foreign Affairs, Despatch No. 146/E., 6 April 1939, 2.

45. Stephen Hemsley Longrigg, *Iraq, 1900 to 1950: A Political, Social, and Economic History* (London: Oxford University Press, 1953), 278.

46. Brad Fisk, "Dujaila: Iraq's Pilot Project for Land Settlement," *Economic Geography* 28, no. 4 (1952): 344, https://doi.org/10.2307/141972.

47. Sara Pursley, *Familiar Futures: Time, Selfhood, and Sovereignty in Iraq* (Stanford, CA: Stanford University Press, 2019), 127; see also Fisk, "Dujaila," 349.

48. "Managing Salinity in Iraq's Agriculture: Current State, Causes, and Impacts," Iraq Salinity Assessment (International Center for Agricultural Research in the Dry Areas, 2013), 26.

49. Pursley, *Familiar Futures*, 147.

50. Mitchell, *Carbon Democracy*, 49–54.

51. Doreen Warriner, *Land Reform and Development in the Middle East: A Study of Egypt, Syria, and Iraq*, 2nd ed. (Westport, CT: Greenwood Press, 1975), 118.

52. Mahdi, *State and Agriculture in Iraq*, 123.

53. Mahdi, 141; see also the discussion of this approach in Sassoon, *Economic Policy in Iraq, 1932–1950*, 146–47.

54. TNA, FO 371/45324, British Ambassador, Baghdad to Secretary of State for Foreign Affairs, Despatch No. 187, 3 May 1945, 1.

55. Ibid.

56. Ibid., 2.

57. This was the only time Americans from the Bureau of Reclamation would be involved in Iraqi water projects. For discussion of the tour, see TNA, FO 371/45324, British Ambassador, Baghdad to Secretary of State for Foreign Affairs, Despatch No. 266, 25 June 1945. For more on Bureau-Iraq connections, see Christopher Sneddon, *Concrete Revolution: Large Dams, Cold War Geopolitics, and the US Bureau of Reclamation* (Chicago: University of Chicago Press, 2015), 178–79.

58. Sassoon, *Economic Policy in Iraq, 1932–1950*, 146.

59. Gerke, "The Iraq Development Board and British Policy, 1945–50," 235.

60. Ahmed Sousa, *Fayaḍānāt Baghdād fī al-tārīkh: baḥth fī tārīkh fayaḍānāt anhur al-'Irāq wa-ta'thīruhā bi-al-nisbah li-madīnat Baghdād* (Baghdad: Al-Adib Press, 1963), 557–64.

61. Nazik al-Mala'ika, "The Drowned Cemetery," as quoted in Sousa, 563–64.

62. Like Willcocks, Haigh proposed two large flood escapes that would protect the country from yearly inundation, one on each river. Both projects included provisions for irrigation, with the potential to bring nearly 7.5 million acres of new land into cultivation. The works were estimated to cost $111 million.

63. There was apparently considerable friction between Haigh and Nuri al-Said, who refused in 1949 to renew Haigh's contract. SA, GB 0033 SAD, Walter Crawford,

"Trip to Baghdad from 15/6/49 to 25/6/49," 25 June 1949. For more on what was known about dams and irrigation affecting communities in Egypt, see Jennifer L. Derr, *The Lived Nile: Environment, Disease, and Material Colonial Economy in Egypt* (Stanford, CA: Stanford University Press, 2019).

64. This approach was, of course, representative of Cold War-era ideologies of state modernization. Haigh was not an exception to the engineering approaches of the time. For more on how these ideologies influenced a wide range of state activities in the Middle East, see Nathan J. Citino, *Envisioning the Arab Future: Modernization in U.S.-Arab Relations, 1945–1967* (Cambridge: Cambridge University Press, 2017).

65. The project as envisioned was also expensive: constructing the channel into the depression meant cutting through a rise that set the depression off from the Tigris River, a task that would require moving 70 million cubic meters of earth. Still, Haigh's report noted that the project could not only lower the amount of water in the Tigris at times of flood but also provide storage of floodwater for use during the summer growing season. The movement of water could also conceivably be converted to electrical power if designed appropriately, such that the Wadi Tharthar project could serve flood control, irrigation storage, and hydroelectric purposes. See C. Voûte, "Contributions of Photo-Interpretation to Engineering Projects in Various Stages of Execution," *Photogrammetria* 19 (1962): 179–91, https://doi.org/10.1016/S0031-8663(62)80093-3.

66. During the Ottoman period, work on the flood escape had started in 1913 based on William Willcocks's plans; World War I had interrupted those efforts. In 1932, the Iraqi government resurrected the project with a contract for work signed in 1939, but war once again interceded.

67. Frank Fraser Haigh, "Report on the Control of the Rivers of Iraq and the Utilization of Their Waters" (Baghdad: Baghdad Press, 1951), 122–23, Institution of Civil Engineers, London, UK. Haigh intended to evacuate the salt in stages to keep Euphrates River salinity at 80 parts per 100,000. However, in 1951, experts at the International Bank of Reconstruction and Development thought this too high for irrigation purposes and found the Euphrates already at or above 70 parts per 100,000 in the autumn at al-Samawah before the introduction of any additional salinity from other projects. See "The Economic Development of Iraq," 186 and 214.

68. Haigh, "Report on the Control of the Rivers," 142.

69. Haigh, 146. Haigh's report in this way gave some credence to the earlier idea of "regeneration"; the concept simply needed a different language of revenue and development to put in place.

70. Haigh, 175.

71. Haigh, 176–77.

72. On the one hand, flood control could reduce the periodic inundations that helped support the high water table. On the other hand, the extension of flow irrigation could do the opposite by introducing additional water into new areas. Moreover, storage of water would allow greater flows in the summer, increasing seepage into the water table at precisely the time of higher evaporation. Haigh, 178.

73. Haigh, 178.

74. In 1945, British Foreign Secretary Ernest Bevin established the British Middle East Office (BMEO) to promote and coordinate British technical assistance work in the region. Paul W. T. Kingston, *Britain and the Politics of Modernization in the Middle East, 1945–1958* (Cambridge: Cambridge University Press, 2002), https://doi.org/10.1017/CBO9780511563539.

75. TNA, FO 371/61621, H. R. Stewart, Memorandum on Haigh's proposal, 28 November 1946, 1.

76. TNA, FO 371/61621, British Middle East Office to British Embassy, Baghdad, No. 117, 20 December 1946.

77. TNA, FO 371/61621, E. D. Pridie, Memorandum to W. F. Crawford, 29 November 1946.

78. Bevin could draw from several precedents for transforming a plan for hydraulic control into a program targeting other social and economic factors, precedents with rather mixed results. The United States government had advertised its transformation of the Tennessee River Valley and sought to export this expertise around the world. By the 1940s, the British had implemented several aspects of the Gezira scheme along the Blue Nile in Sudan, including an administration that Victoria Bernal argues, "had much more to do with control over Sudanese farmers than with control over water." TNA, FO 371/61621, "Note for the Secretary of State's Talk with the Iraqi Foreign Minister about the Development of Irrigation in Iraq," 1 February 1947; Victoria Bernal, "Colonial Moral Economy and the Discipline of Development: The Gezira Scheme and 'Modern' Sudan," *Cultural Anthropology* 12, no. 4 (November 1997): 462, https://doi.org/10.1525/can.1997.12.4.447. See also Matthew David Owen, "For the Progress of Man: The TVA, Electric Power, and the Environment, 1939–1969" (PhD diss., Vanderbilt University, 2014). For more on the Tennessee Valley Authority's international appeal, see David Ekbladh, "'Mr. TVA': Grass-Roots Development, David Lilienthal, and the Rise and Fall of the Tennessee Valley Authority as a Symbol for U.S. Overseas Development, 1933–1973," *Diplomatic History* 26, no. 3 (July 1, 2002): 335–74, https://doi.org/10.1111/1467-7709.00315. On the TVA's appeal in the Middle East in the 1950s: Bochenski and Diamond, "TVA's in the Middle East."

79. TNA, FO 371/61621, British Embassy, Baghdad to Secretary of State for Foreign Affairs, Despatch No. 110, 3 April 1947, 2.

80. TNA, FO 371/61621, "Record of Meeting held at the British Embassy, Baghdad," 22 April 1947.

81. The treaty negotiated the exit of the British forces that had been occupying the country since 1941 but also created a joint British and Iraqi defense board, renewed British control over its air bases at Habbaniyya and Shu'eiba, and gave the United Kingdom rights to transit troops and warships across Iraqi territory. The period of major demonstrations against the regime is known as al-Wathbah. See Batatu, *The Old Social Classes*, 554–57.

82. For more on the protests and discourse during this period, see Orit Bashkin, *The Other Iraq: Pluralism and Culture in Hashemite Iraq* (Stanford, CA: Stanford University Press, 2009), 87–123.

83. In October 1948, a new Iraqi Prime Minister, Muzahim al-Pachachi, created a version of a Development Board to encourage a deal, but not in the format agreed by Salih Jabr. This version did not have an independent staff or budget, and the British embassy refused to support it. See Gerke, "The Iraq Development Board and British Policy, 1945–50," 239.

84. Even after the law was passed, the government dithered in bringing it about. Only after Nuri al-Said returned to the prime ministry for the sixth time in September 1950 was the Board founded with the mandate to put Iraq's growing oil wealth toward social and economic development. Taufiq as-Suwaidi, *My Memoirs: Half a Century of the History of Iraq and the Arab Cause*, trans. Nancy Roberts (Boulder, CO: Lynne Rienner Publishers, 2013), 409.

85. Mahmud Al-Habib, "The Iraqi Development Board," *The Southwestern Social Science Quarterly* 36, no. 2 (1955): 185–90; Stanley John Habermann, "The Iraq Development Board: Administration and Program," *Middle East Journal* 9, no. 2 (1955): 179–86.

86. These studies at least mention salinization and drainage in connection with the cultivator's plight, though some, such as Abbas Alnasrawi, classify drainage as providing only an indirect benefit to sharecroppers. While it is true sharecroppers did not own the land, it is hard to see how an overall increase in the quantity of food would not have been a direct benefit to individuals living near subsistence levels. See Alnasrawi, *The Economy of Iraq*, 26. Carl Iversen in his report on monetary policy in Iraq referred to cultivator's standard of living as "deplorably low," which then led to an "agricultural labor force ... very inefficiently utilized." See Carl Iversen, *A Report on Monetary Policy in Iraq* (Baghdad: National Bank of Iraq, 1954). Kamil Mahdi follows Iversen's lead, noting that "poor health and nutrition probably weakened a cultivator's physical abilities, while ignorance and superstitious tradition must have clouded his comprehension of an increasingly complex and changing environment." The comment about ignorance also echoes the rhetoric of development experts of the time. Mahdi, *State and Agriculture in Iraq*, 133.

87. "The Economic Development of Iraq," 17.

88. Fuad Baali, *Relation of the People to the Land in Southern Iraq* (Gainesville: University of Florida Press, 1966), 62.

89. Several books about Iraq's economic situation cover this topic, but two good examples of this discourse include: Doreen Warriner, *Land and Poverty in the Middle East* (London: Royal Institute of International Affairs, 1948), 116–19; Edith Tilton Penrose and E. F. Penrose, *Iraq: International Relations and National Development* (Boulder, CO: Westview Press, 1978), 174–77.

90. As cited in Marion Farouk-Sluglett and Peter Sluglett, *Iraq since 1958: From Revolution to Dictatorship* (London: I.B. Tauris, 1990), 32.

91. Atheel Al-Jomard, "Internal Migration in Iraq," in *The Integration of Modern Iraq*, ed. Abbas Kelidar (New York: St. Martin's Press, 1979), 111–22.

92. Doris G. Phillips, "Rural-to-Urban Migration in Iraq," *Economic Development and Cultural Change* 7, no. 4 (1959): 405–21.

93. Iversen, *A Report on Monetary Policy in Iraq*, 77–78. The Food and Agriculture Organization found in a study that spanned 1958–1959 that wheat was grown on only 38 percent of the area in the lower basin dedicated to grain. See *FAO Mediterranean Development Project, Iraq*, III-39.

94. Simmons attributes this shift to the agrarian reform laws passed after the 1958 coup, but reading between the lines shows ecological impacts of drought and salinization. John L. Simmons, "Agricultural Development in Iraq: Planning and Management Failures," *Middle East Journal* 19, no. 2 (1965): 131.

95. John Anthony Allan, as quoted in Kaitlin Stack Whitney and Kristoffer Whitney, "John Anthony Allan's 'Virtual Water': Natural Resources Management in the Wake of Neoliberalism," *Arcadia*, no. 11 (Spring 2018), https://doi.org/10.5282/rcc/8316.

96. Tony Allan coined the term, "virtual water," at a seminar in London in 1993, but he used another term, "embedded water," before that when describing the phenomenon of imported water. John Anthony Allan and others, "Fortunately There Are Substitutes for Water Otherwise Our Hydro-Political Futures Would Be Impossible," *Priorities for Water Resources Allocation and Management* 13, no. 4 (1993): 26.

97. See, for example, John Robert McNeill, *Something New under the Sun: An Environmental History of the Twentieth-Century World* (New York: W.W. Norton & Company, 2000).

98. Elliot reproduces the British view of events, wherein the flood justified the Development Board and by extension British water management. Matthew Elliot, *"Independent Iraq": British Influence from 1941 to 1958* (London: I.B. Tauris, 1996), 111, https://doi.org/10.5040/9780755612307.

99. Karol Sorby, "The 1952 Uprising in Iraq and Regent's Role in Its Crushing (Iraq from al-Watba to al-Intifāda: 1949–1952)," *Asian and African Studies* 12 (2003): 177; 183–87.

100. Yitzhak Nakash, *The Shi'is of Iraq* (Princeton, NJ: Princeton University Press, 1994), 128–32.

101. "Statement of the Socialist Nation Party, 20 August 1954," as quoted in Safwa, *Ṣāliḥ Jabr: sīrah siyāsīyah*, 496.

102. Elliot, *Independent Iraq*, 111.

103. CAC, SALT 1/21, John Boyd-Carpenter to Lord Salter, 21 January 1954.

104. Sousa, *Fayaḍānāt Baghdād fī al-tārīkh*, 576–77.

105. Naṣīr al-Chādirchī, *Mudhakkirāt Naṣīr al-Chādirchī: ṭufūlah mutanāqiḍah, shabāb mutamarrid, ṭarīq al-matā'ib* (Bayrūt: al-Madá lil-I'lām wa-al-Thaqāfah wa-al-Funūn, 2017).

106. Sousa, *Fayaḍānāt Baghdād fī al-tārīkh*, 577.

107. As quoted in Sousa, *Fayaḍānāt Baghdād fī al-tārīkh*, 582–83.

108. CAC, SALT 1/21, "Appeal by Lord Salter for the Flood Victims of Baghdad." See also Sousa, 590–92.

109. Baali, *Relation of the People to the Land in Southern Iraq*, 49.

110. Sousa and the other published sources I inspected on the 1954 flood do not mention any casualty figures. Sousa, *Fayaḍānāt Baghdād fī al-tārīkh*, 589–90. See also CAC, SALT 1/21, "Appeal by Lord Salter for the Flood Victims of Baghdad."

111. Muhammad Bahjat al-Athari, "Baghdad and the Flood," as quoted in Sousa, 595–99.

112. The great flood of 1954 barely earns a mention in most English-language histories tracing the political and social life of Iraq in the 1950s.

113. The so-called "Third River" project was the name given to the outfall drain during the 1950s, though not much progress was made. Apparently, those building the project had not heard of the Iraq Petroleum Company's use of the title to describe oil pipelines. Masour Askari, "Iraq's Ecological Disaster," *International Review*, February 12, 2003.

114. Mukhalad Abdullah et al., "Soil Salinity of Mesopotamia and the Main Drains," *Journal of Earth Sciences and Geotechnical Engineering* 10, no. 4 (2020): 224.

115. The marshes also held oil fields, making them a strategic prize that Iran hoped would force Iraq to the negotiating table. Pierre Razoux, *The Iran-Iraq War*, trans. Nicholas Elliott (Cambridge, MA: The Belknap Press of Harvard University Press, 2015), 261–62.

116. Partow, Hassan, "The Mesopotamian Marshlands: Demise of an Ecosystem" (Nairobi: UNEP, 2001), viii.

117. Partow, Hassan, ix.

118. Abbas Alnasrawi, "Iraq: Economic Sanctions and Consequences, 1990–2000," *Third World Quarterly* 22, no. 2 (April 2001): 214, https://doi.org/10.1080/01436590120037036.

119. There was also the issue of graft and kickbacks as powerful actors siphoned funds. Mark Califano and Jeffrey Meyer, *Good Intentions Corrupted: The Oil for Food Scandal and the Threat to the UN* (New York: PublicAffairs, 2009).

120. Alissa J. Rubin and Bryan Denton, "A Climate Warning from the Cradle of Civilization," *The New York Times*, July 29, 2023. Gale Academic OneFile (accessed May 27, 2025).

121. Tom Westcott, "Iraq: Fishermen Fear Shrinking Lake Razzaza Spells End to Their Livelihoods," *Middle East Eye*, January 30, 2022, https://www.middleeasteye.net/news/iraq-lake-razzaza-milh-shrinking-dying-fishing-trade.

122. Khayyun Amtair Rahi and Todd Halihan, "Salinity Evolution of the Tigris River," *Regional Environmental Change* 18 (October 2018): 2117–27, http://dx.doi.org/10.1007/s10113-018-1344-4.

123. Rubin and Denton, "A Climate Warning from the Cradle of Civilization."

124. Bushraa R Yaseen et al., "Environmental Impacts of Salt Tide in Shatt Al-Arab-Basra/Iraq," *IOSR Journal of Environmental Science, Toxicology and Food Technology* 10, no. 1 (January 2016): 35–43.

125. Azhar Al-Rubaie, "From Palm Trees to Homes: Iraqi Agricultural Land Lost to Desert," Al Jazeera, May 26, 2022, https://www.aljazeera.com/news/2022/5/26/climate-change-ravages-iraq-as-palm-trees-make-way-for-desert.

126. Achref Chibani, "Sand and Dust Storms in the MENA Region: A Problem Awaiting Mitigation," Arab Center Washington DC, July 299, 2024, https://arabcenterdc.org/resource/sand-and-dust-storms-in-the-mena-region-a-problem-awaiting-mitigation/.

127. Amirhossein Montazeri et al., "Effects of Upstream Activities of Tigris-Euphrates River Basin on Water and Soil Resources of Shatt al-Arab Border River," *Science of The Total Environment* 858 (February 2023): 159751, https://doi.org/10.1016/j.scitotenv.2022.159751.

128. Mina Aldroubi, "Iraq Could Have No Rivers by 2040, Government Report Warns," *The National*, December 2, 2021; see also Safaa A. R. Al-Asadi, "The Future of Freshwater in Shatt Al-Arab River (Southern Iraq)," *Journal of Geography and Geology* 9, no. 2 (May 27, 2017): 24, https://doi.org/10.5539/jgg.v9n2p24.

129. Zeinab Shuker, "Water, Oil and Iraq's Climate Future," *Middle East Report* 306 (Spring 2023).

Chapter Three

1. There are karst regions in Iraq as well, primarily in the western and southern desert. Eric Gilli, "Karst Areas of Turkey," in *Caves and Karst of Turkey–Volume 2: Geology, Hydrogeology and Karst*, ed. Gültekin Günay et al. (Cham: Springer International Publishing, 2022), 55–65, https://doi.org/10.1007/978-3-030-95361-4_7. Varoujan K. Sissakian, Dhiya'a Al-Deen K. Ajar, and Maher T. Zaini, "Karstification Influence on the Drainage System, Examples from the Iraqi Southern Desert," *Iraqi Bulletin of Geology and Mining* 8, no. 2 (2012): 99–115.

2. Nasrat Adamo, Nadhir Al-Ansari, and Varoujan K. Sissakian, "How Dams Can Affect Freshwater Issues in the Euphrates-Tigris Basins," *Journal of Earth Sciences and Geotechnical Engineering* 10, no. 1 (2020): 61–62; see also Faisal Husain, "Sediment of the Tigris and Euphrates Rivers: An Early Modern Perspective," *Water History* 13 (April 1, 2021): 13–32, https://doi.org/10.1007/s12685-020-00256-2.

3. These are the Turkish names. The Karasu ("black water" in Turkish) is known merely as the Euphrates in Armenian (*Ephrāt*). The Murat is Çemê Muradê in Kurdish and Aratsani in Armenian.

4. Refik Akarun, *Zor ve Sorunlu Temel Üzerinde Yapılan Bir Büyük Baraj Keban Barajı / A Large Dam on Difficult Foundation: Keban Dam* (Yapı Teknik Engineering and Consultancy Co., 1999), 1–2; Mine Orhon, Sibel Esendal, and M. A. Kazak, *Türkiye'deki Barajlar / Dams in Turkey* (Ankara: Devlet Su İşleri Genel Müdürlüğü, 1991), 382.

5. Kerem Öktem, "When Dams Are Built on Shaky Grounds: Policy Choice and Social Performance of Hydro-Project Based Development in Turkey," *Erdkunde* 56, no. 3 (2002): 315.

6. Nancy Y. Reynolds, "Building the Past: Rockscapes and the Aswan High Dam in Egypt," in *Water on Sand: Environmental Histories of the Middle East and North Africa*, ed. Alan Mikhail (Oxford: Oxford University Press, 2012), 185.

7. My definition of "environmental imaginary" derives from Diana K. Davis's work: "[T]he constellation of ideas that groups of humans develop about a given landscape, usually local or regional, that commonly includes assessments about that environment as well as how it came to be in its current state." Diana K. Davis and Edmund Burke III, eds., *Environmental Imaginaries of the Middle East and North Africa* (Athens: Ohio University Press, 2011), 3.

8. In anthropology, a literature reflects on the "poetics of infrastructure," but the emphasis is less on linguistic forms and more on aesthetics, desire and fantasy. See, for example, Brian Larkin, "The Politics and Poetics of Infrastructure," *Annual Review of Anthropology* 42 (2013): 327–43.

9. Şükrü Kacar, "Unutulmaz Bir Ani," *Elazığ Hakimiyet Haber*, May 31, 2016, https://www.elazighakimiyethaber.com/yazi/sukru-kacar/unutulmaz-bir-ani/1563/.

10. "Yeni Fırat." *Yeni Fırat* 1, no. 1 (1962), 3. More information on *Yeni Fırat*'s editor may be found at Gülda Çetindağ Süme and Selamı Çakmakcı, "Yolcu, Fikret Memişoğlu," in *Türk Edebiyatı İsimler Sözlüğü*, June 14, 2019, https://teis.yesevi.edu.tr/madde-detay/yolcu-fikret-memisoglu.

11. Abdullah Şengül, "Arif Nihat Asya," in *Türk Edebiyatı İsimler Sözlüğü*, July 17, 2018, https://teis.yesevi.edu.tr/madde-detay/asya-arif-nihat.

12. Much like the American lawyer Francis Scott Key's 1814 poem, "The Star-Spangled Banner," the imagery of the poem is martial. Instead of rockets and bombs, Asya evokes the "eagle of war" and martyrdom. Soner Yalçın, "Unutulmaz 'Bayrak' şairinin hazin hikâyesi," *Hürriyet*, June 6, 2010, https://www.hurriyet.com.tr/unutulmaz-bayrak-sairinin-hazin-hik-yesi-14944225.

13. Jacob M. Landau, "Ultra-Nationalist Literature in the Turkish Republic: A Note on the Novels of Hüseyin Nihâl Atsız," *Middle Eastern Studies* 39, no. 2 (April 2003): 205 and 208–9, https://doi.org/10.1080/714004510; see also Umut Uzer, "Racism in Turkey: The Case of Huseyin Nihal Atsiz," *Journal of Muslim Minority Affairs* 22, no. 1 (April 2002): 119–30, https://doi.org/10.1080/13602000220124863.

14. Jacob Landau further compares the epic and romantic qualities of Atsız's novel to Walter Scott and the *Niebelungenlied*. Landau, "Ultra-Nationalist Literature in the Turkish Republic," 207 and 209.

15. Mitat Durmuş, "Niyazi Yıldırım Gençosmanoğlu," in *Türk Edebiyatı İsimler Sözlüğü*, February 17, 2019, https://teis.yesevi.edu.tr/madde-detay/gencosmanoglu-niyazi-yildirim; Ahmet Turan Sinan and Fatma Döner Doğan, "Niyazi Yıldırım Gençosmanoğlu'nun Şiirlerinde Mekân: Harput ve Palu," *Uluslararası Palu Sempozyumu Bildiriler Kitabı* 1 (December 2018): 1–18.

16. Arif Nihat Asya, "Fırat," *Elazığ Gazetesi*, June 11, 1963.

17. Connecting national vitality to mothers, particularly in the concept of *anavatan* (motherland), and to the act of birth has a longer history in Turkish nationalism. Carol

Delaney notes how Mustafa Kemal rewrote the last lines of a Namik Kemal poem that connects motherhood to the nation: "The last two lines of Namik Kemal's poem read: the foe thrusts his knife into the heart of the land/there was none to save our ill-fated mother. Mustafa Kemal who imagined himself as the land's savior changed the last line to read 'but yes, one is found to save our ill-fated mother.'" Carol Delaney, "Father State, Motherland, and the Birth of Modern Turkey," in *Naturalizing Power: Essays in Feminist Cultural Analysis*, ed. Sylvia Yanagisako and Carol Delaney (New York: Routledge, 1995), 186.

18. Gençosmanoğlu, Yıldırım N. "Fırat'la Hesaplaşma," *Elazığ Gazetesi*, July 13, 1963.

19. See, for example, Erik J. Zürcher, "The Ottoman Legacy of the Turkish Republic: An Attempt at a New Periodization," *Die Welt Des Islams* 32, no. 2 (1992): 237–53, https://doi.org/10.2307/1570835; Nathan J. Citino, "The Ottoman Legacy in Cold War Modernization," *International Journal of Middle East Studies* 40, no. 4 (2008): 579–97.

20. Scholar Muhammet Özcan collected 93 of the poet's unpublished poems, which had remained in notebooks and diaries in the care of Dökmeci's daughter. Muhammet Özcan, "Cenani Dökmeci'nin şiirlerinde yapı ve tema / Structure and style of poetry in Cenani Dökmeci" (Master's thesis, Fırat Üniversitesi, 2015), vi and 3.

21. Cenani Dökmeci, "Fırat'la Söyleşme," *Yeni Fırat* 28 (1966): 7. Turkish Muslims believe Urfa was once Ur of the Chaldeans and the birthplace of the prophet Abraham. For more, see Elif Batuman, "The Sanctuary: The World's Oldest Temple and the Dawn of Civilization," *New Yorker*, December 19, 2011.

22. Senemoğlu, Bahattin. "Fırat ve Baraj." *Elazığ Gazetesi*, July 19, 1963. Begüm Adalet, *Hotels and Highways: The Construction of Modernization Theory in Cold War Turkey* (Stanford, CA: Stanford University Press, 2018), 85–120.

23. "A river flowed out of Eden to water the garden, and there it divided and became four rivers. . . . And the name of the third river is the Tigris, which flows east of Assyria. And the fourth river is the Euphrates." Gen. 2:10–14 ESV.

24. Edwin T. Layton, *The Revolt of the Engineers: Social Responsibility and the American Engineering Profession* (Cleveland, OH: Press of Case Western Reserve University, 1971), 3.

25. John Black, "The Military Influence on Engineering Education in Britain and India, 1848–1906," *The Indian Economic & Social History Review* 46, no. 2 (April 2009): 211–39, https://doi.org/10.1177/001946460904600203.

26. Several studies in the latter third of the twentieth century focused on tools and technology, rather than on technologists. See, for instance, Daniel R. Headrick, *The Tools of Empire: Technology and European Imperialism in the Nineteenth Century* (New York: Oxford University Press, 1981). Several good studies now exist tracing the work of engineers and their environmental and social imaginings. See, for example, Rudolf Mrázek, *Engineers of Happy Land: Technology and Nationalism in a Colony* (Princeton, NJ: Princeton University Press, 2002); Fredrik Meiton, *Electrical Palestine: Capital and Technology from Empire to Nation* (Oakland: University of California Press, 2019);

Isacar Bolaños, "Water, Engineers, and French Environmental Imaginaries of Ottoman Iraq, 1868–1908," *Environmental History* 27, no. 4 (October 1, 2022): 772–98, https://doi.org/10.1086/721180.

27. Aristotle writes that a poet must act as "an imitator, like a painter or any other artist, [and] must of necessity imitate one of three objects—things as they were or are, things as they are said or thought to be, or things as they ought to be." Aristotle, "Poetics," The Internet Classics Archive, accessed July 25, 2024, http://classics.mit.edu/Aristotle/poetics.mb.txt.

28. Eduardo Kohn, *How Forests Think: Toward an Anthropology beyond the Human* (Berkeley: University of California Press, 2013).

29. Bruno Latour, *Science in Action: How to Follow Scientists and Engineers Through Society* (Cambridge, MA: Harvard University Press, 1987).

30. The intellectual path here is thus more Jacques Derrida, science as art—and less Bruno Latour, science as social construction.

31. Howard Nemerov, "Poetry." *Encyclopedia Britannica*, https://www.britannica.com/art/poetry.

32. James C. Scott, *Seeing like a State: How Certain Schemes to Improve the Human Condition Have Failed* (New Haven, CT: Yale University Press, 2008).

33. Gabrielle Hecht, *The Radiance of France: Nuclear Power and National Identity after World War II* (Cambridge, MA: MIT Press, 1998).

34. Aristotle, "Poetics."

35. A. Sönmez, "The Re-Emergence of the Idea of Planning and the Scope and Targets of the 1963–1967 Plan," in *Planning in Turkey*, ed. S. İlkin and E. İnanç (Ankara: Orta Doğu Teknik Üniversitesi, 1967), 34; see also Vedat Milor, "The Genesis of Planning in Turkey," *New Perspectives on Turkey* 4 (1990): 1–30.

36. Forms of development planning may be traced to colonial regimes in 1900s Africa, and like engineering in colonial contexts, was oriented toward the extension of public works infrastructure. See Albert Waterston et al., *Development Planning: Lessons of Experience* (Baltimore, MD: Johns Hopkins Press, 1965), 28–44.

37. James Ferguson, *The Anti-Politics Machine: "Development," Depoliticization, and Bureaucratic Power in Lesotho* (Minneapolis: University of Minnesota Press, 1994).

38. Sinan Yıldırmaz, *Politics and the Peasantry in Post-War Turkey: Social History, Culture and Modernization*, Library of Ottoman Studies 46 (London, I.B. Tauris, 2017), 25.

39. Max Weston Thornburg, Graham Spry, and George Henry Soule, *Turkey: An Economic Appraisal* (New York: The Twentieth Century Fund, 1949), 26–27; see also Dale Stahl, "The Two Rivers: Water, Development and Politics in the Tigris-Euphrates Basin, 1920–1975" (PhD diss., Columbia University, 2014), 74–84.

40. Erik Jan Zürcher, *Turkey: A Modern History*, 3rd ed. (London: I.B. Tauris, 2004), 197–98.

41. Preliminary engineering studies of the Euphrates River at Keban began in 1938, and support for the project had built within the Turkish bureaucracy over the 1950s.

İbrahim Deriner, "Keban Barajı ve Hidroelektrik Santralı Hakkında Bazı Bilgiler," *Türkiye Mühendislik Haberleri* 8, no. 92 (1962). See also Stahl, "The Two Rivers: Water, Development and Politics in the Tigris-Euphrates Basin, 1920–1975" (PhD diss., Columbia University), 76–86.

42. Mehmet Turgut, *Gap'ın Sahipleri* (Istanbul: Boğazici Yayınları, 1995), 38–40.

43. "The Constitution of Republic of Turkey (1961)," *Islamic Studies* 2, no. 4 (1963): 477.

44. *First Five-Year Development Plan 1963–1967* (Ankara: State Planning Organization, 1963), iii.

45. Begüm Adalet, "Agricultural Infrastructures: Land, Race, and Statecraft in Turkey," *Environment and Planning D: Society and Space* 40, no. 6 (December 2022): 981, https://doi.org/10.1177/02637758221124139.

46. For more on Russo-Ottoman competition, see Michael A. Reynolds, *Shattering Empires: The Clash and Collapse of the Ottoman and Russian Empires, 1908–1918* (Cambridge: Cambridge University Press, 2011). On the Hamidian regiments, Janet Klein, *The Margins of Empire: Kurdish Militias in the Ottoman Tribal Zone* (Stanford, CA: Stanford University Press, 2011). There is a large literature on the Armenian Genocide. A useful reference with multiple perspectives is Ronald Grigor Suny, Norman M. Naimark, and Fatma Müge Göçek, eds., *A Question of Genocide: Armenians and Turks at the End of the Ottoman Empire* (New York: Oxford University Press, 2015).

47. Uğur Ümit Üngör, *The Making of Modern Turkey: Nation and State in Eastern Anatolia, 1913–1950* (Oxford: Oxford University Press, 2011). For more on how infrastructure related to these processes, see Adalet, *Hotels and Highways*, 121–57.

48. Robert W. Olson, *The Emergence of Kurdish Nationalism and the Sheikh Said Rebellion, 1880–1925* (Austin: University of Texas Press, 1991).

49. The law referred to Kurds as "nomadic individuals of non-Turkish origin" and allowed the government "to deport nomads who do not share the Turkish culture outside the national boundaries." As quoted in Soner Çağaptay, "Reconfiguring the Turkish Nation in the 1930s," *Nationalism and Ethnic Politics* 8, no. 2 (June 2002): 72–73, https://doi.org/10.1080/13537110208428662.

50. Martin Bruinessen, "Genocide in Kurdistan? The Suppression of the Dersim Rebellion in Turkey (1937–38) and the Chemical War against the Iraqi Kurds (1988)," in *Genocide: Conceptual and Historical Dimensions*, ed. George J. Andreopoulos (Philadelphia: University of Pennsylvania Press, 1997), 141–70.

51. For his efforts to highlight the plight of Kurds in eastern Anatolia, Beşikci's books were banned and he was imprisoned for 17 years. İsmail Beşikçi, *Doğu Anadolu'nun düzeni: sosyo-ekonomik ve etnik temeller*, Yurt ve dünya sorunları, 1 (Erzurum, Turkey: E. Yayinlari, 1969).

52. Dale J. Stahl, "A Technopolitical Frontier: The Keban Dam Project and Southeastern Anatolia," in *Transforming Socio-Natures in Turkey: Landscapes, State and Environmental Movements*, ed. Onur İnal and Ethemcan Turhan (New York: Routledge, 2020), 31–51.

53. This approach had its roots in the economic program of the Demokrat Parti, which came to power in 1950 in Turkey's first multi-party elections. Mesut Yeğen,

"The Kurdish Question in Turkish State Discourse," *Journal of Contemporary History* 34, no. 4 (1999): 564–65.

54. Turgut, *Gap'ın Sahipleri*, 41–42.

55. Christopher Sneddon, *Concrete Revolution: Large Dams, Cold War Geopolitics, and the US Bureau of Reclamation* (Chicago: University of Chicago Press, 2015).

56. Millet Meclisi Tutanak Dergisi, *Keban Barajı ve Aşağı Fırat Havzası Kalkınma Projesi Hakkında Millet Meclisi Adına Araştırma Komisyonu* (Ankara: Millet Meclisi, 1962), 359.

57. John P. Callahan, "New Fields Tested by Bond and Share," *New York Times*, September 2 1945, 45.

58. *Elazığ Gazetesi*, "Keban Barajı Projelerini inşa edecek firmanın temsilcisi Amerikalı Sibentin bugün şehrimize geliyor," February 15, 1963, 1; "Keban'da Yeni Sondajlar," March 12, 1963, 1; "Ardıçoğlu: Baraj Alanında Tetkiklerde Bulundu," May 26, 1963, 1.

59. Transportation Research Board and National Research Council, "U.S. Federal and Corps of Engineers Water Resources Planning Guidelines," in *Inland Navigation System Planning: The Upper Mississippi River-Illinois Waterway* (Washington, DC: The National Academies Press, 2001), https://doi.org/10.17226/10072.

60. Erin A. Cech asserts, "Engineers and scientists were called upon in the 1920s to help instill technocratic decision-making procedures into public policymaking.... While the technological skepticism of the 1950s–1970s challenged the notion that technocratic leadership was possible or desirable, the ideology of depoliticization remained essentially intact." Erin A. Cech, "The (Mis)Framing of Social Justice: Why Ideologies of Depoliticization and Meritocracy Hinder Engineers' Ability to Think About Social Injustices," in *Engineering Education for Social Justice*, ed. Juan Lucena, vol. 10 (Dordrecht: Springer Netherlands, 2013), 71–72, https://doi.org/10.1007/978-94-007-6350-0_4.

61. "Report to Electric Power Resources Survey and Development Administration for Engineering and Economic Feasibility of Keban Dam and Hydroelectric Project" (New York: EBASCO Services Inc., 1963), I-5, Charles River Watershed Surveys, Reports, and Maps, 1962–1970; Records of the Office of the Chief of Engineers, 1789–1999, Record Group 77, U.S. National Archives Boston.

62. İsmail Beşikçi, *Devletlerarası Sömürge Kürdistan* (İstanbul: Alan Yayıncılık, 1990), 17.

63. This analysis owes much to Timothy Mitchell's observations about the work of economists in "making" the economy. See Timothy Mitchell, "The Work of Economics: How a Discipline Makes Its World," *European Journal of Sociology / Archives Européennes de Sociologie / Europäisches Archiv Für Soziologie* 46, no. 2 (2005): 297–320.

64. "Report to Electric Power Resources Survey and Development Administration for Engineering and Economic Feasibility of Keban Dam and Hydroelectric Project", Charles River Watershed Surveys, Reports, and Maps, 1962–1970; Records of the

Office of the Chief of Engineers, 1789–1999, Record Group 77, U.S. National Archives, Boston, II-7.

65. "Report to Electric Power Resources Survey and Development Administration," Appendix E.

66. The other two sections are labeled "foreign trade," though the narrative focuses primarily on trade deficits, and "foreign investment," which is mainly about foreign loans.

67. "Report to Electric Power Resources Survey and Development Administration," II-8 and II-27.

68. Michel Callon, "Introduction: The Embeddedness of Economic Markets in Economics," in *The Laws of the Markets*, ed. Michel Callon (Oxford: Blackwell Publishers, 1998), 35.

69. Adalet, "Agricultural Infrastructures," 978.

70. Land reform and settlement in Turkey followed patterns similar to the Dujayla Settlement in Iraq. See Adalet, 983.

71. "Report to Electric Power Resources Survey and Development Administration," II-29.

72. "Report to Electric Power Resources Survey and Development Administration," II-31.

73. The discussion of the five offices occupies nearly forty pages and each is represented with multiple organizational charts and tables of employees and budgets. "Report to Electric Power Resources Survey and Development Administration," II-29–68.

74. "Report to Electric Power Resources Survey and Development Administration," III-1.

75. Some Keban project electricity was meant for Urfa and Antep. Even so, there was no provision in the initial phases of the Keban project for agricultural development, despite the assertion of agricultural growth contained within the commission report to parliament. An irrigation water pumping project, Uluova Eyüpbağları, using Keban reservoir water was eventually constructed in 1987; another project, Kuzuova, opened in 1999. Yaşar Baş, "Elazığ İli, Sarıyakup Köyü Tarihi," *Bingöl University Social Sciences Institute Journal* 6 (2016): 133–135.

76. "Report to Electric Power Resources Survey and Development Administration," IV-24–28. For more on the power of mapping regions of Kurdistan, see Karen Culcasi, "Cartographically Constructing Kurdistan within Geopolitical and Orientalist Discourses," *Political Geography* 25, no. 6 (August 1, 2006): 680–706, https://doi.org/10.1016/j.polgeo.2006.05.008.

77. "Report to Electric Power Resources Survey and Development Administration," IV-8–9.

78. "Report to Electric Power Resources Survey and Development Administration," IV-12.

79. Rob Nixon, *Slow Violence and the Environmentalism of the Poor* (Cambridge, MA: Harvard University Press, 2011), 150–74, https://doi.org/10.2307/j.ctt2jbsgw.

80. Michel-Rolph Trouillot, *Silencing the Past: Power and the Production of History* (Boston: Beacon Press, 1995), 70–107.

81. Keban was known from the period of the Hittites (1500–1200 B.C.) as a rich source of minerals. The Iron Age civilization of the Urartu, centered around Lake Van in modern Turkey's far east, mined gold and silver in the area. Keban and the surrounding communities were later subject to a series of invasions from virtually every direction: Arabs from the south, Byzantine Greeks from the west, Seljuk Turks and Mongolians from the north and east. These invasions left a rich archaeological history in the surrounding hills. See *Doomed by the Dam: A Survey of the Monuments Threatened by the Creation of the Keban Dam Flood Area, Elazığ 18–29 October 1966*, Orta Doğu Teknik Üniversitesi (Ankara, Turkey) (Ankara: METU Faculty of Architecture, Dept. of Restoration, 1967).

82. Interestingly, the dam itself is hardly named in the archaeological reports. Laurent Dissard, "Submerged Stories from the Sidelines of Archaeological Science: The History and Politics of the Keban Dam Rescue Project (1967–1975) in Eastern Turkey" (PhD diss., University of California, Berkeley, 2011), 69–76.

83. Joseph D. Lombardo, "In the Kingdom of Dams: Water, Governance, and the Keban Dam Project in Eastern Anatolia, 1961–1974" (PhD diss., The New School, 2018), 91 and 106.

84. "Keban Baraj Gölü Altında Kalacak Köy ve Mahallerin Anket Çalışması Sonucu" (Ankara: İmar ve İskan Bakanlığı, 1966); Turan Ersoy, *Keban Iskan Problemi* (Ankara: Devlet Planlama Teşkilatı, 1968); Turan Ersoy, "Keban Barajının İnşası Dolayısıyla Açıkta Kalan Halkın İskan ve İstihdamı," *Sosyal Siyaset Konferansları Dergisi*, no. 21 (1970): 23–38; Oya Silier, *Keban Köylerinde Sosyo Ekonomik Yapı ve Yeniden Yerleşim Sorunları* (Ankara: Orta Doğu Teknik Üniversitesi, 1976).

85. Lombardo, "In the Kingdom of Dams," 94–134.

86. "KEBAN BARAJI 50 YAŞINDA!," Devlet Su İşleri Genel Müdürlüğü, September 9, 2024, https://www.dsi.gov.tr/Haber/Detay/13356.

87. Doğan Gürpınar, "Anatolia's Eternal Destiny Was Sealed: Seljuks of Rum in the Turkish National(Ist) Imagination from the Late Ottoman Empire to the Republican Era," *European Journal of Turkish Studies*, May 2, 2012, https://doi.org/10.4000/ejts.4547.

88. Carole Hillenbrand, *Turkish Myth and Muslim Symbol: The Battle of Manzikert* (Edinburgh: Edinburgh University Press, 2007), 196–225.

89. Refik Akarun, *A Large Dam on Difficult Foundation: Keban Dam* (Istanbul: Yapı Teknik Engineering and Consultancy Co., 1999), II-7 to II-11.

90. Two recent earthquakes have struck the region. In March 2010, an earthquake of magnitude 6.1 with its epicenter halfway between the towns of Elazığ and Bingöl killed 42 people. In January 2020, an even larger earthquake at 6.7 magnitude, struck the area, with an epicenter very close to the Karakaya Dam, located downriver of the Keban Dam. The two dams showed little to no damage, though ruptures below the surface of the reservoir released gas bubbles that could be seen on the surface of the

water. Cetin Kemal Onder et al., "Geotechnical Aspects of Reconnaissance Findings after 2020 January 24th, M6.8 Sivrice–Elazig–Turkey Earthquake," *Bulletin of Earthquake Engineering* 19 (July 2021): 3415–59, https://doi.org/10.1007/s10518-021-01112-1. For more on the gas explosions see, M. Ali Özdemir and Nusret Özgen, "Keban Barajından Su Kaçakları ve Sunduğu Doğal Potansiyel," *Afyon Kocatepe Üniversitesi Sosyal Bilimler Dergisi* 6, no. 1 (June 2004): 65–86.

91. Akarun, *Zor ve Sorunlu Temel Üzerinde Yapılan Bir Büyük Baraj Keban Barajı / A Large Dam on Difficult Foundation: Keban Dam*; Özdemir and Özgen, "Keban Barajından Su Kaçakları ve Sunduğu Doğal Potansiyel."

92. These facets of the Keban Dam's construction may help in connecting the ways geologists and historians see the past. Dale J. Stahl, "The Dam as Catastrophe: Connecting Geological Models to Modern History," *Water History* 13, June 21, 2021, https://doi.org/10.1007/s12685-021-00278-4.

93. Hasan Tosun, "Earthquakes and Dams," in *Earthquake Engineering*, ed. Abbas Moustafa (Rijeka: IntechOpen, 2015), https://doi.org/10.5772/59372; Rick Gore, "Anatolia—A History Forged by Disaster," *National Geographic Magazine*, accessed September 6, 2022, https://www.nationalgeographic.com/science/article/anatolian-history.

94. "Sular Çekildi Tarlalar Ortaya Çıktı," *Günışığı Gazetesi*, April 23, 2021, https://www.gunisigigazetesi.net/elazig-guncel/sular-cekildi-tarlalar-ortaya-cikti-h83523.html.

Chapter Four

1. Daniel Hillel, *Rivers of Eden: The Struggle for Water and the Quest for Peace in the Middle East* (New York: Oxford University Press, 1994), 103; Vakur Sümer, "Handle with Care! The Tragedy of the Tabqa Dam," ORSAM-Center for Middle Eastern Studies, n.d., https://www.orsam.org.tr/en/handle-with-care-the-tragedy-of-the-tabqa-dam/.

2. The study noted that "large dams in Mediterranean climates exert the strongest influence" on climate near the reservoir. Ahmed Mohamed Degu et al., "The Influence of Large Dams on Surrounding Climate and Precipitation Patterns," *Geophysical Research Letters* 38, no. 4 (2011), https://doi.org/10.1029/2010GL046482.

3. Gülşen Kum, "The Influence of Dams on Surrounding Climate: The Case of Keban Dam / Barajların Çevre İklime Etkisi: Keban Barajı Örneği," *Gaziantep University Journal of Social Sciences* 15 (2016): 193–204.

4. Yves T. Prairie et al., "Greenhouse Gas Emissions from Freshwater Reservoirs: What Does the Atmosphere See?" *Ecosystems (New York, N.Y.)* 21 (2018): 1058–71, https://doi.org/10.1007/s10021-017-0198-9.

5. Bridget R. Deemer et al., "Greenhouse Gas Emissions from Reservoir Water Surfaces: A New Global Synthesis," *BioScience* 66, no. 11 (November 1, 2016): 949–64, https://doi.org/10.1093/biosci/biw117.

6. *Tufan fi Balad al-Ba'th [A Flood in Ba'th Country]* (AMIP–ARTE France, 2003).

7. Gökçe Şencan, "For Hasankeyf the Bell Tolls," *International Rivers* (blog), February 12, 2020, https://www.internationalrivers.org/news/for-hasankeyf-the-bell-tolls/.

8. Timothy Mitchell warns against falling into the trap of techno-politics, which "is a particular form of manufacturing, a certain way of organizing the amalgam of human and nonhuman, things and ideas, so that the human, the intellectual, the realm of intentions and ideas seems to come first and to control and organize the nonhuman." Timothy Mitchell, *Rule of Experts: Egypt, Techno-Politics, Modernity* (Berkeley: University of California Press, 2002), 43.

9. Richard White, *The Organic Machine* (New York: Hill and Wang, 1995).

10. Donna Haraway, "The Cyborg Manifesto: Science, Technology, and Socialist-Feminism in the Late Twentieth Century," in *Simians, Cyborgs, and Women: The Reinvention of Nature* (New York: Routledge, 1991), 69.

11. Sara B. Pritchard, *Confluence: The Nature of Technology and the Remaking of the Rhône*, Harvard Historical Studies 172 (Cambridge, MA: Harvard University Press, 2011), 16.

12. Pritchard, 17.

13. I use this term for its dual meaning, both to describe the stilling of water captured by the reservoirs and the use of reservoir water for human settlements, some of which were constructed to restrict the movement of human groups.

14. Environmental studies have analyzed various aspects of these visions in history, from control over water to the development of governmental and economic systems for designating and managing natural resources. See, for instance, Patrick McCully, *Silenced Rivers: The Ecology and Politics of Large Dams* (London: Zed Books, 1996); Richard Drayton, *Nature's Government: Science, Imperial Britain, and the "Improvement" of the World* (New Haven, CT: Yale University Press, 2000); John F. Richards, *The Unending Frontier: An Environmental History of the Early Modern World* (Berkeley: University of California Press, 2005); David Blackbourn, *The Conquest of Nature: Water, Landscape, and the Making of Modern Germany* (New York: Norton, 2007); John Bellamy Foster, Brett Clark, and Richard York, *The Ecological Rift: Capitalism's War on the Earth* (New York: Monthly Review Press, 2010).

15. Building from observations by historian Kate Brown, this section examines the implications of these similarities. See Kate Brown, "Gridded Lives: Why Kazakhstan and Montana Are Nearly the Same Place," *The American Historical Review* 106, no. 1 (2001): 17–48; Kate Brown, *Plutopia: Nuclear Families, Atomic Cities, and the Great Soviet and American Plutonium Disasters* (New York: Oxford University Press, 2013).

16. *Film Muhawala 'an Sadd al-Furat [Film Essay on the Euphrates Dam]*, 1970; *al-Hayat al-Yawmiyah fi Qarya Suriyya [Everyday Life in a Syrian Village]* (Mu'assasah al-'Āmmah lil-Sīnimā, 1974); *Tufan fi Balad al-Ba'th [A Flood in Ba'th Country]*.

17. 'Abd al-Salām al-'Ujaylī, *al-Maghmūrūn: riwāyah*, Ṭab'ah khāṣṣah (Ṭarābulus, al-Jamāhīrīyah al-'Arabīyah al-Lībīyah al-Sha'bīyah al-Ishtirākīyah: al-Munsha'ah al-'Āmmah, 1984).

18. As quoted in H. Turgut, *Demirel'in Dunyasi*, 49.
19. As quoted in Hulûsi Turgut, *Demirel'in dünyası* (İstanbul: ABC Ajansı Yayınları, 1992), 56.
20. Turgut, 54–55.
21. Turgut, 49.
22. M. Şükrü Hanioğlu, *Atatürk: An Intellectual Biography* (Princeton, NJ: Princeton University Press, 2011).
23. See Nilüfer Göle, *Mühendisler ve İdeoloji: Öncü Devrimcilerden Yenilikçi Seçkinlere* (İstanbul: Metis Yayınları, 1998).
24. Nilüfer Göle, *Mühendisler ve İdeoloji: Öncü Devrimcilerden Yenilikçi Seçkinlere*, 2nd ed. (İstanbul: Metis Yayınları, 1998), 199–207.
25. As quoted in Murat Arslan, *Süleyman Demirel* (İstanbul: İletişim Yayınları, 2019), 32–33.
26. In 1986, Demirel gave a series of interviews that were published in *Yeni Nesil* newspaper and, with material from *Köprü* magazine, collected in a volume on "The Spiritual Side of Development." Süleyman Demirel, *Kalkınmanın Manevi Yönü* (İstanbul: Yeni Asya Yayınları, 1987).
27. Hulûsi Turgut, *GAP ve Demirel: 50 yıl* (ABC Basın Ajansı, 2000), 58.
28. As quoted in Turgut, *Demirel'in dünyası*, 109.
29. "The Earth Mover," *Time*, May 3, 1954.
30. Murat Arslan, *Süleyman Demirel* (İstanbul: İletişim Yayınları, 2019), 57–58.
31. These included the Demirköprü, Kemer, and Hirfanlı dams. H. Turgut, *Demirel'in Dünyası*, 118. For more on the Bureau of Reclamation, see Christopher Sneddon, *Concrete Revolution: Large Dams, Cold War Geopolitics, and the US Bureau of Reclamation* (Chicago: University of Chicago Press, 2015); Donald J. Pisani, *Water and American Government: The Reclamation Bureau, National Water Policy, and the West, 1902–1935* (Berkeley: University of California Press, 2002); and Donald Worster, *Rivers of Empire: Water, Aridity, and the Growth of the American West* (New York: Pantheon Books, 1985).
32. Turgut, *Demirel'in dünyası*, 121.
33. Max Weber, *The Theory of Social and Economic Organization* (London: Free Press of Glencoe, 1947).
34. Donald Worster, *Rivers of Empire: Water, Aridity, and the Growth of the American West* (New York: Pantheon Books, 1985).
35. Erik Jan Zürcher, *The Young Turk Legacy and Nation Building: From the Ottoman Empire to Atatürk's Turkey* (London: I. B. Tauris, 2010), 136–50.
36. Arslan, *Süleyman Demirel*, 39.
37. Turgut, *Demirel'in dünyası*, 136–37.
38. Turgut, 150–51.
39. Arslan, *Süleyman Demirel*, 42.
40. Demirel would use this understanding to great effect in parliamentary political battles. See Süleyman Demirel, *Nereden geldik, nereye gidiyoruz: Türkiye'nin siyasî iktisadî panoraması* (Yeni Asya Yayınevi, 1978).

41. Rostow originally published *The Stages of Growth* in 1960. W. W. Rostow, *The Stages of Economic Growth: A Non-Communist Manifesto*, 3rd ed. (Cambridge: Cambridge University Press, 1991), https://doi.org/10.1017/CBO9780511625824.

42. Süleyman Demirel, *Büyük Türkiye* (İstanbul: Dergâh Yayınları, 1975), 334.

43. Demirel, 339.

44. Demirel, 338.

45. Demirel, 346.

46. Demirel, *Kalkınmanın Manevi Yönü*, 19.

47. Adalet, *Hotels and Highways*, 14.

48. Turkey plans to build a canal connecting the Black Sea and the Sea of Marmara for a price tag of $65 billion. Bethan McKernan, "'I Get Nightmares': Turks Fear Impact of Erdoğan's $65bn Istanbul Canal," *The Guardian*, June 26, 2021. https://www.theguardian.com/world/2021/jun/26/i-get-nightmares-turks-fear-impact-of-erdogans-65bn-istanbul-canal.

49. Michael M. Gunter and M. Hakan Yavuz, "Turkish Paradox: Progressive Islamists versus Reactionary Secularists," *Critique: Critical Middle Eastern Studies* 16, no. 3 (January 1, 2007): 289–301, https://doi.org/10.1080/10669920701616633.

50. Several scholars have pursued these flows in considering the construction of major infrastructure. Andrew Needham, *Power Lines: Phoenix and the Making of the Modern Southwest* (Princeton, NJ: Princeton University Press, 2014); Deborah Cowen, *The Deadly Life of Logistics: Mapping Violence in Global Trade* (Minneapolis: University of Minnesota Press, 2014), 53–90; Timothy Mitchell, *Carbon Democracy: Political Power in the Age of Oil* (London; New York: Verso, 2011).

51. Discussion about the dams in the Cold War context can be found in Monib El-Khatib, "The Syrian Tabqa Dam: Its Development and Impact," *The Geographical Bulletin* 26 (November 1, 1984): 19–28. For more on Cold War competition in the Middle East, see Rashid Khalidi, *Sowing Crisis: The Cold War and American Dominance in the Middle East* (Boston: Beacon Press, 2009); David W. Lesch, ed., *The Middle East and the United States: A Historical and Political Reassessment*, 4th ed. (Boulder, CO: Westview Press, 2007).

52. US Central Intelligence Agency records contain multiple secret reports about al-Tabqa Dam, including a video smuggled out of the country, testifying to the Cold War national security interest in the project. See "Euphrates River Development," Current Intelligence Weekly Special Report (Central Intelligence Agency, May 13, 1966), Freedom of Information Act Electronic Reading Room.

53. On infrastructure as a site where power and ontology coalesce, see Rudolf Mrázek, *Engineers of Happy Land: Technology and Nationalism in a Colony* (Princeton, NJ: Princeton University Press, 2002); and Jane Bennett, *Vibrant Matter: A Political Ecology of Things* (Durham, NC: Duke University Press, 2010).

54. Brown, "Gridded Lives: Why Kazakhstan and Montana Are Nearly the Same Place," 21.

55. Brown, 21.

56. Brown, 47.

57. Laurent Dissard, "Submerged Stories from the Sidelines of Archaeological Science: The History and Politics of the Keban Dam Rescue Project (1967–1975) in Eastern Turkey" (Berkeley: University of California Press, 2011); China Sajadian, "The Drowned and the Displaced: Afterlives of Agrarian Developmentalism across the Lebanese-Syrian Border," *Mashriq & Mahjar: Journal of Middle East & North African Migration Studies* 10, no. 1 (March 21, 2023), https://doi.org/10.24847/v10i12023.347.

58. Christian Velud, "Une expérience d'administration régionale en Syrie durant le Mandat Français: conquête, colonisation et mise en valeur de la Gazira, 1920–1936" (PhD diss., l'Université Lumière Lyon 2, 1991); Dale J. Stahl, "A Technopolitical Frontier: The Keban Dam Project and Southeastern Anatolia," in *Transforming Socio-Natures in Turkey: Landscapes, State and Environmental Movements*, ed. Onur İnal and Ethemcan Turhan (New York: Routledge, 2020), 31–51.

59. Kurdish Human Rights Project, "The Impact of Large-Scale Dam Construction on Regional Security in the Kurdish Regions of Turkey" (Alternative Water Forum, Istanbul, Turkey, March 21, 2009).

60. Behrooz Morvaridi, "Resettlement, Rights to Development and the Ilisu Dam, Turkey," *Development and Change* 35, no. 4 (2004): 719–41, https://doi.org/10.1111/j.0012-155X.2004.00377.x; Isabel Zhang, "Submerging Kurdish History in Turkey: A Case Study of the Ilısu Dam," *The Middle East International Journal for Social Sciences* 3, no. 1 (March 2021): 1–8.

61. Jordi Tejel, *Syria's Kurds: History, Politics and Society*, trans. Emily Welle and Jane Welle, Routledge Advances in Middle East and Islamic Studies 16 (London: Routledge, 2009), 60–62.

62. Any number of studies on Turkish economic development largely exclude military support and industries in their assessments. See, for instance, Max Weston Thornburg, Graham Spry, and George Henry Soule, *Turkey: An Economic Appraisal* (New York: The Twentieth Century Fund, 1949); Zvi Yehuda Hershlag, *Turkey: The Challenge of Growth* (Leiden, Netherlands: E. J. Brill, 1968); Berch Berberoglu, *Turkey in Crisis: From State Capitalism to Neocolonialism* (London: Zed Books, 1982); Bent Hansen, *Egypt and Turkey*, A World Bank Comparative Study (New York: Oxford University Press, 1991).

63. Richard P. Tucker and Edmund Russell, eds., *Natural Enemy, Natural Ally: Toward an Environmental History of Warfare* (Corvallis: Oregon State University Press, 2004); Jacob Darwin Hamblin, *Arming Mother Nature: The Birth of Catastrophic Environmentalism* (New York: Oxford University Press, 2013); Emmanuel Kreike, *Scorched Earth: Environmental Warfare as a Crime against Humanity and Nature*, Human Rights and Crimes against Humanity (Princeton, NJ: Princeton University Press, 2021).

64. For more on this period, see John M. VanderLippe, *The Politics of Turkish Democracy* (Albany: State University of New York Press, 2005), 137–87; William M. Hale, *Turkish Foreign Policy, 1774–2000* (London: Frank Cass, 2000), 109–39.

65. A full list of claims may be found at TNA, FO 371/101874/1152/1.

66. Helms to Eden, January 14, 1952; M. Zorlu's speech, January 10, 1952, TNA, FO 371/101874/1152/12. See also FO 371/101874/1152/16 for the French communiqué.

67. The UK Treasury decided not to immediately publish a Turkish default and chose instead to tell the British parliament that Turkey had "discontinued service on these loans." Brief by Sir Herbert Brittain, January 29, 1952, TNA, FO 371/101874/1152/17.

68. Christelow to Egger, March 18, 1952, TNA, FO 371/101874/1152/24. The British later worked to obtain American support through the US Ambassador to Ankara, George McGhee. See Helms to McGhee, May 12, 1952, and May 13, 1952; and Fox to Cheetham, May 22, 1952 in TNA, FO 371/101874/1152/37 and 1152/45. The Treasury Department's tactic alarmed the British Foreign Office as well, which responded that "the course proposed by the Treasury may well gain nothing, but may merely intensely irritate one of our best friends, particularly as the Seyhan dam project is apparently a very good one and worth support on merit." Strang to Bridges, March 20, 1952, TNA, FO 371/101874/1152/21.

69. Telegram from Washington to Foreign Office, April 28, 1952, TNA, FO 371/101874/1152/29.

70. Telegram from Ankara to Foreign Office, May 18, 1952, TNA, FO 371/101874/1152/38.

71. Telegram from Washington to Foreign Office, June 19, 1952, TNA, FO 371/101874/1152/61; see also 1152/52, telegram from Ankara to Foreign Office, June 12, 1952, for Ankara's demand and 1152/54, Scott-Fox to Cheetham, June 9, 1952, for American reaction. Throughout the remainder of the year, the British and Turkish governments negotiated the terms of a loan settlement. The British ended up settling for a mere £3 million out of a requested £16 million.

72. Baran Tuncer, "External Financing of the Turkish Economy and Its Foreign Policy Implications," in *Turkey's Foreign Policy in Transition, 1950–1974* (Leiden, Netherlands: E. J. Brill, 1975), 208–12.

73. Selin M. Bölme, "The Politics of Incirlik Air Base," *Insight Turkey* 9, no. 3 (2007): 84–86.

74. The term "military-industrial complex" was famously coined by US President Dwight D. Eisenhower in his 1961 farewell address, and the concept continues to animate debates about the US military establishment and American involvement in Middle Eastern conflicts. See Charles J. Dunlap, "The Military-Industrial Complex," *Daedalus* 140, no. 3 (2011): 135–47.

75. "Memorandum for Phlips Talbot from Robert Komer, February 19, 1963, John F. Kennedy Library, quoted in Bruce R. Kunihulm, "Turkey and the West Since World War II," in *Turkey Between East and West*, ed. Vojtech Mastny and R. Craig Nation (Boulder, CO: Westview Press, 1997), 59.

76. John White, "A Pledge for Development," III-5, January 4, 1966, TNA, FO 957/273.

77. The connection was evident even in the membership of the OECD consortium created to negotiate development aid to Turkey. Only three states in the consortium were not also members of NATO: Austria, Sweden, and Switzerland.

78. The nature of foreign assistance changed in the 1960s, requiring a commensurate shift in Turkey's approach to aid. Donors became more apt to fund projects as opposed to broad programs, even as Turkish institutions were generating the kinds of development plans that had earlier worked to secure aid. The Five-Year Development Plans were difficult to finance externally, so the Turkish government returned to applying for aid from international organizations instead of foreign governments. In 1961, Turkey requested a NATO aid consortium. When that was not forthcoming, Ankara turned toward the Organisation for Economic Co-operation and Development (OECD) but struggled to reach appropriate terms. Ferenc A. Váli, *Bridge across the Bosporus: The Foreign Policy of Turkey* (Baltimore, MD: Johns Hopkins Press, 1971), 331.

79. The total aid for the project totaled some $135 million, of which the United States provided $40 million.

80. See Chapter 3 and, "Milletvekilimiz Nurettin Ardıçoğlu Keban Barajı mevzuu için Cemal Gürsel'e bir muhtıra verdi," *Elazığ Gazetesi*, December 24, 1963, 1.

81. SCI-Impregilo was a Franco-Italian consortium made up of Compagnie de Constructions Internationales, Impresit-Girola-Lodigiani, Compagnie Français d'Entreprises, and ARI Construction. Elena Calandri, "Italy's Foreign Assistance Policy, 1959–1969," *Contemporary European History* 12, no. 4 (2003): 521–22.

82. Tabitha Petran, *Syria* (New York: Praeger, 1972), 215; for more on this period, see Patrick Seale, *The Struggle for Syria: A Study of Post-War Arab Politics, 1945–1958* (New Haven, CT: Yale University Press, 1987).

83. The following volume contains both the 1922 report and related contemporary essays. Charles Héraud, *Une Mission de Reconnaissance de L'Euphrate En 1922* (Damas: Institut Français de Damas, 1995). For additional background on nineteenth-century French engineers and visions of Syria, see Isacar Bolaños, "Water, Engineers, and French Environmental Imaginaries of Ottoman Iraq, 1868–1908," *Environmental History* 27, no. 4 (October 1, 2022): 772–98, https://doi.org/10.1086/721180.

84. Walter Ferguson Crawford, "Visit of Sir A. Overton and W.F. Crawford to Syria and Lebanon," November 30, 1946, SA, GB 0033 SAD, 504/5/1. For the Iraq case, see Chapter 2.

85. Walter Ferguson Crawford, "Syria Jan 1952," Trek Notes—Syria, SA, GB 0033 SAD, 504/5/1.

86. Crawford, "Syria Jan 1952."

87. The Bank declared that additional study of the site was necessary to determine both irrigation possibilities and power demands. Walter Ferguson Crawford, "Syrian Development," Trek Notes—Syria, June 27, 1950, SA, GB 0033 SAD, 504/5/1; Jean R. de Fargues and Neil Bass, "Preliminary Report on the Youssef Pasha Dam for Production of Electrical Power and for Irrigation of the Euphrates Valley" (International

Bank for Reconstruction and Development, April 1956). For more on the relationship between agriculture, peasant farmers and politics in Syria, see Hanna Batatu, *Syria's Peasantry, the Descendants of Its Lesser Rural Notables, and Their Politics* (Princeton, NJ: Princeton University Press, 1999); Raymond A. Hinnebusch, *Peasant and Bureaucracy in Ba'thist Syria: The Political Economy of Rural Development* (Boulder, CO: Westview Press, 1989).

88. The statement was published in *Pravda*, April 17, 1955. Translation quoted from Sabrina P. Ramet, *The Soviet-Syrian Relationship since 1955: A Troubled Alliance*, Westview softcover ed, (Boulder, CO: Westview Press, 1990), 15–16.

89. Syria signed a trade agreement with the Soviet Union in November 1955. The International Fair in Damascus, started in 1954, featured several countries from the Soviet sphere of influence and eventually led to Syria signing trade agreements with many. For more on the Suez Crisis of 1956, see Guy Laron, *Origins of the Suez Crisis: Postwar Development Diplomacy and the Struggle over Third World Industrialization, 1945–1956* (Baltimore, MD: Johns Hopkins University Press, 2013).

90. The Soviet study suggested that this site held greater potential for irrigation, nearly 1.5 million acres, and increased electricity production.

91. In addition, Cairo, which dominated Syria through the UAR, was not at this time friendly in its relations to the Soviet Union. Central Intelligence Agency, "Euphrates River Development," Current Intelligence Weekly, May 13, 1966, 1. Declassified Documents Reference System (CK3100337117).

92. According to Guy Laron, a water project was "a shield with which the Syrians defended themselves against Nasser's attempts to meddle in their affairs." In 1959 Israel announced the creation of the National Water Carrier, which would divert Sea of Galilee waters to the Negev desert. Syrian Ba'thist leaders wanted Nasser to wage a guerilla campaign against the diversion; he refused. When those leaders rebelled against the UAR, they published transcripts to show how Nasser had undermined Syrian interests and betrayed the Palestinian cause. Guy Laron, *The Six-Day War: The Breaking of the Middle East* (New Haven, CT: Yale University Press, 2017), 34.

93. See Massimiliano Trentin, "Modernization as State Building: The Two Germanies in Syria, 1963–1972," *Diplomatic History* 33, no. 3 (June 2009): 487–505, https://doi.org/10.1111/j.1467-7709.2009.00782.x.

94. In February 1963, West Germany offered the Syrian government a smaller loan of 350 million marks ($87.5 million). By September, though, only 20 million marks had been forwarded for a feasibility study. Damascus, upset at German procrastination, threatened to send a delegation to Moscow to see about reviving Soviet interest. Jean Fleury, dispatch no. 1295, "Projets allemands en Syrie," September 26, 1963, CADN, 188PO/B/2.

95. Germany tried to lure the Syrians back, announcing that the credit offered for construction of the dam was "still considered valid on the German side." Paul-Henry Manière, dispatch no. 950/AL, "Participation allemande à la Foire de Damas," September 16, 1965, CADN, 188PO/B/2. For American conversations about Germany's

recognition of Israel, see "Memorandum of Conversation Between President Johnson and Chancellor Erhard," June 12, 1964, *Foreign Relations of the United States, 1964–68*, vol. XV, doc. 49, 111–115.

96. Damascus entered negotiations with East German representatives on an array of commercial and technical agreements, which were signed in Damascus on August 2. The GDR opened a consulate general in Damascus on August 23 and signed a credit agreement of $25 million on October 23. François Charles-Roux, dispatch no. 763/AL, "Activités allemandes en Syrie," July 22, 1965, CADN, 188PO/B/2. For more on East German involvement in Syrian affairs, see Massimiliano Trentin, *Engineers of Modern Development: East German Experts in Ba'thist Syria, 1965–1972* (CLEUP, 2010).

97. François Charles-Roux, dispatch no. 1053/AL, "Déclarations syriennes et allemandes au sujet du barrage de l'Euphrate," October 21, 196 and dispatch no. 1322/AL, "Relations entre la République Fédérale, Israêl et le monde arabe," December 28, 1965, CADN, 188PO/B/2.

98. In March 1965, the Lyndon Johnson administration approved the sale of 210 tanks to Israel and 100 to Jordan, and in 1966 sold 48 warplanes to Israel and 36 to Jordan. Laron, *The Six-Day War*, 188–89.

99. The new leaders of Syria released imprisoned Communists and authorized the party's newspaper, *Nidal al-Sha'b*. Martin Seymour, "The Dynamics of Power in Syria since the Break with Egypt," *Middle Eastern Studies* 6, no. 1 (1970): 41–43.

100. François Charles-Roux, dispatch no. 61/AL, "Financement du barrage de l'Euphrate: ouvertures syriennes en direction de l'U.R.S.S.," January 20, 1966; dispatch no. 103/AL, "Projet de barrage sur l'Euphrate," February 27, 1966; dispatch no. 461/AL, "Publication du Protocole d'accord syro-soviétique sur le barrage de l'Euphrate," May 5, 1966, CADN, 188PO/B/2.

101. François Charles-Roux, dispatch no. 500/AL, "La RFA et le barrage sur l'Euphrate," May 12, 1966, CADN, 108PO/B/2.

102. Zohurul Bari, "Syrian-Iraqi Dispute Over the Euphrates Waters," *International Studies* 16, no. 2 (April 1, 1977): 227–44, https://doi.org/10.1177/002088177701600203.

103. Brynjar Lia, "The Islamist Uprising in Syria, 1976–82: The History and Legacy of a Failed Revolt," *British Journal of Middle Eastern Studies* 43 (2016): 541–59, http://dx.doi.org/10.1080/13530194.2016.1139442; Raphaël Lefèvre, *Ashes of Hama: The Muslim Brotherhood in Syria* (Oxford: Oxford University Press, 2013).

104. Mary Beth D Nikitin, Paul K Kerr, and Andrew Feickert, "Syria's Chemical Weapons: Issues for Congress" (Washington, DC: Congressional Research Service, September 30, 2013).

105. David McDowall, *A Modern History of the Kurds* (London: I. B. Tauris, 1997), 418–44.

106. James Ron, *Weapons Transfers and Violations of the Laws of War in Turkey* (New York: Human Rights Watch, 1995), https://archive.hrw.org/legacy/summaries/s.turkey95n.html.

107. Ryan Browne and Jennifer Hansler, "US Believes Reports Turkey Misused US-Supplied Weapons in Syria Incursion Are 'Credible,'" CNN, November 6, 2019, https://www.cnn.com/2019/11/06/politics/us-turkey-weapon-misuse/index.html.

108. John F. Devlin, "The Baath Party: Rise and Metamorphosis," *The American Historical Review* 96, no. 5 (1991): 1402–4, https://doi.org/10.2307/2165277.

109. Lisa Wedeen, *Ambiguities of Domination: Politics, Rhetoric, and Symbols in Contemporary Syria* (Chicago: University of Chicago Press, 1999).

110. "Ma'rid 'an sadd al-Furat fi madinat ath-Thawra," (Exhibition about the Euphrates Dam in the city of ath-Thawra), *al-Ba'th*, July 1, 1973, 8.

111. Omar Amiralay, *Film Essay on the Euphrates Dam*. Syrian Arab Television, 1970. See also Rasha Salti, "The Cruel Sea: A Conversation with Omar Amiralay," *Bidoun*, Fall 2008, https://www.bidoun.org/articles/pulp-the-archive.

112. *al-Hayat al-Yawmiyah fi Qarya Suriyya [Everyday Life in a Syrian Village]*.

113. Samuel Sweeney, "An Overlooked Syrian Writer," *Dappled Things*, 2021, https://www.dappledthings.org/reviews/an-overlooked-syrian-writer.

114. The dam flooded 59 villages inhabited by 62,226 people. Hans Meliczek, "Land Settlement in the Euphrates Basin of Syria," *Ekistics* 53, no. 318/319 (1986): 202–12.

115. China Sajadian's ethnography of the children of the displaced who were made refugees in Lebanon by the twenty-first century Syrian Civil War points to this relocation as part of a longer process of Syrian government policies. Several of the families she interviewed were relocated from the Euphrates Valley to Qamishli, a town on the Turkish border over 250 km from the Euphrates River. Sajadian, "The Drowned and the Displaced." For more on the creation of an "Arab Belt" along Syria's frontiers, see Tejel, *Syria's Kurds*.

116. Samuel Sweeney suggests that al-'Ujayli shies away from criticizing the politics of relocation, but my reading of the text is that the author shows something worse: the state cares so little for the people of the flooded valley that no reason beyond "government interest" is necessary. Obedience is the only thing the state requires, such that even minor efforts to explain the state's policies are unimportant. See Sweeney, "An Overlooked Syrian Writer."

117. al-'Ujaylī, *al-Maghmūrūn*, 164.

118. al-'Ujaylī, 304.

119. al-'Ujaylī, 315.

120. al-'Ujaylī, 348–49.

121. al-'Ujaylī, 23–24.

122. Mariam Karouny, "Buried Hopes: Burst Dam Swamps Syrian Villages," *The Guardian*, June 6, 2002. https://www.theguardian.com/world/2002/jun/06/1.

123. *Tufan fi Balad al-Ba'th [A Flood in Ba'th Country]*.

Conclusion

1. In Syria, the Tishrin Dam, built near the old Yusuf Pasha dam site investigated by the French during the Mandate period, became operational in 1999. Iraq had already

built large dams on Tigris River tributaries flowing from the Zagros Mountains: the Dukan Dam (1959) on the Little Zab and the Darbandikhan Dam (1961) on the Diyala River. A year after the opening of the Mosul Dam on the Tigris River, Iraq unveiled the Haditha Dam in 1987 on the Euphrates River.

2. Dursun Yıldız notes a series of reports written in the 1960s exploring first the Euphrates and then the Tigris as creating the foundational engineering knowledge required for the water infrastructure on the rivers. Dursun Yıldız, *GAP bölgede ekonomik, stratejik ve siyasal gelişmeler* (İstanbul: Truva Yayınları, 2009), 17–18.

3. Mehmet Özdoğan, "Aşağı Fırat Havzası 1977 Yüzey Araştırmaları," Aşağı Fırat Projesi Yayınları (İstanbul: Orta Doğu Teknik Üniversitesi, 1977); *The Southeastern Anatolia Project Master Plan Study, Volume 2 Master Plan*, 4 vols. (Ankara: State Planning Organization, 1989).

4. "GAP Regional Development Administration," gap.gov.tr, n.d., http://www.gap.gov.tr/en/what-is-gap-page-1.html.

5. The Silvan Dam on the Batman River, a tributary of the Tigris River, is one of those still under construction; it will be second only to the Atatürk Dam in irrigation provision. "Minister Yumaklı: 'Contract Signed for Electricity Production at Silvan Dam and HEPP,'" (In Turkish) April 25, 2024, https://www.tarimorman.gov.tr/Haber/6270/Bakan-Yumakli-Silvan-Baraji-Ve-Heste-Elektrik-Uretimi-Icin-Sozlesme-Imzalandi.

6. Leila M. Harris, "Irrigation, Gender, and Social Geographies of the Changing Waterscapes of Southeastern Anatolia," *Environment and Planning D: Society and Space* 24, no. 2 (2006): 204, https://doi.org/10.1068/d03k.

7. Matthew Brunwasser, "Zeugma After the Flood," *Archaeology Magazine*, December 2012, https://archaeology.org/issues/november-december-2012/features/features-zeugma-after-the-flood/.

8. Martin van Bruinessen, "The Kurds in Turkey," *MERIP Reports*, no. 121 (1984): 7, https://doi.org/10.2307/3011035.

9. Wolfgang Wohlwend, "'Our Heads Did Not Accept It'–Development and Nostalgia in Southeastern Anatolia," *Zeitschrift Für Ethnologie* 140, no. 2 (2015): 221.

10. *The Ascending Place of Light: GAP Southeast Anatolia Region* (Fersa Ofset Tesisleri, 2007), 7.

11. Cevdet Yılmaz, ed., *Güneydoğu Anadolu Mutfağı* (GAP Eylem Planı, 2011), 3.

12. Arda Bilgen, "The Southeastern Anatolia Project (GAP) in Turkey: An Alternative Perspective on the Major Rationales of GAP," *Journal of Balkan and Near Eastern Studies* 21, no. 5 (August 21, 2018): 532–52 https://doi.org/10.1080/19448953.2018.1506287.

13. Harris, "Irrigation, Gender, and Social Geographies of the Changing Waterscapes of Southeastern Anatolia," 194.

14. Mehmet Yesilnacar and Sinan Uyanik, "Investigation of Water Quality of the World's Largest Irrigation Tunnel System, the Sanliurfa Tunnels in Turkey," *Fresenius Environmental Bulletin* 14 (January 1, 2005): 300.

15. "GAP General Presentation," 2021, http://www.gap.gov.tr/en/gap-general-presentation-page-31.html.

16. Leila M. Harris, "Contested Sustainabilities: Assessing Narratives of Environmental Change in Southeastern Turkey," *Local Environment* 14, no. 8 (September 2009): 699–720, https://doi.org/10.1080/13549830903096452.

17. Dogan Altinbilek and Cecilia Tortajada, "The Atatürk Dam in the Context of the Southeastern Anatolia (GAP) Project," in *Impacts of Large Dams: A Global Assessment*, ed. Asit K Biswas, Dogan Altinbilek, and Cecilia Tortajada (Berlin: Springer, 2012), 186–87.

18. Osman Çopur, "Cotton Production in Turkey," in *Book of Abstracts*, 2016, 327–31; "Cotton: World Markets and Trade" (Foreign Agricultural Service, United States Department of Agriculture, March 2024).

19. Ali Volkan Bilgili et al., "Post-Irrigation Degradation of Land and Environmental Resources in the Harran Plain, Southeastern Turkey," *Environmental Monitoring and Assessment* 190, no. 660 (November 2018): 2–4, https://doi.org/10.1007/s10661-018-7019-2. On fertilizer production, see Yunhu Gao and André Cabrera Serrenho, "Greenhouse Gas Emissions from Nitrogen Fertilizers Could Be Reduced by up to One-Fifth of Current Levels by 2050 with Combined Interventions," *Nature Food* 4 (February 2023): 170–78, https://doi.org/10.1038/s43016-023-00698-w.

20. Çağlar Erdoğan, "Cotton and Products Annual" (United States Department of Agriculture, March 24, 2020), 2.

21. As quoted in Yıldız, *GAP bölgede ekonomik, stratejik ve siyasal gelişmeler*, 23.

22. Süleyman Demirel, *Büyük Türkiye* (İstanbul: Dergâh Yayınları, 1975), 253–54.

23. Diana K. Davis, *Resurrecting the Granary of Rome: Environmental History and French Colonial Expansion in North Africa* (Athens: Ohio University Press, 2007); Diana K. Davis, *The Arid Lands: History, Power, Knowledge*, History for a Sustainable Future (Cambridge, MA: MIT Press, 2016).

24. Bilgen, "The Southeastern Anatolia Project (GAP) in Turkey," 535–36.

25. Timothy Mitchell, *Rule of Experts: Egypt, Techno-Politics, Modernity* (Berkeley: University of California Press, 2002), 52.

26. Karen Barad, *Meeting the Universe Halfway: Quantum Physics and the Entanglement of Matter and Meaning* (Durham, NC: Duke University Press, 2007), 185.

27. Leila M. Harris, "Water and Conflict Geographies of the Southeastern Anatolia Project," *Society & Natural Resources* 15, no. 8 (September 2002): 743–59, https://doi.org/10.1080/08941920290069326; see also John F. Kolars and William Mitchell, *The Euphrates River and the Southeast Anatolia Development Project* (Carbondale: Southern Illinois University Press, 1991), 30–34.

28. "Governments Withdraw Financial Support for Ilisu Dam in Turkey," Amnesty International, July 7, 2009, https://www.amnesty.org/en/latest/news/2009/07/governments-withdraw-financial-support-ilisu-dam-turkey-20090707/; Jacques Leslie, "In a Major Reversal, the World Bank Is Backing Mega Dams," Yale E360, December 19, 2024, https://e360.yale.edu/features/world-bank-hydro-dams.

29. Harris, "Contested Sustainabilities"; Ahmet Ertugrul Akyol, "Sustainability and Inclusivity of the Southeastern Anatolia Project (GAP) in Southeastern Turkey: The Case of Şanlıurfa Province" (master's thesis, University of Amsterdam, 2013); Arda Bilgen, "Turkey's Southeastern Anatolia Project (GAP): A Qualitative Review of the Literature," *British Journal of Middle Eastern Studies* 47, no. 4 (November 23, 2018): 652–71, https://doi.org/10.1080/13530194.2018.1549978; see also Matthias Leese and Simon Meisch, "Securitising Sustainability? Questioning the 'Water, Energy and Food-Security Nexus'," *Water Alternatives* 8, no. 1 (2015): 15.

30. Bağış earned a PhD from the London School of Economics in 1974 and was a faculty member in international relations at Hacettepe University in Ankara. Ali İhsan Bağış, *Southeastern Anatolia Project: The Cradle of Civilisation Regenerated* (İstanbul: Interbank, 1989), 9.

31. Bağış, 5–7.

32. Such a comment corresponds well to what Diana K. Davis writes of European perceptions of arid environments: "It is not widely recognized that Western thinking about deserts has been tightly bound up with European notions of forests and their 'proper' form . . . [leading] to the development of the widely influential theory of desiccation that identified humans as the primary agents in the process later named desertification." Davis, *The Arid Lands*, 20–21.

33. Barad writes of an "intertwining of ethics, knowing, and being," which I have somewhat modified here. Barad, *Meeting the Universe Halfway*, 185.

34. We might add some additional terms for Williams's list from the histories in this book—the injustice of nature, and the civilizing of nature. Raymond Williams, "Ideas of Nature," in *Problems in Materialism and Culture* (London: Verso, 1980), 84.

35. Osama Esber, "Iraq and the Arab World on the Edge of the Abyss," *The Markaz Review* (blog), January 14, 2021, https://themarkaz.org/iraq-and-the-arab-world-on-the-edge-of-the-abyss/.

36. Khālid Sulaymān, *Ḥurrās al-miyāh: al-jafāf wa-al-taghayyur al-manākhī fī al-ʿIrāq* (Dimashq: Dār al-Madá lil-Iʻlām wa-al-Thaqāfah wa-al-Funūn, 2020), 48–49.

37. "Nature Iraq Organization," International Union for Conservation of Nature, n.d., https://iucn.org/our-union/members/iucn-members/nature-iraq-organization.

38. "Iraq: Environmentalists Face Retaliation" Human Rights Watch, February 23, 2023, https://www.hrw.org/news/2023/02/23/iraq-environmentalists-face-retaliation.

39. Stephen C. Lonergan and Jassim Al-Asadi, *The Ghosts of Iraq's Marshes: A History of Conflict, Tragedy, and Restoration* (Cairo: The American University in Cairo Press, 2024).

40. This, he says, is an important difference between his work and "Iraqi writing [which] is largely focused on urban elements in daily life—the café, the street, the home, and so on." Khudayyir also notes how little modern Iraqi literature deals with the floods on the Tigris and Euphrates. Muhammad Khudayyir, as quoted in Sulaymān, *Ḥurrās al-miyā*, 42.

41. Muḥammad Khuḍayyir, "Qiṣṣatān Min Taḥt Al-Arḍ," *Tāmarrā*, no. 1 (2017): 66.

42. Sulaymān, *Ḥurrās al-miyā*, 38.

43. Jeremy S. Pal and Elfatih A. B. Eltahir, "Future Temperature in Southwest Asia Projected to Exceed a Threshold for Human Adaptability," *Nature Climate Change* 6 (February 2016): 197–200, https://doi.org/10.1038/nclimate2833.

44. Dipesh Chakrabarty, *The Climate of History in a Planetary Age* (Chicago: University of Chicago Press, 2021).

45. And, in this way, is allied with other such efforts in environmental history, such as, Emily O'Gorman, *Wetlands in a Dry Land: More-than-Human Histories of Australia's Murray-Darling Basin*, (Seattle: University of Washington Press, 2021).

Bibliography

Archives Consulted

INDIA
National Archives of India (NAI)

FRANCE
Centre des Archives diplomatiques de Nantes (CADN)
Centre des Archives diplomatiques du ministère des Affaires étrangères

TURKEY
Atatürk Kitaplığı
Başbakanlık Osmanlı Arşivi (BOA)
Başbakanlık Cumhuriyet Arşivi
Beyazıt Devlet Kütüphanesi
Devlet Su İşleri Kütüphanesi
Türkiye Büyük Millet Meclisi Kütüphanesi

UNITED KINGDOM
The British Library (BL)
Churchill Archive Center, Cambridge University (CAC)
Institution of Civil Engineers, London
The National Archives (TNA)
Middle East Centre Archive, Oxford University (MECA)
Sudan Archive, Durham University (SAD)

UNITED STATES
National Archives and Record Administration (NARA)
Library of Congress

Periodicals and Journals

IN TURKISH:
Elazığ Gazetesi
Yeni Fırat

IN ARABIC:
Al-Baʿth

IN ENGLISH:
Times of Mesopotamia

Other Published Sources

Abdullah, Mukhalad, Nadhir al-Ansari, Nasrat Adamo, Varoujan Sissakian, and Jan Laue. "Soil Salinity of Mesopotamia and the Main Drains." *Journal of Earth Sciences and Geotechnical Engineering* 10, no. 4 (2020): 221–30.

Adalet, Begüm. "Agricultural Infrastructures: Land, Race, and Statecraft in Turkey." *Environment and Planning D: Society and Space* 40, no. 6 (December 2022): 975–93. https://doi.org/10.1177/02637758221124139.

———. *Hotels and Highways: The Construction of Modernization Theory in Cold War Turkey*. Stanford, CA: Stanford University Press, 2018.

Adamo, Nasrat, Nadhir Al-Ansari, and Varoujan K. Sissakian. "How Dams Can Affect Freshwater Issues in the Euphrates-Tigris Basins." *Journal of Earth Sciences and Geotechnical Engineering* 10, no. 1 (2020): 43–76.

Ahmad, Feroz. "The Agrarian Policy of the Young Turks, 1908–1918." In *From Empire to Republic: Essays on the Late Ottoman Empire and Modern Turkey*, Vol. 1. Istanbul: İstanbul Bilgi University Press, 2008.

Akarun, Refik. *Zor ve Sorunlu Temel Üzerinde Yapılan Bir Büyük Baraj Keban Barajı / A Large Dam on Difficult Foundation: Keban Dam*. Yapı Teknik Engineering and Consultancy Co., 1999.

Akpınar, Deniz. *Osmanlı'da su projeleri Hindiye barajı*. Istanbul: Arı Sanat, 2017.

Akyol, Ahmet Ertugrul. "Sustainability and Inclusivity of the Southeastern Anatolia Project (GAP) in Southeastern Turkey: The Case of Şanlıurfa Province." MSc Thesis, University of Amsterdam, 2013.

Al-Ansari, Nadhir, Nasrat Adamo, and Varoujan K Sissakian. "Hydrological Characteristics of the Tigris and Euphrates Rivers," n.d.

Al-Ansari, Nadhir, Issa E Issa, Varoujan Sissakian, and Nasrat Adamo. "Mystery of Mosul Dam the Most Dangerous Dam in the World: The Project." *Journal of Earth Sciences and Geotechnical Engineering* 5, no. 3 (2015).

Aldroubi, Mina. "Iraq Could Have No Rivers by 2040, Government Report Warns." *The National*, December 2, 2021. https://www.thenationalnews.com/mena/iraq/2021/12/02/iraq-could-have-no-rivers-by-2040-government-report-warns/.

Al-Habib, Mahmud. "The Iraqi Development Board." *The Southwestern Social Science Quarterly* 36, no. 2 (1955): 185–90.

al-Hayat al-Yawmiyah fi Qarya Suriyya [Everyday Life in a Syrian Village]. Mu'assasah al-'Āmmah lil-Sīnimā, 1974.
Al-Jomard, Atheel. "Internal Migration in Iraq." In *The Integration of Modern Iraq*, edited by Abbas Kelidar, 111–22. New York: St. Martin's Press, 1979.
Allan, John Anthony. et al. "Fortunately There Are Substitutes for Water Otherwise Our Hydro-Political Futures Would Be Impossible." *Priorities for Water Resources Allocation and Management* 13, no. 4 (1993): 26.
Al-Monitor. "Floods Cause Severe Damage in Southern Iraq." May 9, 2013. https://www.al-monitor.com/originals/2013/05/floods-iraq-rain-damages-deaths.html.
Alnasrawi, Abbas. "Iraq: Economic Sanctions and Consequences, 1990–2000." *Third World Quarterly* 22, no. 2 (April 2001): 205–18. https://doi.org/10.1080/01436590120037036.
———. *The Economy of Iraq: Oil, Wars, Destruction of Development and Prospects, 1950–2010*. Contributions in Economics and Economic History 154. Westport, CT: Greenwood Press, 1994.
Al-Rubaie, Azhar. "Climate Change Ravages Iraq as Palm Trees Make Way for Desert." Al Jazeera, May 26, 2022. https://www.aljazeera.com/news/2022/5/26/climate-change-ravages-iraq-as-palm-trees-make-way-for-desert.
Al-Shabībī, Muḥammad Riḍā. *Dīwān Al-Shabībī*. Cairo: Maṭba'at Lajnat al-Ta'līf wa-al-Tarjamah wa-al-Nashr, 1940.
Altinbilek, Dogan, and Cecilia Tortajada. "The Atatürk Dam in the Context of the Southeastern Anatolia (GAP) Project." In *Impacts of Large Dams: A Global Assessment*, edited by Asit K Biswas, Dogan Altinbilek, and Cecilia Tortajada, 171–99. Berlin: Springer, 2012.
Alwash, Azzam. "The Mosul Dam: Turning a Potential Disaster into a Win-Win Solution." *Viewpoints* 98 (2016): 1–2.
Amnesty International. "Governments Withdraw Financial Support for Ilisu Dam in Turkey," July 7, 2009. https://www.amnesty.org/en/latest/news/2009/07/governments-withdraw-financial-support-ilisu-dam-turkey-20090707/.
Anderson, Benedict R. *Imagined Communities: Reflections on the Origin and Spread of Nationalism*. New York: Verso, 1983.
Appadurai, Arjun, ed. *The Social Life of Things: Commodities in Cultural Perspective*. Cambridge: Cambridge University Press, 1986.
Aqrawi, A. A. M., and G. Evans. "Sedimentation in the Lakes and Marshes (Ahwar) of the Tigris-Euphrates Delta, Southern Mesopotamia." *Sedimentology* 41, no. 4 (August 1994): 755–76. https://doi.org/10.1111/j.1365-3091.1994.tb01422.x.
Aristotle. "Poetics." The Internet Classics Archive. Accessed July 25, 2024. http://classics.mit.edu/Aristotle/poetics.mb.txt.
Arslan, Murat. *Süleyman Demirel*. İstanbul: İletişim Yayınları, 2019.
Ashkenas, Jeremy, Archie Tse, Derek Watkins, and Karen Yourish. "A Rogue State Along Two Rivers." *The New York Times*, July 3, 2014. https://www.nytimes.com/interactive/2014/07/03/world/middleeast/syria-iraq-isis-rogue-state-along-two-rivers.html.

Askari, Masour. "Iraq's Ecological Disaster." *International Review*, February 12, 2003.

Baali, Fuad. *Relation of the People to the Land in Southern Iraq*. Gainesville: University of Florida Press, 1966.

Bağış, Ali İhsan. *Southeastern Anatolia Project: The Cradle of Civilisation Regenerated*. İstanbul: Interbank, 1989.

Barad, Karen. *Meeting the Universe Halfway: Quantum Physics and the Entanglement of Matter and Meaning*. Durham, NC: Duke University Press, 2007.

Bari, Zohurul. "Syrian-Iraqi Dispute Over the Euphrates Waters." *International Studies* 16, no. 2 (April 1, 1977): 227–44. https://doi.org/10.1177/002088177701600203.

Batatu, Hanna. *Syria's Peasantry, the Descendants of Its Lesser Rural Notables, and Their Politics*. Princeton, NJ: Princeton University Press, 1999.

———. *The Old Social Classes and the Revolutionary Movements of Iraq: A Study of Iraq's Old Landed and Commercial Classes and of Its Communists, Ba'thists and Free Officers*. Princeton, NJ: Princeton University Press, 1989.

Batuman, Elif. "The Sanctuary: The World's Oldest Temple and the Dawn of Civilization." *New Yorker*, December 19, 2011.

Bennett, Jane. *Vibrant Matter: A Political Ecology of Things*. Durham, NC: Duke University Press, 2010.

Berberoglu, Berch. *Turkey in Crisis: From State Capitalism to Neocolonialism*. London: Zed Books, 1982.

Bernal, Victoria. "Colonial Moral Economy and the Discipline of Development: The Gezira Scheme and 'Modern' Sudan." *Cultural Anthropology* 12, no. 4 (November 1997): 447–79. https://doi.org/10.1525/can.1997.12.4.447.

Beşikçi, İsmail. *Devletlerarası Sömürge Kürdistan*. İstanbul: Alan Yayıncılık, 1990.

———. *Doğu Anadolu'nun düzeni: sosyo-ekonomik ve etnik temeller*. Yurt ve dünya sorunları, 1. Erzurum, Turkey: E. Yayinlari, 1969.

Bet-Shlimon, Arbella. *City of Black Gold: Oil, Ethnicity, and the Making of Modern Kirkuk*. Stanford, CA: Stanford University Press, 2019.

Bilgen, Arda. "The Southeastern Anatolia Project (GAP) in Turkey: An Alternative Perspective on the Major Rationales of GAP." *Journal of Balkan and Near Eastern Studies* 21, no. 5 (August 21, 2018): 532–52. https://doi.org/10.1080/19448953.2018.1506287.

———. "Turkey's Southeastern Anatolia Project (GAP): A Qualitative Review of the Literature." *British Journal of Middle Eastern Studies* 47, no. 4 (November 23, 2018): 652–71. https://doi.org/10.1080/13530194.2018.1549978.

Bilgili, Ali Volkan, İrfan Yeşilnacar, Kotera Akihiko, Takanori Nagano, Aydın Aydemir, Hüseyin Sefa Hızlı, and Ayşin Bilgili. "Post-Irrigation Degradation of Land and Environmental Resources in the Harran Plain, Southeastern Turkey." *Environmental Monitoring and Assessment* 190, no. 11 (November 2018): 660. https://doi.org/10.1007/s10661-018-7019-2.

Biswas, Asit K. *International Waters of the Middle East: From Euphrates-Tigris to Nile*. Bombay: Oxford University Press, 1994.

Black, John. "The Military Influence on Engineering Education in Britain and India, 1848–1906." *The Indian Economic & Social History Review* 46, no. 2 (April 2009): 211–39. https://doi.org/10.1177/001946460904600203.

Blackbourn, David. *The Conquest of Nature: Water, Landscape, and the Making of Modern Germany*. New York: Norton, 2007.

Bochenski, Feliks, and William Diamond. "TVA's in the Middle East." *Middle East Journal* 4, no. 1 (1950): 52–82.

Bolaños, Isacar. "Water, Engineers, and French Environmental Imaginaries of Ottoman Iraq, 1868–1908." *Environmental History* 27, no. 4 (October 1, 2022): 772–98. https://doi.org/10.1086/721180.

Bölme, Selin M. "The Politics of Incirlik Air Base." *Insight Turkey* 9, no. 3 (2007): 82–91.

Brown, Bill. "Thing Theory." *Critical Inquiry* 28, no. 1 (2001): 1–22.

Brown, Kate. "Gridded Lives: Why Kazakhstan and Montana Are Nearly the Same Place." *The American Historical Review* 106, no. 1 (2001): 17–48.

———. *Plutopia: Nuclear Families, Atomic Cities, and the Great Soviet and American Plutonium Disasters*. New York: Oxford University Press, 2013.

Browne, Ryan, and Jennifer Hansler. "US Believes Reports Turkey Misused US-Supplied Weapons in Syria Incursion Are 'Credible.'" CNN, November 6, 2019. https://www.cnn.com/2019/11/06/politics/us-turkey-weapon-misuse/index.html.

Bruinessen, Martin. "Genocide in Kurdistan? The Suppression of the Dersim Rebellion in Turkey (1937–38) and the Chemical War against the Iraqi Kurds (1988)." In *Genocide: Conceptual and Historical Dimensions*, edited by George J. Andreopoulos, 141–70. Philadelphia: University of Pennsylvania Press, 1997.

Bruinessen, Martin van. "The Kurds in Turkey." *MERIP Reports*, no. 121 (1984): 6–12, 14. https://doi.org/10.2307/3011035.

Brunwasser, Matthew. "Zeugma After the Flood." *Archaeology Magazine*, December 2012. https://archaeology.org/issues/november-december-2012/features/features-zeugma-after-the-flood/.

Bulloch, John, and Ādil Darwīsh. *Water Wars: Coming Conflicts in the Middle East*. Repr. London: Gollancz, 1996.

Buringh, Pieter. *Soils and Soil Conditions in Iraq*. Wageningen, Netherlands: H. Veenman & Zonen N. V., 1960.

Burns, Norman. "Development Projects in Iraq: 1. The Dujaylah Land Settlement." *Middle East Journal* 5, no. 3 (Summer 1951): 362–66.

Çağaptay, Soner. "Reconfiguring the Turkish Nation in the 1930s." *Nationalism and Ethnic Politics* 8, no. 2 (June 2002): 67–82. https://doi.org/10.1080/13537110208428662.

Calandri, Elena. "Italy's Foreign Assistance Policy, 1959–1969." *Contemporary European History* 12, no. 4 (2003): 509–25.

Califano, Mark, and Jeffrey Meyer. *Good Intentions Corrupted: The Oil for Food Scandal and the Threat to the UN*. New York: PublicAffairs, 2009.

Callon, Michel. "Introduction: The Embeddedness of Economic Markets in Economics." In *The Laws of the Markets*, edited by Michel Callon, 1–57. Oxford: Blackwell Publishers, 1998.

Cech, Erin A. "The (Mis)Framing of Social Justice: Why Ideologies of Depoliticization and Meritocracy Hinder Engineers' Ability to Think About Social Injustices." In *Engineering Education for Social Justice*, vol. 10, edited by Juan Lucena, 67–84. Dordrecht: Springer Netherlands, 2013. https://doi.org/10.1007/978-94-007-6350-0_4.

Çetinsaya, Gökhan. *The Ottoman Administration of Iraq, 1890–1908*. New York: Routledge, 2006. https://doi.org/10.4324/9780203332467.

Ceylan, Ebubekir. *The Ottoman Origins of Modern Iraq: Political Reform, Modernization and Development in the Nineteenth-Century Middle East*. London: I.B. Tauris, 2011.

Chibani, Achref. "Sand and Dust Storms in the MENA Region: A Problem Awaiting Mitigation." Arab Center. Washington, DC, July 29, 2022. https://arabcenterdc.org/resource/sand-and-dust-storms-in-the-mena-region-a-problem-awaiting-mitigation/.

Christensen, Peter. *The Decline of Iranshahr: Irrigation and Environments in the History of the Middle East, 500 B.C. to A.D. 1500*. Copenhagen: Museum Tusculanum Press, University of Copenhagen, 1993.

Churchill, Winston S. "'My Grandfather Invented Iraq.'" *Wall Street Journal*, March 10, 2003.

Citino, Nathan J. *Envisioning the Arab Future: Modernization in U.S.-Arab Relations, 1945–1967*. Global and International History. Cambridge: Cambridge University Press, 2017.

———. "The Ottoman Legacy in Cold War Modernization." *International Journal of Middle East Studies* 40, no. 4 (2008): 579–97.

Cohen, Stuart. *British Policy in Mesopotamia, 1903–1914*. London: Ithaca Press for the Middle East Centre, St. Antony's College, Oxford, 1976.

Cole, Camille. "Controversial Investments: Trade and Infrastructure in Ottoman-British Relations in Iraq, 1861–1918." *Middle Eastern Studies* 54, no. 5 (September 3, 2018): 744–68. https://doi.org/10.1080/00263206.2018.1462164.

Cole, Camille Lyans. "Empire on Edge: Land, Law, and Capital in Gilded Age Basra." PhD diss., Yale University, 2020.

Coole, Diana. "Agentic Capacities and Capacious Historical Materialism: Thinking with New Materialisms in the Political Sciences." *Millennium: Journal of International Studies* 41, no. 3 (June 2013): 451–69. https://doi.org/10.1177/0305829813481006.

Çopur, Osman. "Cotton Production in Turkey." In *Book of Abstracts*, 327–31, 2016.

"Cotton: World Markets and Trade." Foreign Agricultural Service, United States Department of Agriculture, March 2024.

Cowen, Deborah. *The Deadly Life of Logistics: Mapping Violence in Global Trade*. Minneapolis: University of Minnesota Press, 2014.

Cronon, William. *Nature's Metropolis: Chicago and the Great West*. New York: Norton, 1992.
Culcasi, Karen. "Cartographically Constructing Kurdistan within Geopolitical and Orientalist Discourses." *Political Geography* 25, no. 6 (August 1, 2006): 680–706. https://doi.org/10.1016/j.polgeo.2006.05.008.
Davis, Diana K. *Resurrecting the Granary of Rome: Environmental History and French Colonial Expansion in North Africa*. Athens: Ohio University Press, 2007.
———. *The Arid Lands: History, Power, Knowledge*. History for a Sustainable Future. Cambridge, MA: MIT Press, 2016.
Davis, Diana K., and Edmund Burke III, eds. *Environmental Imaginaries of the Middle East and North Africa*. Athens: Ohio University Press, 2011.
Davis, Eric. *Memories of State: Politics, History, and Collective Identity in Modern Iraq*. Berkeley: University of California Press, 2008.
Deemer, Bridget R., John A. Harrison, Siyue Li, Jake J. Beaulieu, Tonya DelSontro, Nathan Barros, José F. Bezerra-Neto, Stephen M. Powers, Marco A. dos Santos, and J. Arie Vonk. "Greenhouse Gas Emissions from Reservoir Water Surfaces: A New Global Synthesis." *BioScience* 66, no. 11 (November 1, 2016): 949–64. https://doi.org/10.1093/biosci/biw117.
Degu, Ahmed Mohamed, Faisal Hossain, Dev Niyogi, Roger Pielke Sr., J. Marshall Shepherd, Nathalie Voisin, and Themis Chronis. "The Influence of Large Dams on Surrounding Climate and Precipitation Patterns." *Geophysical Research Letters* 38, no. 4 (2011). https://doi.org/10.1029/2010GL046482.
Delaney, Carol. "Father State, Motherland, and the Birth of Modern Turkey." In *Naturalizing Power: Essays in Feminist Cultural Analysis*, edited by Sylvia Yanagisako and Carol Delaney. New York: Routledge, 1995.
Demirel, Süleyman. *Büyük Türkiye*. İstanbul: Dergâh Yayınları, 1975.
———. *Kalkınmanın Manevi Yönü*. İstanbul: Yeni Asya Yayınları, 1987.
———. *Nereden geldik, nereye gidiyoruz: Türkiye'nin siyasî iktisadî panoraması*. İstanbul: Yeni Asya Yayınevi, 1978.
Deriner, İbrahim. "Keban Barajı ve Hidroelektrik Santralı Hakkında Bazı Bilgiler." *Türkiye Mühendislik Haberleri* 8, no. 92 (1962): 3–5.
Derr, Jennifer. "Drafting a Map of Colonial Egypt: The 1902 Aswan Dam, Historical Imagination, and the Production of Agricultural Geography." In *Environmental Imaginaries of the Middle East and North Africa*, edited by Diana K. Davis and Edmund Burke III, 136–57. Athens: Ohio University Press, 2011.
Derr, Jennifer L. *The Lived Nile: Environment, Disease, and Material Colonial Economy in Egypt*. Stanford, CA: Stanford University Press, 2019.
Devlet Su İşleri Genel Müdürlüğü. "KEBAN BARAJI 50 YAŞINDA!," September 9, 2024. https://www.dsi.gov.tr/Haber/Detay/13356.
Devlin, John F. "The Baath Party: Rise and Metamorphosis." *The American Historical Review* 96, no. 5 (1991): 1396–1407. https://doi.org/10.2307/2165277.

Dieleman, P.J. *Reclamation of Salt Affected Soils in Iraq: Soil Hydrological and Agricultural Studies*. Wageningen, Netherlands: International Institute for Land Reclamation and Improvement, 1963. https://edepot.wur.nl/59923.

Dissard, Laurent. "Submerged Stories from the Sidelines of Archaeological Science: The History and Politics of the Keban Dam Rescue Project (1967–1975) in Eastern Turkey." PhD diss., University of California, Berkeley, 2011.

Dodge, Toby. *Inventing Iraq: The Failure of Nation-Building and a History Denied*. New York: Columbia University Press, 2003.

Dökmeci, Cenani. "Fırat'la Söyleşme." *Yeni Fırat* 28 (1966): 7.

Dolbee, Samuel. *Locusts of Power: Borders, Empire, and Environment in the Modern Middle East*. Cambridge: Cambridge University Press, 2023.

Doomed by the Dam: A Survey of the Monuments Threatened by the Creation of the Keban Dam Flood Area, Elazığ 18–29 October 1966. Orta Doğu Teknik Üniversitesi. Ankara: METU Faculty of Architecture, Dept. of Restoration, 1967.

Drayton, Richard. *Nature's Government: Science, Imperial Britain, and the "Improvement" of the World*. New York: Yale University Press, 2000.

Dunlap, Charles J. "The Military-Industrial Complex." *Daedalus* 140, no. 3 (2011): 135–47.

Durmuş, Mitat. "Niyazi Yıldırım Gençosmanoğlu." In *Türk Edebiyatı İsimler Sözlüğü*, February 17, 2019. https://teis.yesevi.edu.tr/madde-detay/gencosmanoglu-niyazi-yildirim.

Ekbladh, David. "'Mr. TVA': Grass-Roots Development, David Lilienthal, and the Rise and Fall of the Tennessee Valley Authority as a Symbol for U.S. Overseas Development, 1933–1973." *Diplomatic History* 26, no. 3 (July 1, 2002): 335–74. https://doi.org/10.1111/1467-7709.00315.

———. *The Great American Mission: Modernization and the Construction of an American World Order*. America in the World. Princeton, NJ: Princeton University Press, 2010.

Ekici, Cevat, ed. *Osmanlı Döneminde Irak: Plan, Fotoğraf ve Belgelerle*. İstanbul: Osmanlı Arşivi Daire Başkanlığı, 2006.

El-Khatib, Monib. "The Syrian Tabqa Dam: Its Development and Impact." *The Geographical Bulletin* 26 (November 1, 1984): 19–28.

Elliot, Matthew. *"Independent Iraq": British Influence from 1941 to 1958*. London: I.B. Tauris, 1996. https://doi.org/10.5040/9780755612307.

Erdoğan, Çağlar. "Cotton and Products Annual." United States Department of Agriculture, March 24, 2020.

Ersoy, Turan. "Keban Barajının İnşası Dolayısıyla Açıkta Kalan Halkın İskan ve İstihdamı." *Sosyal Siyaset Konferansları Dergisi*, no. 21 (1970): 23–38.

———. *Keban İskan Problemi*. Ankara: Devlet Planlama Teşkilatı, 1968.

Esber, Osama. "Iraq and the Arab World on the Edge of the Abyss." *The Markaz Review*, January 14, 2021. https://themarkaz.org/iraq-and-the-arab-world-on-the-edge-of-the-abyss/.

Escobar, Arturo. *Encountering Development: The Making and Unmaking of the Third World*. Princeton, NJ: Princeton University Press, 2012.

Eugene Rogan. *The Arabs: A History*. New York: Basic Books, 2009.

"Euphrates River Development." Current Intelligence Weekly Special Report. Central Intelligence Agency, May 13, 1966. Freedom of Information Act Electronic Reading Room, https://www.cia.gov/readingroom/document/cia-rdp79-00927a005300020002-2

FAO Mediterranean Development Project, Iraq: Country Report. Rome, Italy: Food and Agriculture Organization of the United Nations, 1959.

Fargues, Jean R. de, and Neil Bass. "Preliminary Report on the Youssef Pasha Dam for Production of Electrical Power and for Irrigation of the Euphrates Valley." International Bank for Reconstruction and Development, April 1956.

Farouk-Sluglett, Marion, and Peter Sluglett. *Iraq since 1958: From Revolution to Dictatorship*. London: I.B. Tauris, 1990.

Ferguson, James. *The Anti-Politics Machine: "Development," Depoliticization, and Bureaucratic Power in Lesotho*. Minneapolis: University of Minnesota Press, 1994.

Fiege, Mark. *The Republic of Nature: An Environmental History of the United States*. Seattle: University of Washington Press, 2012.

Film Muhawala 'an Sadd al-Furat [Film Essay on the Euphrates Dam], 1970.

Fisk, Brad. "Dujaila: Iraq's Pilot Project for Land Settlement." *Economic Geography* 28, no. 4 (1952): 343–54. https://doi.org/10.2307/141972.

Foster, John Bellamy, Brett Clark, and Richard York. *The Ecological Rift: Capitalism's War on the Earth*. New York: Monthly Review Press, 2010.

Gao, Yunhu, and André Cabrera Serrenho. "Greenhouse Gas Emissions from Nitrogen Fertilizers Could Be Reduced by up to One-Fifth of Current Levels by 2050 with Combined Interventions." *Nature Food* 4 (February 2023): 170–78. https://doi.org/10.1038/s43016-023-00698-w.

"GAP General Presentation," 2021. http://www.gap.gov.tr/en/gap-general-presentation-page-31.html.

gap.gov.tr. "GAP Regional Development Administration," n.d. http://www.gap.gov.tr/en/what-is-gap-page-1.html.

Gaud, William S. "The Green Revolution: Accomplishments and Apprehensions." Presented at the Society for International Development, Washington, DC, March 8, 1968. http://www.agbioworld.org/biotech-info/topics/borlaug/borlaug-green.html.

Gerges, Fawaz. *Making the Arab World: Nasser, Qutb, and the Clash That Shaped the Middle East*. Princeton, NJ: Princeton University Press, 2018.

Gerke, Gerwin. "The Iraq Development Board and British Policy, 1945–50." *Middle Eastern Studies* 27, no. 2 (April 1991): 231–55. https://doi.org/10.1080/00263209108700858.

Ghalib, Ali. *Malaria and Malaria in Iraq*. Facts and Prospects in Iraq 5. Jerusalem: The New Publishers Iraq, 1944.

Gilli, Eric. "Karst Areas of Turkey." In *Caves and Karst of Turkey–Volume 2: Geology, Hydrogeology and Karst*, edited by Gültekin Günay, Koray Törk, İsmail Noyan GÜNER, and Eric Gilli, 55–65. Cham, Switzerland: Springer International Publishing, 2022. https://doi.org/10.1007/978-3-030-95361-4_7.

Gilmartin, David. *Blood and Water: The Indus River Basin in Modern History*. Oakland: University of California Press, 2015.

Göle, Nilüfer. *Mühendisler ve İdeoloji: Öncü Devrimcilerden Yenilikçi Seçkinlere*. 2nd ed. İstanbul: Metis Yayınları, 1998.

Gore, Rick. "Anatolia—A History Forged by Disaster." *National Geographic Magazine*. Accessed September 6, 2022. https://www.nationalgeographic.com/science/article/anatolian-history.

Gratien, Chris. *The Unsettled Plain: An Environmental History of the Late Ottoman Frontier*. Stanford, CA: Stanford University Press, 2022.

Günışığı Gazetesi. "Sular Çekildi Tarlalar Ortaya Çıktı." April 23, 2021. https://www.gunisigigazetesi.net/elazig-guncel/sular-cekildi-tarlalar-ortaya-cikti-h83523.html.

Gunter, Michael M., and M. Hakan Yavuz. "Turkish Paradox: Progressive Islamists versus Reactionary Secularists." *Critique: Critical Middle Eastern Studies* 16, no. 3 (January 1, 2007): 289–301. https://doi.org/10.1080/10669920701616633.

Gürpınar, Doğan. "Anatolia's Eternal Destiny Was Sealed: Seljuks of Rum in the Turkish National(ist) Imagination from the Late Ottoman Empire to the Republican Era." *European Journal of Turkish Studies*, May 2, 2012. https://doi.org/10.4000/ejts.4547.

Habermann, Stanley John. "The Iraq Development Board: Administration and Program." *Middle East Journal* 9, no. 2 (1955): 179–86.

Haider, Saleh. "Land Problems of Iraq." PhD diss., London School of Economics, 1942.

Haigh, Frank Fraser. "Report on the Control of the Rivers of Iraq and the Utilization of Their Waters." Institution of Civil Engineers, London, UK. Baghdad: Baghdad Press, 1951.

Haj, Samira. *The Making of Iraq, 1900–1963: Capital, Power, and Ideology*. Albany: State University of New York Press, 1997.

Hale, William M. *Turkish Foreign Policy, 1774–2000*. London: Frank Cass, 2000.

Hamblin, Jacob Darwin. *Arming Mother Nature: The Birth of Catastrophic Environmentalism*. New York: Oxford University Press, 2013.

Ḥamdānī, Maḥmūd Shawqī. *Lamaḥāt min taṭawwur al-rayy fī al-ʿIrāq*. Baghdad: al-Maṭbaʿa al-Saʿdūn, 1984.

Hanioğlu, M. Şükrü. *Atatürk: An Intellectual Biography*. Princeton, NJ: Princeton University Press, 2011.

Hansen, Bent. *Egypt and Turkey*. New York: Oxford University Press, 1991.

Haraway, Donna. "The Cyborg Manifesto: Science, Technology, and Socialist-Feminism in the Late Twentieth Century." In *Simians, Cyborgs, and Women: The Reinvention of Nature*, 292–324. New York: Routledge, 1991.

Harris, Leila M. "Contested Sustainabilities: Assessing Narratives of Environmental Change in Southeastern Turkey." *Local Environment* 14, no. 8 (September 2009): 699–720. https://doi.org/10.1080/13549830903096452.

———. "Irrigation, Gender, and Social Geographies of the Changing Waterscapes of Southeastern Anatolia." *Environment and Planning D: Society and Space* 24, no. 2 (2006): 187–213. https://doi.org/10.1068/d03k.

———. "Water and Conflict Geographies of the Southeastern Anatolia Project." *Society & Natural Resources* 15, no. 8 (September 2002): 743–59. https://doi.org/10.1080/08941920290069326.

Haydar, Rustum, and Najda Fathi Safwa. *Mudhakkirāt Rustum Haydar*. Beirut: al-Dār al-'Arabīyah lil-Mawsū'āt, 1988.

Headrick, Daniel R. *The Tools of Empire: Technology and European Imperialism in the Nineteenth Century*. New York: Oxford University Press, 1981.

Hecht, Gabrielle, ed. *Entangled Geographies: Empire and Technopolitics in the Global Cold War*. Cambridge, MA: MIT Press, 2011.

———. *The Radiance of France: Nuclear Power and National Identity after World War II*. Cambridge, MA: MIT Press, 1998.

Héraud, Charles. *Une Mission de Reconnaissance de L'Euphrate En 1922*. Damas: Institut Français de Damas, 1995.

Hillel, Daniel. *Rivers of Eden: The Struggle for Water and the Quest for Peace in the Middle East*. New York: Oxford University Press, 1994.

Hillenbrand, Carole. *Turkish Myth and Muslim Symbol: The Battle of Manzikert*. Edinburgh: Edinburgh University Press, 2007.

Hinnebusch, Raymond A. *Peasant and Bureaucracy in Ba'thist Syria: The Political Economy of Rural Development*. Boulder, CO: Westview Press, 1989.

Hodder, Ian. *Entangled: An Archaeology of the Relationships Between Humans and Things*. Hoboken, NJ: John Wiley & Sons, Inc., 2012.

Hodge, Joseph Morgan. *Triumph of the Expert: Agrarian Doctrines of Development and the Legacies of British Colonialism*. Athens: Ohio University Press, 2007.

Hommes, Lena, Rutgerd Boelens, and Harro Maat. "Contested Hydrosocial Territories and Disputed Water Governance: Struggles and Competing Claims over the Ilisu Dam Development in Southeastern Turkey." *Geoforum* 71 (May 1, 2016): 9–20. https://doi.org/10.1016/j.geoforum.2016.02.015.

Husain, Faisal. *Rivers of the Sultan: The Tigris and Euphrates in the Ottoman Empire*. New York: Oxford University Press, 2021.

———. "Sediment of the Tigris and Euphrates Rivers: An Early Modern Perspective." *Water History* 13 (April 1, 2021): 13–32. https://doi.org/10.1007/s12685-020-00256-2.

International Union for Conservation of Nature. "Nature Iraq Organization," n.d. https://iucn.org/our-union/members/iucn-members/nature-iraq-organization.

"Iraq: Environmentalists Face Retaliation | Human Rights Watch," February 23, 2023. https://www.hrw.org/news/2023/02/23/iraq-environmentalists-face-retaliation.

Irving, Clive. "Gertrude of Arabia, the Woman Who Invented Iraq." *The Daily Beast*, June 17, 2014.
Iversen, Carl. *A Report on Monetary Policy in Iraq*. Baghdad: National Bank of Iraq, 1954.
Jackson, John. "Engineering Problems of Mesopotamia and the Euphrates Valley." *Empire Review* 29 (1915): 193–99.
Jacobs, Matthew F. *Imagining the Middle East: The Building of an American Foreign Policy, 1918–1967*. Chapel Hill: University of North Carolina Press, 2011.
Jacobsen, Thorkild, and Robert M. Adams. "Salt and Silt in Ancient Mesopotamian Agriculture." *Science* 128, no. 1251–1258 (November 21, 1958): 9.
John M. VanderLippe. *The Politics of Turkish Democracy*. Albany: State University of New York Press, 2005.
Kacar, Şükrü. "Unutulmaz Br Ani." *Elazığ Hakimiyet Haber*, May 31, 2016. https://www.elazighakimiyethaber.com/yazi/sukru-kacar/unutulmaz-bir-ani/1563/.
Kadhim, Abbas. *Reclaiming Iraq: The 1920 Revolution and the Founding of the Modern State*. Austin: University of Texas Press, 2012.
Karouny, Mariam. "Buried Hopes: Burst Dam Swamps Syrian Villages." *The Guardian*, June 6, 2002. https://www.theguardian.com/world/2002/jun/06/1.
"Keban Baraj Gölü Altında Kalacak Köy ve Mahallerin Anket Çalışması Sonucu." Ankara: İmar ve İskan Bakanlığı, 1966.
Khadduri, Majid. *Independent Iraq: A Study in Iraqi Politics since 1932*. London: Oxford University Press, 1951.
Khalidi, Rashid. *Sowing Crisis: The Cold War and American Dominance in the Middle East*. Boston: Beacon Press, 2009.
Khuḍayyir, Muḥammad. "Qiṣṣatān Min Taḥt Al-Arḍ." *Tāmarrā*, no. 1 (2017): 66–69.
Kibaroglu, Ayşegül, and Sezin Iba Gürsoy. "Water–Energy–Food Nexus in a Transboundary Context: The Euphrates–Tigris River Basin as a Case Study." *Water International* 40, no. 5–6 (September 19, 2015): 824–38. https://doi.org/10.1080/02508060.2015.1078577.
Kingston, Paul W. T. *Britain and the Politics of Modernization in the Middle East, 1945–1958*. Cambridge Middle East Studies, no. 4. Cambridge: Cambridge University Press, 2002. https://doi.org/10.1017/CBO9780511563539.
Klein, Janet. *The Margins of Empire: Kurdish Militias in the Ottoman Tribal Zone*. Stanford, CA: Stanford University Press, 2011.
Kliot, Nurit. *Water Resources and Conflict in the Middle East*. London: Routledge, 1993.
Kloppenburg, Jack Ralph. *First the Seed: The Political Economy of Plant Biotechnology*. Madison: University of Wisconsin Press, 2005.
Kohn, Eduardo. *How Forests Think: Toward an Anthropology beyond the Human*. Berkeley: University of California Press, 2013.
Kolars, John F., and William Mitchell. *The Euphrates River and the Southeast Anatolia Development Project*. Carbondale: Southern Illinois University Press, 1991.

Koselleck, Reinhart. *Futures Past: On the Semantics of Historical Time*. Translated by Keith Tribe. New York: Columbia University Press, 2004.

———. *The Practice of Conceptual History: Timing History, Spacing Concepts*. Translated by Todd Samuel Presner. Stanford, CA: Stanford University Press, 2002.

Kreike, Emmanuel. *Scorched Earth: Environmental Warfare as a Crime against Humanity and Nature*. Princeton, NJ: Princeton University Press, 2021.

Kubba, Sam. *The Iraqi Marshlands and the Marsh Arabs: The Ma'dan, Their Culture and the Environment*. New York: Ithaca Press, 2011.

Kum, Gülşen. "The Influence of Dams on Surrounding Climate: The Case of Keban Dam / Barajların Çevre İklime Etkisi: Keban Barajı Örneği." *Gaziantep University Journal of Social Sciences* 15 (2016): 193–204.

Kuniholm, Bruce R. "Turkey and the West Since World War II." In *Turkey Between East and West*, edited by Vojtech Mastny and R. Craig Nation. Boulder, CO: Westview Press, 1997.

Kurdish Human Rights Project. "The Impact of Large-Scale Dam Construction on Regional Security in the Kurdish Regions of Turkey." Presented at the Alternative Water Forum, Istanbul, Turkey, March 21, 2009.

Landau, Jacob M. "Ultra-Nationalist Literature in the Turkish Republic: A Note on the Novels of Hüseyin Nihâl Atsız." *Middle Eastern Studies* 39, no. 2 (April 2003): 204–10. https://doi.org/10.1080/714004510.

Larkin, Brian. "The Politics and Poetics of Infrastructure." *Annual Review of Anthropology* 42 (2013): 327–43.

Laron, Guy. *Origins of the Suez Crisis: Postwar Development Diplomacy and the Struggle over Third World Industrialization, 1945–1956*. Baltimore, MD: Johns Hopkins University Press, 2013.

———. *The Six-Day War: The Breaking of the Middle East*. New Haven, CT: Yale University Press, 2017.

Latham, Michael E. *The Right Kind of Revolution: Modernization, Development, and U.S. Foreign Policy from the Cold War to the Present*. Ithaca, NY: Cornell University Press, 2010.

Latour, Bruno. *Reassembling the Social: An Introduction to Actor-Network-Theory*. Oxford: Oxford University Press, 2007.

———. *Science in Action: How to Follow Scientists and Engineers Through Society*. Cambridge, MA: Harvard University Press, 1987.

Layton, Edwin T. *The Revolt of the Engineers: Social Responsibility and the American Engineering Profession*. Cleveland, OH: Press of Case Western Reserve University, 1971.

Leese, Matthias, and Simon Meisch. "Securitising Sustainability? Questioning the 'Water, Energy and Food-Security Nexus'." *Water Alternatives* 8, no. 1 (2015): 15.

Lefèvre, Raphaël. *Ashes of Hama: The Muslim Brotherhood in Syria*. Oxford: Oxford University Press, 2013.

Lemke, Thomas. "Varieties of Materialism." *BioSocieties* 10 (December 1, 2015): 490–95. https://doi.org/10.1057/biosoc.2015.41.

Lesch, David W., ed. *The Middle East and the United States: A Historical and Political Reassessment*. 4th ed. Boulder, CO: Westview Press, 2007.

Leslie, Jacques. "In a Major Reversal, the World Bank Is Backing Mega Dams." Yale E360, December 19, 2024. https://e360.yale.edu/features/world-bank-hydro-dams.

Lewis, Bernard. *The Emergence of Modern Turkey*. London: Oxford University Press, 1961.

Li, Darryl. "A Jihadism Anti-Primer." *MERIP*, no. 276 (Fall 2015). https://merip.org/2015/12/a-jihadism-anti-primer/.

Lia, Brynjar. "The Islamist Uprising in Syria, 1976–82: The History and Legacy of a Failed Revolt." *British Journal of Middle Eastern Studies* 43, no. 4 (2016): 541–59. http://dx.doi.org/10.1080/13530194.2016.1139442.

Lombardo, Joseph D. "In the Kingdom of Dams: Water, Governance, and the Keban Dam Project in Eastern Anatolia, 1961–1974." PhD diss., The New School, 2018.

Lonergan, Stephen C., and Jassim Al-Asadi. *The Ghosts of Iraq's Marshes: A History of Conflict, Tragedy, and Restoration*. Cairo: The American University in Cairo Press, 2024.

Longrigg, Stephen Hemsley. *Iraq, 1900 to 1950: A Political, Social, and Economic History*. London: Oxford University Press, 1953.

Lorenz, Frederick M., and Edward J. Erickson. *The Euphrates Triangle: Security Implications of the Southeastern Anatolia Project*. Washington, DC: National Defense University Press, 1999.

Low, Michael Christopher. *Imperial Mecca: Ottoman Arabia and the Indian Ocean Hajj*. Columbia University Press, 2020.

Mahdi, Kamil A. *State and Agriculture in Iraq: Modern Development, Stagnation and the Impact of Oil*. Reading, UK: Ithaca Press, 2000.

"Managing Salinity in Iraq's Agriculture: Current State, Causes, and Impacts." Iraq Salinity Assessment. International Center for Agricultural Research in the Dry Areas, 2013.

Manela, Erez. *The Wilsonian Moment: Self-Determination and the International Origins of Anticolonial Nationalism*. Oxford: Oxford University Press, 2009.

Mapping Militant Organizations. "The Islamic State." Accessed April 21, 2023. https://cisac.fsi.stanford.edu/mappingmilitants/profiles/islamic-state.

Marr, Phebe Ann. *The Modern History of Iraq*. 2nd ed. Boulder, CO: Westview Press, 2004.

Mason, Kenneth. "Notes on the Canal System and Ancient Sites of Babylonia in the Time of Xenophon." *The Geographical Journal* 56, no. 6 (1920): 468–81. https://doi.org/10.2307/1780469.

McCully, Patrick. *Silenced Rivers: The Ecology and Politics of Large Dams*. London: Zed Books, 1996.

McDowall, David. *A Modern History of the Kurds*. London: I. B. Tauris, 1997.
McKernan, Bethan. "'I Get Nightmares': Turks Fear Impact of Erdoğan's $65bn Istanbul Canal." *The Guardian*, June 26, 2021. https://www.theguardian.com/world/2021/jun/26/i-get-nightmares-turks-fear-impact-of-erdogans-65bn-istanbul-canal.
McNeill, John Robert. *Something New under the Sun: An Environmental History of the Twentieth-Century World*. New York: W.W. Norton & Company, 2000.
Meiton, Fredrik. *Electrical Palestine: Capital and Technology from Empire to Nation*. Oakland: University of California Press, 2019.
Meliczek, Hans. "Land Settlement in the Euphrates Basin of Syria." *Ekistics* 53, no. 318/319 (1986): 202–12.
Mikhail, Alan. *Nature and Empire in Ottoman Egypt: An Environmental History*. Cambridge: Cambridge University Press, 2011.
Milor, Vedat. "The Genesis of Planning in Turkey." *New Perspectives on Turkey* 4 (1990): 1–30.
"Minister Yumaklı: 'Contract Signed for Electricity Production at Silvan Dam and HEPP,'" April 25, 2024. https://www.tarimorman.gov.tr/Haber/6270/Bakan-Yumakli-Silvan-Baraji-Ve-Heste-Elektrik-Uretimi-Icin-Sozlesme-Imzalandi.
Mitchell, Timothy. *Carbon Democracy: Political Power in the Age of Oil*. London: Verso, 2011.
———. *Rule of Experts: Egypt, Techno-Politics, Modernity*. Berkeley: University of California Press, 2002.
———. "The Work of Economics: How a Discipline Makes Its World." *European Journal of Sociology / Archives Européennes de Sociologie / Europäisches Archiv Für Soziologie* 46, no. 2 (2005): 297–320.
Montazeri, Amirhossein, Mehdi Mazaheri, Saeed Morid, and Mohammad Reza Mosaddeghi. "Effects of Upstream Activities of Tigris-Euphrates River Basin on Water and Soil Resources of Shatt al-Arab Border River." *Science of The Total Environment* 858 (February 2023): 159751. https://doi.org/10.1016/j.scitotenv.2022.159751.
Morozova, Galina S. "A Review of Holocene Avulsions of the Tigris and Euphrates Rivers and Possible Effects on the Evolution of Civilizations in Lower Mesopotamia." *Geoarchaeology* 20, no. 4 (2005): 401–23. https://doi.org/10.1002/gea.20057.
Morton, Michael Quentin. "River of Oil: Early Oil Exploration in Iraq." *GEO ExPro* 12, no. 1 (February 2015).
Morvaridi, Behrooz. "Resettlement, Rights to Development and the Ilisu Dam, Turkey." *Development and Change* 35, no. 4 (2004): 719–41. https://doi.org/10.1111/j.0012-155X.2004.00377.x.
Mrázek, Rudolf. *Engineers of Happy Land: Technology and Nationalism in a Colony*. Princeton, NJ: Princeton University Press, 2002.
Nakash, Yitzhak. *The Shi'is of Iraq*. Princeton, NJ: Princeton University Press, 1994.
———. *The Shi'is of Iraq*. 4th ed. Princeton, NJ: Princeton University Press, 1996.

Naṣīr al-Chādirchī. *Mudhakkirāt Naṣīr al-Chādirchī: ṭufūlah mutanāqiḍah, shabāb mutamarrid, ṭarīq al-matāʿib*. Bayrūt: al-Madá lil-Iʿlām wa-al-Thaqāfah wa-al-Funūn, 2017.

Needham, Andrew. *Power Lines: Phoenix and the Making of the Modern Southwest*. Princeton, NJ: Princeton University Press, 2014.

Nikitin, Mary Beth D, Paul K Kerr, and Andrew Feickert. "Syria's Chemical Weapons: Issues for Congress." Congressional Research Service, September 30, 2013.

Nixon, Rob. *Slow Violence and the Environmentalism of the Poor*. Cambridge, MA: Harvard University Press, 2011. https://doi.org/10.2307/j.ctt2jbsgw.

Nunn, Wilfrid. *Tigris Gunboats: The Forgotten War in Iraq 1914–1917*. London: Chatham Publishing, 2007.

Nuttall, Sarah. *Entanglement: Literary and Cultural Reflections on Post Apartheid*. Johannesburg: Wits University Press, 2009.

O'Gorman, Emily. *Wetlands in a Dry Land: More-than-Human Histories of Australia's Murray-Darling Basin*. Seattle: University of Washington Press, 2021.

Öktem, Kerem. "When Dams Are Built on Shaky Grounds: Policy Choice and Social Performance of Hydro-Project Based Development in Turkey." *Erdkunde* 56, no. 3 (2002): 310–25.

Olson, Robert W. *The Emergence of Kurdish Nationalism and the Sheikh Said Rebellion, 1880–1925*. Austin: University of Texas Press, 1991.

Onder, Cetin Kemal, Cakir Elife, Ilgac Makbule, Can Gizem, Soylemez Berkan, et al. "Geotechnical Aspects of Reconnaissance Findings after 2020 January 24th, M6.8 Sivrice–Elazig–Turkey Earthquake." *Bulletin of Earthquake Engineering* 19, no. 9 (July 2021): 3415–59. https://doi.org/10.1007/s10518-021-01112-1.

Orhon, Mine, Sibel Esendal, and M. A. Kazak. *Türkiye'deki Barajlar / Dams in Turkey*. Ankara: Devlet Su İşleri Genel Müdürlüğü, 1991.

Orit Bashkin. *The Other Iraq: Pluralism and Culture in Hashemite Iraq*. Stanford, CA: Stanford University Press, 2009.

Orta-Doğu'da Su Sorunu. EIUK. Ankara: Dişişleri Bakanlığı Bölgesel ve Sınıraşan Sular Dairesi, 1994.

Owen, Matthew David. "For the Progress of Man: The TVA, Electric Power, and the Environment, 1939–1969." PhD diss., Vanderbilt University, 2014.

Owen, Roger. *The Middle East in the World Economy, 1800–1914*. London: Methuen, 1981.

Özcan, Muhammet. "Cenani Dökmeci'nin şiirlerinde yapı ve tema / Structure and style of poetry in Cenani Dökmeci." Fırat Üniversitesi, 2015. https://openaccess.firat.edu.tr/xmlui/handle/11508/15268.

Özdemir, M. Ali, and Nusret Özgen. "Keban Barajından Su Kaçakları ve Sunduğu Doğal Potansiyel." *Afyon Kocatepe Üniversitesi Sosyal Bilimler Dergisi* 6, no. 1 (June 2004): 65–86.

Özdoğan, Mehmet. "Aşağı Fırat Havzası 1977 Yüzey Araştırmaları." Aşağı Fırat Projesi Yayınları. İstanbul: Orta Doğu Teknik Üniversitesi, 1977.

Pal, Jeremy S., and Elfatih A. B. Eltahir. "Future Temperature in Southwest Asia Projected to Exceed a Threshold for Human Adaptability." *Nature Climate Change* 6 (February 2016): 197–200. https://doi.org/10.1038/nclimate2833.

Partow, Hassan. "The Mesopotamian Marshlands: Demise of an Ecosystem." Nairobi: UNEP, 2001.

Pedersen, Susan. "Getting Out of Iraq—in 1932: The League of Nations and the Road to Normative Statehood." *The American Historical Review* 115, no. 4 (October 1, 2010): 975–1000. https://doi.org/10.1086/ahr.115.4.975.

Penrose, Edith Tilton, and E. F. Penrose. *Iraq: International Relations and National Development*. Boulder, CO: Westview Press, 1978.

Petran, Tabitha. *Syria*. New York: Praeger, 1972.

Philipps, Dave, Azmat Khan, and Eric Schmitt. "A Dam in Syria Was on a 'No-Strike' List. The U.S. Bombed It Anyway." *International New York Times*, January 22, 2022. Gale Academic OneFile (accessed May 27, 2025).

Phillips, Doris G. "Rural-to-Urban Migration in Iraq." *Economic Development and Cultural Change* 7, no. 4 (1959): 405–21.

Prairie, Yves T., Jukka Alm, Jake Beaulieu, Nathan Barros, Tom Battin, Jonathan Cole, Paul del Giorgio, et al. "Greenhouse Gas Emissions from Freshwater Reservoirs: What Does the Atmosphere See?" *Ecosystems* 21 (2018): 1058–71. https://doi.org/10.1007/s10021-017-0198-9.

Pritchard, Sara B. *Confluence: The Nature of Technology and the Remaking of the Rhône*. Harvard Historical Studies 172. Cambridge, MA: Harvard University Press, 2011.

Pursley, Sara. *Familiar Futures: Time, Selfhood, and Sovereignty in Iraq*. Stanford Studies in Middle Eastern and Islamic Societies and Cultures. Stanford, CA: Stanford University Press, 2019.

——. "'Lines Drawn on an Empty Map': Iraq's Borders and the Legend of the Artificial State." *Jadaliyya*, June 2, 2015. https://www.jadaliyya.com/Details/32140.

"Qānūn Al-Rayy Wa-al-Sidād Sanat 1923." Baghdad: Maṭbaʿat al-Ḥukūmah, n.d.

Rahi, Khayyun Amtair, and Todd Halihan. "Salinity Evolution of the Tigris River." *Regional Environmental Change* 18 (October 2018): 2117–27. http://dx.doi.org/10.1007/s10113-018-1344-4.

Ramet, Sabrina P. *The Soviet-Syrian Relationship since 1955: A Troubled Alliance*. Westview softcover ed. Boulder, CO: Westview Press, 1990.

Raydan, Noam. "If The Daura Refinery Could Speak (Part I)." *The Chokepoint*, August 15, 2022. https://chokepoint.substack.com/p/if-the-daura-refinery-could-speak.

Razoux, Pierre. *The Iran-Iraq War*. Translated by Nicholas Elliott. Cambridge, MA: The Belknap Press of Harvard University Press, 2015.

"Report to Electric Power Resources Survey and Development Administration for Engineering and Economic Feasibility of Keban Dam and Hydroelectric Project." New York: EBASCO Services Inc., 1963. Charles River Watershed Surveys, Reports, and Maps, 1962–1970; Records of the Office of the Chief of Engineers, 1789–1999, Record Group 77. U.S. National Archives Boston.

Reynolds, Michael A. *Shattering Empires: The Clash and Collapse of the Ottoman and Russian Empires, 1908–1918*. Cambridge, MA: Cambridge University Press, 2011.

Reynolds, Nancy Y. "Building the Past: Rockscapes and the Aswan High Dam in Egypt." In *Water on Sand: Environmental Histories of the Middle East and North Africa*, edited by Alan Mikhail. Oxford: Oxford University Press, 2012.

Richards, John F. *The Unending Frontier: An Environmental History of the Early Modern World*. Berkeley: University of California Press, 2005.

Riḍwān, Walīd. *Muškilat al-miyāh baina Sūriyā wa-Turkiyā: (asbāb al-muškila-al-mašāri' al-mā'iya as-sūriya-āfāq al-ḥall)*. Ṭab'a 1. Bairūt: Šarikat al-Maṭbū'āt li-t-Tauzī' wa-'n-Našr, 2006.

Ron, James. *Weapons Transfers and Violations of the Laws of War in Turkey*. New York: Human Rights Watch, 1995. https://archive.hrw.org/legacy/summaries/s.turkey95n.html.

Rostow, W. W. *The Stages of Economic Growth: A Non-Communist Manifesto*. 3rd ed. Cambridge: Cambridge University Press, 1991. https://doi.org/10.1017/CBO9780511625824.

Rubin, Alissa J., and Bryan Denton. "A Climate Warning from the Cradle of Civilization." *The New York Times*, July 29, 2023. Gale Academic OneFile.

Russell, Edmund. *Evolutionary History: Uniting History and Biology to Understand Life on Earth*. Cambridge: Cambridge University Press, 2011.

Rutledge, Ian. *Enemy on the Euphrates: The British Occupation of Iraq and the Great Arab Revolt, 1914–1921*. London: Saqi Books, 2014.

Safwa, Najda Fathi. *Ṣāliḥ Jabr: sīrah siyāsīyah*. Bayrūt, Lubnān: Dār al-Sāqī, 2016.

Sajadian, China. "The Drowned and the Displaced: Afterlives of Agrarian Developmentalism across the Lebanese-Syrian Border." *Mashriq & Mahjar: Journal of Middle East & North African Migration Studies* 10, no. 1 (March 21, 2023). https://doi.org/10.24847/v10i12023.347.

Salti, Rasha. "The Cruel Sea: A Conversation with Omar Amiralay." *Bidoun*, Fall 2008. https://www.bidoun.org/articles/pulp-the-archive.

Saraiva, Tiago. *Fascist Pigs: Technoscientific Organisms and the History of Fascism*. Cambridge, MA: MIT Press, 2018.

Sassoon, Joseph. *Economic Policy in Iraq, 1932–1950*. London: F. Cass, 1987.

Satia, Priya. "'A Rebellion of Technology': The British Arabian Imaginary." In *Environmental Imaginaries of the Middle East and North Africa*, edited by Diana K. Davis and Edmund Burke III. Athens: Ohio University Press, 2011.

———. "Developing Iraq: Britain, India and the Redemption of Empire and Technology in the First World War." *Past & Present* 197, no. 1 (November 1, 2007): 211–55. https://doi.org/10.1093/pastj/gtm008.

Sayce, Archibald H., John Jackson, L. W. King, F. R. Maunsell, and William Willcocks. "The Garden of Eden and Its Restoration: Discussion." *The Geographical Journal* 40, no. 2 (August 1912): 145–48. https://doi.org/10.2307/1778460.

Schayegh, Cyrus. *The Middle East and the Making of the Modern World*. Cambridge, MA: Harvard University Press, 2017.

Schouwenburg, Hans. "Back to the Future?" March 28, 2015. https://doi.org/10.18352/22130624-00301003.

Scott, James C. *Seeing like a State: How Certain Schemes to Improve the Human Condition Have Failed*. New Haven, CT: Yale University Press, 2008.

Seale, Patrick. *The Struggle for Syria: A Study of Post-War Arab Politics, 1945–1958*. New Haven, CT: Yale University Press, 1987.

Şen, Zekâi. *Sınır aşan sularımız*. İstanbul: Su Vakfı Yayınları, 2002.

Şencan, Gökçe. "For Hasankeyf the Bell Tolls." *International Rivers*, February 12, 2020. https://www.internationalrivers.org/news/for-hasankeyf-the-bell-tolls/.

Şengül, Abdullah. "Arif Nihat Asya." In *Türk Edebiyatı İsimler Sözlüğü*, July 17, 2018. https://teis.yesevi.edu.tr/madde-detay/asya-arif-nihat.

Seymour, Martin. "The Dynamics of Power in Syria since the Break with Egypt." *Middle Eastern Studies* 6, no. 1 (1970): 35–47.

Shapland, Greg. *Rivers of Discord: International Water Disputes in the Middle East*. London: Hurst, 1997.

Shawkat, Naji. *Sīrah wa-dhikrayāt: thamānīn ʿāman 1894–1974*. Baghdad: Manshūrāt Maktabat al-Yaqẓah al-ʿArabīyah, 1974.

Shiva, Vandana. *The Violence of the Green Revolution: Third World Agriculture, Ecology and Politics*. London: Zed Books, 1991.

Shuker, Zeinab. "Water, Oil and Iraq's Climate Future." *Middle East Report* 306 (Spring 2023).

Sick, Gary. "Foreword." In *The Creation of Iraq, 1914–1921*, edited by Reeva S. Simon and Eleanor Harvey Tejirian. New York: Columbia University Press, 2004.

Silier, Oya. *Keban Köylerinde Sosyo Ekonomik Yapı ve Yeniden Yerleşim Sorunları*. Ankara: Orta Doğu Teknik Üniversitesi, 1976.

Simmons, John L. "Agricultural Development in Iraq: Planning and Management Failures." *Middle East Journal* 19, no. 2 (1965): 129–40.

Simon, Reeva S. *Iraq between the Two World Wars: The Militarist Origins of Tyranny*. New York: Columbia University Press, 2004.

Sinan, Ahmet Turan, and Fatma Döner Doğan. "Niyazi Yıldırım Gençosmanoğlu'nun Şiirlerinde Mekân: Harput ve Palu." *Uluslararası Palu Sempozyumu Bildiriler Kitabı* 1 (December 2018): 1–18.

Sissakian, Varoujan K, Nasrat Adamo, Nadhir Al-Ansari, Mukhalad Abdullah, and Jan Laue. "Sea Level Changes in the Mesopotamian Plain and Limits of the Arabian Gulf: A Critical Review." *Journal of Earth Sciences and Geotechnical Engineering* 10, no. 4 (2020): 87–110.

Sissakian, Varoujan K., Dhiya'a Al-Deen K. Ajar, and Maher T. Zaini. "Karstification Influence on the Drainage System, Examples from the Iraqi Southern Desert." *Iraqi Bulletin of Geology and Mining* 8, no. 2 (2012): 99–115.

Sluglett, Peter. *Britain in Iraq: Contriving King and Country, 1914–1932*. New York: Columbia University Press, 2007.
Sneddon, Christopher. *Concrete Revolution: Large Dams, Cold War Geopolitics, and the US Bureau of Reclamation*. Chicago: University of Chicago Press, 2015.
Sohrabi, Nader. *Revolution and Constitutionalism in the Ottoman Empire and Iran*. New York: Cambridge University Press, 2011.
Sönmez, A. "The Re-Emergence of the Idea of Planning and the Scope and Targets of the 1963–1967 Plan." In *Planning in Turkey*, edited by S. İlkin and E. İnanç, 28–43. Ankara: Orta Doğu Teknik Üniversitesi, 1967.
Sorby, Karol. "The 1952 Uprising in Iraq and Regent's Role in Its Crushing (Iraq from al-Watba to al-Intifāda: 1949–1952)." *Asian and African Studies* 12 (2003): 166–93.
Sousa, Ahmed. *Fayaḍānāt Baghdād fī al-tārīkh: baḥth fī tārīkh fayaḍānāt anhur al-ʿIrāq wa-taʾthīruhā bi-al-nisbah li-madīnat Baghdād*. Baghdad: Al-Adib Press, 1963.
———. *Irrigation in Iraq: Its History and Development*. Jerusalem: New Publishers Iraq, 1945.
———. *Taṭawwur al-rayy fī al-ʿIrāq*. Baghdad: Maṭbaʿat al-Maʿārif, 1946.
———. *Wādī al-Furāt wa-mashrūʿ saddat al-Hindīyah*. Al-Ṭabʿah 1. Baghdad: Maṭbaʿat al-Maʿārif, 1944.
Stack Whitney, Kaitlin, and Kristoffer Whitney. "John Anthony Allan's 'Virtual Water': Natural Resources Management in the Wake of Neoliberalism." *Arcadia*, no. 11 (Spring 2018). https://doi.org/10.5282/rcc/8316.
Stahl, Dale J. "A Technopolitical Frontier: The Keban Dam Project and Southeastern Anatolia." In *Transforming Socio-Natures in Turkey: Landscapes, State and Environmental Movements*, edited by Onur İnal and Ethemcan Turhan, 31–51. New York: Routledge, 2020.
———. "The Dam as Catastrophe: Connecting Geological Models to Modern History." *Water History*, June 21, 2021. https://doi.org/10.1007/s12685-021-00278-4.
———. "The Two Rivers: Water, Development and Politics in the Tigris-Euphrates Basin, 1920–1975." PhD diss., Columbia University, 2014.
Stork, Joe. "Oil and the Penetration of Capitalism in Iraq." In *Oil and Class Struggle*, edited by Petter Nore and Terisa Turner, 172–98. London: Zed Books, 1980.
Sulaymān, Khālid. *Ḥurrās al-miyāh: al-jafāf wa-al-taghayyur al-manākhī fī al-ʿIrāq*. Dimashq: Dār al-Mada lil-Iʿlām wa-al-Thaqāfah wa-al-Funūn, 2020.
Süme, Gülda Çetindağ, and Selamı Çakmakcı. "Yolcu, Fikret Memişoğlu." In *Türk Edebiyatı İsimler Sözlüğü*, June 14, 2019. https://teis.yesevi.edu.tr/madde-detay/yolcu-fikret-memisoglu.
Sümer, Vakur. "Handle with Care! The Tragedy of the Tabqa Dam." ORSAM-Center for Middle Eastern Studies, n.d. https://www.orsam.org.tr/en/handle-with-care-the-tragedy-of-the-tabqa-dam/.
Summitt, April R. *Contested Waters: An Environmental History of the Colorado River*. Boulder: University Press of Colorado, 2019.
Suny, Ronald Grigor, Norman M. Naimark, and Fatma Müge Göçek, eds. *A Question of Genocide: Armenians and Turks at the End of the Ottoman Empire*. New York: Oxford University Press, 2015.

Suwaidī, Taufīq as-. *My Memoirs: Half a Century of the History of Iraq and the Arab Cause*. Translated by Nancy Roberts. Boulder, CO: Lynne Rienner Publishers, 2013.
Sweeney, Samuel. "An Overlooked Syrian Writer." *Dappled Things*, 2021. https://www.dappledthings.org/reviews/an-overlooked-syrian-writer.
Tarbush, Mohammad A. *The Role of the Military in Politics: A Case Study of Iraq to 1941*. London: KPI Limited, 1982.
Tejel, Jordi. *Syria's Kurds: History, Politics and Society*. Translated by Emily Welle and Jane Welle. Routledge Advances in Middle East and Islamic Studies 16. London; New York: Routledge, 2009.
The Ascending Place of Light: GAP Southeast Anatolia Region. Fersa Ofset Tesisleri, 2007.
"The Constitution of Republic of Turkey (1961)." *Islamic Studies* 2, no. 4 (1963): 467–519.
"The Economic Development of Iraq." Washington, D.C: International Bank for Reconstruction and Development, 1952.
The Southeastern Anatolia Project Master Plan Study, Volume 2 Master Plan. Ankara: State Planning Organization, 1989.
The Third River. British Petroleum, 1955. https://www.bpvideolibrary.com/record/385.
Thesiger, Wilfred. *The Marsh Arabs*. London: Penguin Classics, 2008.
Thomas, Nicholas. *Entangled Objects: Exchange, Material Culture, and Colonialism in the Pacific*. Cambridge, MA: Harvard University Press, 1991.
Thornburg, Max Weston, Graham Spry, and George Henry Soule. *Turkey: An Economic Appraisal*. New York: The Twentieth Century Fund, 1949.
Tilley, Helen. *Africa as a Living Laboratory: Empire, Development, and the Problem of Scientific Knowledge, 1870–1950*. Chicago: University of Chicago Press, 2013.
Time. "The Earth Mover." May 3, 1954.
Tosun, Hasan. "Earthquakes and Dams." In *Earthquake Engineering*, edited by Abbas Moustafa. Rijeka: IntechOpen, 2015. https://doi.org/10.5772/59372.
Townshend, Charles. *Desert Hell: The British Invasion of Mesopotamia*. Cambridge, MA: Belknap Press of Harvard University Press, 2011.
Trentin, Massimiliano. *Engineers of Modern Development: East German Experts in Ba'thist Syria, 1965–1972*. Padova, Italy: CLEUP, 2010.
———. "Modernization as State Building: The Two Germanies in Syria, 1963–1972." *Diplomatic History* 33, no. 3 (June 2009): 487–505. https://doi.org/10.1111/j.1467-7709.2009.00782.x.
Tripp, Charles. *A History of Iraq*. 3rd ed. Cambridge: Cambridge University Press, 2010.
Trouillot, Michel-Rolph. *Silencing the Past: Power and the Production of History*. Boston: Beacon Press, 1995.
Tsing, Anna Lowenhaupt. *The Mushroom at the End of the World: On the Possibility of Life in Capitalist Ruins*. First paperback printing. Princeton, NJ: Princeton University Press, 2017.

Tucker, Richard P., and Edmund Russell, eds. *Natural Enemy, Natural Ally: Toward an Environmental History of Warfare.* 1st ed. Corvallis: Oregon State University Press, 2004.
Tufan fi Balad al-Ba'th [A Flood in Ba'th Country]. AMIP-ARTE France, 2003.
Tuncer, Baran. "External Financing of the Turkish Economy and Its Foreign Policy Implications." In *Turkey's Foreign Policy in Transition, 1950-1974.* Edited by Kemal H. Karpat, 206-24. Leiden, Netherlands: E. J. Brill, 1975.
Turgut, Hulûsi. *Demirel'in dünyası.* İstanbul: ABC Ajansı Yayınları, 1992.
——. *GAP ve Demirel: 50 yıl.* İstanbul: ABC Basın Ajansı, 2000.
Turgut, Mehmet. *Gap'ın Sahipleri.* Istanbul: Boğazici Yayınları, 1995.
Tvedt, Terje. *The River Nile in the Age of the British: Political Ecology and the Quest for Economic Power.* London: I.B. Tauris, 2004.
Ujaylī, 'Abd al-Salām al-'. *al-Maghmūrūn: riwāyah.* Ṭab'ah khāṣṣah. Ṭarābulus, al-Jamāhīrīyah al-'Arabīyah al-Lībīyah al-Sha'bīyah al-Ishtirākīyah: al-Munsha'ah al-'Āmmah, 1984.
Üngör, Uğur Ümit. *The Making of Modern Turkey: Nation and State in Eastern Anatolia, 1913-1950.* Oxford: Oxford University Press, 2011.
Unlandherm, Frank. "Sir William Willcocks: A Victorian in the Middle East." Senior thesis, Princeton University, 1959.
Uzer, Umut. "Racism in Turkey: The Case of Huseyin Nihal Atsiz." *Journal of Muslim Minority Affairs* 22, no. 1 (April 2002): 119-30. https://doi.org/10.1080/13602000220124863.
Váli, Ferenc A. *Bridge across the Bosporus: The Foreign Policy of Turkey.* Baltimore, MD: Johns Hopkins Press, 1971.
Van de Mieroop, Marc. *A History of the Ancient Near East, ca. 3000-323 B.C.* 2nd ed. Malden, MA: Blackwell Pub, 2007.
Velud, Christian. "Une expérience d'administration régionale en Syrie durant le Mandat Français: conquête, colonisation et mise en valeur de la Gazira, 1920-1936." Doctoral dissertation, l'Université Lumière Lyon 2, 1991.
Voûte, C. "Contributions of Photo-Interpretation to Engineering Projects in Various Stages of Execution." *Photogrammetria* 19 (1962-1964): 179-91. https://doi.org/10.1016/S0031-8663(62)80093-3.
Warriner, Doreen. *Land and Poverty in the Middle East.* London: Royal Institute of International Affairs, 1948.
——. *Land Reform and Development in the Middle East: A Study of Egypt, Syria, and Iraq.* 2d ed. Westport, CT: Greenwood Press, 1975.
Waterston, Albert, Christopher James Martin, August T. Schumacher, and Fritz A. Steuber. *Development Planning: Lessons of Experience.* Baltimore, MD: Johns Hopkins Press, 1965.
Weber, Max. *The Theory of Social and Economic Organization.* London: Free Press of Glencoe, 1947.
Wedeen, Lisa. *Ambiguities of Domination: Politics, Rhetoric, and Symbols in Contemporary Syria.* Chicago: University of Chicago Press, 1999.

Weinthal, Erika. *State Making and Environmental Cooperation: Linking Domestic and International Politics in Central Asia*. Cambridge, MA: MIT Press, 2002.

Westcott, Tom. "Iraq: Fishermen Fear Shrinking Lake Razzaza Spells End to Their Livelihoods." *Middle East Eye*, January 30, 2022. https://www.middleeasteye.net/news/iraq-lake-razzaza-milh-shrinking-dying-fishing-trade.

White, Richard. "Environmental History, Ecology, and Meaning." *The Journal of American History* 76, no. 4 (1990): 1111–16. https://doi.org/10.2307/2936588.

———. *The Organic Machine*. New York: Hill and Wang, 1995.

Willcocks, William. *Irrigation in Mesopotamia*. London: E. & F. N. Spon, Ltd., 1911.

———. "The Garden of Eden and Its Restoration." *The Geographical Journal* 40, no. 2 (August 1912): 129–45. https://doi.org/10.2307/1778459.

———. *The Restoration of the Ancient Irrigation Works on the Tigris, or, The Re-Creation of Chaldea*. Cairo: National Printing Dept., 1903.

———. "Two and a Half Years in Mesopotamia." *Blackwood's Edinburgh Magazine*, March 1916.

Williams, Elizabeth R. *States of Cultivation: Imperial Transition and Scientific Agriculture in the Eastern Mediterranean*. Stanford, CA: Stanford University Press, 2023.

Williams, Raymond. "Ideas of Nature." In *Problems in Materialism and Culture*, 67–85. London: Verso, 1980.

Wilson, Arnold Talbot. *Mesopotamia, 1917–1920; a Clash of Loyalties: A Personal and Historical Record*. London: Oxford University Press, 1931.

Wittfogel, Karl August. *Oriental Despotism: A Study of Total Power*. New Haven, CT: Yale University Press, 1957.

Wohlwend, Wolfgang. "'Our Heads Did Not Accept It'–Development and Nostalgia in Southeastern Anatolia." *Zeitschrift Für Ethnologie* 140, no. 2 (2015): 207–23.

Worster, Donald. *Rivers of Empire: Water, Aridity, and the Growth of the American West*. New York: Pantheon Books, 1985.

Yalçın, Soner. "Unutulmaz 'Bayrak' şairinin hazin hikâyesi." *Hürriyet*, June 6, 2010. https://www.hurriyet.com.tr/unutulmaz-bayrak-sairinin-hazin-hik-yesi-14944225.

Yaseen, Bushraa R, Kadum A Al Asaady, Ali A Kazem, and Miqdam T Chaichan. "Environmental Impacts of Salt Tide in Shatt Al-Arab-Basra/Iraq." *IOSR Journal of Environmental Science, Toxicology and Food Technology* 10, no. 1 (January 2016): 35–43.

Yeğen, Mesut. "The Kurdish Question in Turkish State Discourse." *Journal of Contemporary History* 34, no. 4 (1999): 555–68.

Yergin, Daniel. *The Prize: The Epic Quest for Oil, Money, and Power*. New York: Simon & Schuster, 1991.

Yesilnacar, Mehmet, and Sinan Uyanik. "Investigation of Water Quality of the World's Largest Irrigation Tunnel System, the Sanliurfa Tunnels in Turkey." *Fresenius Environmental Bulletin* 14 (January 1, 2005): 300–306.

Yıldırmaz, Sinan. "Politics and the Peasantry in Post-War Turkey: Social History, Culture and Modernization." Library of Ottoman Studies 46. I.B. Tauris, 2017.

Yıldız, Dursun. *GAP bölgede ekonomik, stratejik ve siyasal gelişmeler*. İstanbul: Truva Yayınları, 2009.

Yılmaz, Cevdet, ed. *Güneydoğu Anadolu Mutfağı*. GAP Eylem Planı, 2011.

Young, Gavin. *Return to the Marshes: Life with the Marsh Arabs of Iraq*. London: Faber & Faber, 2011.

Zhang, Isabel. "Submerging Kurdish History in Turkey: A Case Study of the Ilısu Dam." *The Middle East International Journal for Social Sciences* 3, no. 1 (March 2021): 1–8.

Zubaydī, Muḥammad ʿAbd al-Majīd Ḥassūn. *al-Amn al-māʾī al-ʿIrāqī: dirāsah ʿan sayr al-mufāwaḍāt qassamat al-miyāh al-dawlīyah*. Al-Ṭabʿah 1. Silsilat rasāʾil jāmiʿīyah. Baghdad: Dār al-Shuʾūn al-Thaqāfīyah al-ʿĀmmah, 2008.

Zürcher, Erik J. "The Ottoman Legacy of the Turkish Republic: An Attempt at a New Periodization." *Die Welt Des Islams* 32, no. 2 (1992): 237–53. https://doi.org/10.2307/1570835.

Zürcher, Erik Jan. *The Young Turk Legacy and Nation Building: From the Ottoman Empire to Atatürk's Turkey*. London: I. B. Tauris, 2010.

———. *Turkey: A Modern History*. 3rd ed. London: I. B. Tauris, 2004.

Zvi Yehuda Hershlag. *Turkey: The Challenge of Growth*. Leiden, Netherlands: E. J. Brill, 1968.

ʿAtiyah, Ghassan. *Iraq, 1908–1921: A Socio-Political Study*. Beirut: Arab Institute for Research and Pub, 1973.

Index

Abu Dibs depression, 74–75, 88, 93. *See also* reservoirs
actor-network theory, 11–12, 124. *See also* Latour, Bruno
Adalet, Begüm, 114, 121, 153–54
agency, 10–17, 29, 51, 81, 100, 132–33, 139–40; agential realism, 12; confederate agency, 11; historicity, 13; hybrid, 13, 16, 29, 51; neologisms and grammar of, 13, 132–33; non-human, 12–17; relational concepts of, 10–11; rhythms of change, 81; wildly entangled forms of, 100, 139–40. *See also* entanglement; new materialism; unintention
agriculture, 3, 30, 33, 55, 59–61, 88, 105; as a civilizing instrument, 33; destruction by flood, 88; expansion of, 30, 55, 59–61, 105; origins of, 3. *See also* dams; irrigation; salinization
Ahram, Ariel I., 91
Allan, Tony, 84, 218n96
Alp Arslan, 105, 128–29
Amiralay, Omar, 136, 144, 172–74, 179–80; *Everyday Life in a Syrian Village*, 173–74; *A Film Essay on the Euphrates Dam*, 172–73, 178; *A Flood in Ba'th Country*, 136, 140, 173, 178–79. *See also* Ba'th Party (Syria); Euphrates River
Araz, Melih E., 191
Armenian Genocide, 114–15, 191
Arslan, Murat, 151
al-Asadi, Jassim, 194–96
al-Assad, Hafez, 6, 168, 171–72, 178
Aswan High Dam, 98–99, 166–68, 205n21. *See also* Egypt
Asya, Arif Nihat, 102–3, 178, 221n12
Atatürk Dam, 135, 181, 183, 186, 238n5. *See also* infrastructure; Southeast Anatolia Project
al-Athari, Muhammad Bahjat, 89
Atsız, Hüseyin Nihal, 103

Baba Gurgur, 62, 64
Baghdad, 2, 24–26, 31–32, 34–48, 65, 72–74, 86–89, 93, 97, 198n8, 207n53, 208n75, 213n42; flood protection, 45–48, 74, 208n75; flood vulnerability of, 24, 31, 35, 65, 93, 97, 198n8, 213n42; flooding in and around, 25–26, 34, 39–47, 72–73, 86–89;

267

Baghdad, (cont.)
  founding of, 2; occupation of, 31–32, 207n53. See also floods; Iraq
Baghdad Railway, 29–30
Bahr al-Milh. See Abu Dibs depression
Barad, Karen, 12, 189, 192, 196, 200n35, 240n33. See also agency; entanglement
Basra, 36, 58, 93–94, 195, 206n32
Batatu, Hanna, 26, 56
Battle of Manzikert, 105, 128–29. See also nationalism: Turkish
Baykam, Suphi, 113
Bağış, Ali İhsan, 190–92
Ba'th Party (Syria), 8, 144, 155, 169, 235n92; 1966 coup, 168; critiques of, 173, 178–79; ideas about the Euphrates Dam, 171–72, 187, 192–93; plans for displacement, 157. See also Euphrates River; Syria; al-Tabqa Dam; United Arab Republic
Bennett, Jane, 10–14. See also agency
Bevin, Ernest, 79, 216n74, 216n78. See also Britain
Beşikçi, İsmail, 116, 118–19, 224n51. See also Kurds
Bilgen, Arda, 185, 188–89
Bingöl, Necip, 102
Bird, Hugh Stonehewer, 70
Birecik Dam, 99 (fig.), 183–84
Britain: British Middle East Office, 78, 84, 165, 216n74; claim to inherit Mesopotamian civilization, 20–21, 25, 188, 192, 207n53; conceptions of environment, 27, 34–39, 188, 190, 192, 208n58; concerns about Persian Gulf trade, 29–30, 69, 206n32; engineering and technical expertise, 33–34, 37, 71–73, 76–78, 86–87, 107, 126–27, 165–66, 216n74, 216n78; engineering companies, 5, 30–31, 66–67; in World War I, 32; international relations, with Syria, 165–66; with Turkey, 160–63, 233n67, 233n68, 233n71; with the United States, 71; involvement in irrigation and flood control, 29–30, 33–39, 41–43, 45–51, 70–81, 218n98; involvement in oil sector, 61–64; mandate of Iraq, 20–21, 26, 28, 32, 38–39, 50–51, 60–61, 65, 82, 207n54, 207n83 (see also Iraq: mandate government); relations with post-Mandate Iraq, 56, 65, 79–81, 216n81, 217n83; scientific studies of salinization, 57–60; Suez Crisis, 166; supposed invention of Iraq, 26–28, 50–51, 89, 204n7. See also Iraq Petroleum Company
British Empire. See Britain
British Middle East Office, 78, 84, 165, 216n74
Brown, Kate, 18, 155–57
bunds. See dikes
Bureau of Reclamation (U.S.), 71, 148–49, 214n57
Bury, L. E., 41, 45, 48

calcium carbonate, 97
Callon, Michel, 121
canyons: of the Euphrates River, 1, 4, 95–96, 98, 100–101, 105, 117, 126, 129, 131–32; of the Tigris River, 1, 4, 95–96, 136
Çelikbaş, Fethi, 113
al-Chadirchi, Kamil, 92
al-Chadirchi, Nasir, 87
Chakrabarty, Dipesh, 196
civilization, 10, 14, 17–18, 21–22, 25, 50, 141–43, 145–55, 171, 180, 187–96; ancient (see Mesopotamia); as material and conceptual project, 22, 141–43, 171, 180, 189–90, 193, 196; cradle of, 10, 17, 141, 190; Demirel's conception of, 145–47, 149–55, 188–89;

regeneration of, 18, 21, 25, 50, 187–93. *See also* civilizational dreaming; Mesopotamia; technical civilization

civilizational dreaming: as a Cold War phenomenon, 144, 156, 170–71; as a form of entanglement, 143, 189–90; critique of and resistance to, 22, 143–44, 153, 179–80, 190, 192–93, 196; definition of, 18; through rivers, 141–43, 193. *See also* Mesopotamia; civilization; technical civilization

climate change, 17, 24, 64, 93–94, 132, 135, 181, 187, 194–95; due to evaporation, 135

Cold War, 117, 144, 147, 153, 155–56, 158–61, 164, 166, 169–71, 209n83, 215n64, 231n52; dual layer of imperialism, 158–59

colonialism, 19, 27–28, 32, 38, 44, 141, 166, 192, 223n36; and development planning, 86, 112–13; and environmental engineering, 9, 19, 27, 29, 55, 60, 107; bilateral treaty arrangements, 65; Kurdistan as an example of, 115–16, 118–19, 123

concrete revolution. *See* Sneddon, Christopher

Coole, Diana, 12–13, 201n45

cotton, 186–87, 201n49

Crawford, Walter Ferguson, 165

Cronon, William, 15

cultural turn: relation to new materialism, 11, 200n35

dams, 5–8, 11–12, 16–17, 19, 22, 60, 64, 68, 71, 89, 95–98, 100–102, 106–7, 110–11, 113, 116–33, 139–44, 154–72, 175–76, 178–81, 183–87, 192–93, 196; as Cold War competition, 19, 71, 144, 155; as national or civilizational projects, 6, 8, 22, 106–7, 141–43, 159, 178–79, 192–93 (*see also* technical civilization); as rock and representation, 95, 98, 100; discursive representation of, 100–102, 119–20, 128–29; ecological effects, 60, 64, 89, 97, 130–31, 186–87 (*see also* irrigation; reservoirs; rivers; salinization); entanglement with temporal and natural forces, 100, 128–32, 139–41, 196; financing and foreign aid, 159–71; ideological symbols, 6–8, 113, 116, 172; overview of significance in history, 5–6, 8, 11–12, 16–17; role in state formation, 5–8, 106–7, 143, 154–59, 164, 169–72, 175–76, 179–81; technical reports and imaginaries, 100, 107, 110–11, 117–26 (*see also* engineering; techno-poetics); terrain and geology as agents, 95–97, 101, 129–33; unequal distribution of consequences, 68, 124–28, 183–87 (*see also* displacement). *See also* Aswan High Dam; Atatürk Dam; Birecik Dam; Ilısu Dam; Keban Dam; Mosul Dam; al-Tabqa Dam

date palms: effects of salinization on, 94

Davis, Diana K., 9, 188, 221n7, 240n32

debt. *See* economy

Demirel, Süleyman, 22, 143–54, 160, 180, 188–89; and Kemalism, 146–47; characterization of eastern Anatolia, 188; definition of civilization as technical, 145, 150–51, 153–54, 188; definition of rural life, 145–46; Eisenhower Exchange Fellow, 149; engineer and theorist, 189; *Great Turkey* (*Buyuk Turkiye*), 152–53; in Turkish bureaucracy, 148–52; prime minister of Turkey, 152–54; technical education, 147–48; *The Spiritual Side of Development* (*Kalkınmanın Manevi Yönü*), 153–54; youth, 145–47. *See also* development; engineering; engineers;

Demirel, Süleyman, (cont.)
  Five-Year Development Plans (Turkey); Keban Dam; Southeast Anatolia Project; technical civilization; Turkey
Dersim, 115. *See also* Kurds
determinism, 9, 13, 15, 140, 201n46
development, 10, 14, 21–22, 28–28, 30, 32, 55–57, 70–71, 79–80, 86, 89–90, 122, 155, 192–93; Cold War and geopolitical framing of, 170–71, 180–81; connection to techno-poetics, 108, 118; entanglement of narrative and technical processes, 18, 120–21, 189; planning in Iraq (*see* Development Board (Iraq); Dujayla Land Settlement Project); planning in Syria, 164–66; planning in Turkey (*see* Five-Year Development Plan; Southeast Anatolia Project; Turkey: development planning;); planning, history of, 106, 111–12, 117, 234n78; relation to military aid, 157–63. *See also* engineering; infrastructure; technical civilization
Development Board (Iraq), 81, 86, 89–90, 92, 112, 212n23; adoption of Haigh's plan, 80–81, 218n98; founding of, 79–80, 217n83; in the political ecology of salt, 85. *See also* Iraq Petroleum Company; Irrigation Development Commission (Iraq); political ecology of salt
Devlet Planlama Teşkilatı. *See* State Planning Organization
Devlet Su İşleri. *See* State Hydraulic Works
dikes, 24–25, 31–35, 37, 39, 41–44, 46–47 (figs.), 48, 72, 87–88, 90–91, 208n75. *See also* floods
displacement, 73, 85, 88, 98, 115, 126–28, 132, 136, 140, 157, 170–71, 173, 175–76, 178–80, 184, 186, 192, 237n115; and archaeology, 126–27; in Iraq, 73, 85, 88; in Syria, 136, 157, 173, 175–76, 178–80, 237n115 (*see also* al-'Ujayli, 'Abd al-Salam; Amiralay, Omar); in Turkey, 98, 115, 126–28, 157, 184, 186, 192; of reservoirs, 132, 140. *See also* dams; floods; Keban Dam; Kurds; Southeast Anatolia Project; al-Tabqa Dam
Diyala River. *See* Tigris River: tributaries
Dodge, Toby, 27, 204n16, 209n82, 209n83
drainage, 21, 35, 53–54, 57, 58–60, 67, 70, 76–78, 85, 93, 212n23, 217n86, 219n113; natural forms of, 58; of Iraq's marshlands, 90–92. *See also* irrigation; salinization
drought, 93–94, 132, 140, 145–46, 151, 153, 194, 218n94
Dujayla Land Settlement Project, 67–69, 210n1, 226n70. *See also* irrigation; Kut Barrage; salinization
dust storms, 94
Dökmeci, Cenani, 105, 128–29, 222n20

earthquake, 95, 130, 132, 227n90; Anatolian fault zones and plate tectonics, 95–97. *See also* geology
EBASCO Services, Inc., 117–25, 164; history of, 117. *See also* engineering; engineering companies; Keban Dam; Turkey
economic development. *See* development
economy: narrative imagery and debt as metaphor, 119–22. *See also* dams: development; dams: financing and foreign aid
Egypt, 27, 29, 35–36, 71, 73, 164, 166–67, 170, 174; irrigation experts, 36, 73; relations with Syria, 166–67, 170; Suez Crisis and Aswan High Dam, 166. *See also* Willcocks, William
Elazığ, 96, 102–3, 105–7, 115, 127–28
Elazığ Gazetesi, 102–3

Index

Electric Works Study Administration, 113
embankment. *See* dikes
embodied water. *See* virtual water
engineering: and temporal claims to certainty, 17, 124, 131–32; as a poetics and literary genre, 19, 100–101, 107–8, 110, 119, 124 (*see also* techno-poetics); as civilizational dreams, 18, 141–42, 191; circulation of methods and expertise, 29–30, 117–18, 154–55, 165–66 (*see also* Haigh, Frank Fraser; Willcocks, William); entanglement with narrative, 189; feasibility reports in, 110; history of, 107; imperial ideologies and racial logics of, 35–37, 71, 158–59, 175, 215n64; misconceptions of environment, 57–60, 73, 75–76, 81, 90–93, 187; relation to state formation and power, 7–8, 88–89, 159–60, 170, 178–80, 195; un-engineering as responsive to natural forces, 100, 133, 196. *See also* civilizational dreaming; Demirel, Süleyman; engineers; engineering companies; entanglement
engineering companies, 5; Alexander Gibb and Partners, 165; Balfour, Beatty and Co., Ltd., 213n44; EBASCO Services, Inc., 117–25, 164; Morrison Knudsen, 149; SCI-Impregilo, 164; Sir J. Jackson, Ltd., 31. *See also* engineering; engineers; irrigation
engineers: as a social group, 18; as agents of empire, 31–32, 35–37, 108, 208n58 (*see also* Haigh, Frank Fraser; Willcocks, William; Demirel, Süleyman); as guardians of mutual adaptation, 196; as knowledge producers, 19, 107–8, 110; as part of Islamist movements, 148; as professionals, 107; as storytellers, 100–102, 107, 123–26, 128; intentions and ecological consequences, 56–57, 68, 73–78, 117, 130–32. *See also* engineering; engineering companies; techno-poetics
entanglement: concept and methodology of, 14–17, 121, 126, 192–96, 202n54; historical and geopolitical cases of, 81, 90, 110, 117, 150–51, 155, 158, 170, 187, 189–90; of human and nonhuman forces, 29, 50–51, 56–57, 68–69, 73, 81–90, 100–101, 124–26, 128, 139–41, 159–60, 174–78, 180–81. *See also* agency; civilizational dreaming; engineering; new materialism; technical civilization
enviro-technical analysis, 12
environmental imaginary, 27, 101, 107, 119, 132–33; as an "unimaginary," 126; criticism of, 108. *See also* techno-poetics
Erbakan, Necmettin, 148
Erbil, 93
Erder, Cevat, 126–27
Eroğan, Kadri, 114
Euphrates Dam. *See* al-Tabqa Dam
Euphrates River: conceptual and cultural framings, 19, 22, 37–38, 44, 100–107, 125, 128, 142, 156, 184–85 (*see also* Amiralay, Omar; al-'Ujayli, 'Abd al-Salam; Demirel, Süleyman; Mesopotamia); control, conflict, and contestation, 6–8, 157–59, 166–67, 169–70; ecology and environmental effects, 3, 21, 25, 34, 39, 88, 93, 97, 186–87, 194–95 (*see also* salinization; dust storms; marshlands); geography and regime, 1–3, 23–24, 29, 35–36, 64–65, 96–98, 99 (fig.), 132 (*see also* floods); global and transnational connections, 19–21, 30, 55–56 (*see also* Cold War; dams: financing and foreign aid); historical transformation and continuities, 4, 18, 135–36, 137–38 (figs.), 139–41, 144,

Euphrates River (*cont.*)
154 (*see also* reservoirs; Tigris River; dams); historiographical position, 9–10, 15, 17; technical engineering, 21–22, 30–31, 65–66, 74–76, 90–91, 113. *See also* Hindiyya Barrage; Iraq; Keban Dam; Southeast Anatolia Project; Syria; al-Tabqa Dam; Turkey
evaporation, 52, 54, 59, 74, 94, 134–35, 186, 215n72
Everyday Life in a Syrian Village, 173–74. *See also* Amiralay, Omar

Farhoud, Ibrahim, 168
al-Farsi, Nasrat, 65
Faysal ibn Husayn, 32, 38, 41, 45, 64–65
Fertile Crescent, 56, 78, 141, 187. *See also* Mesopotamia
fertilizer, 55, 94, 187, 212n23
A Film Essay on the Euphrates Dam, 172–73, 178. *See also* Amiralay, Omar; al-Tabqa Dam
fish, 3, 53–54, 93, 119–21
Five-Year Development Plan (Turkey), 112–14, 116, 118, 120. *See also* Turkey
A Flood in Ba'th Country, 136, 140, 173, 178–79. *See also* Amiralay, Omar; Ba'th Party (Syria)
floods, 24–26, 32–37, 39–48, 72, 85–90, 176; characteristics of, 24; debate over causes of, 35–37; in the filling of reservoirs, 176; of 1919, 25–26, 32–34; of 1923, 39; of 1926, 40–48; of 1946, 72; of 1954, 85–90. *See also* Baghdad; Development Board (Iraq); dikes; Euphrates River; Haigh, Frank Fraser; Tigris River; Willcocks, William
France, 38, 103, 127, 161, 188; engineers, 30, 165; Suez Crisis, 166

Gençosmanoğlu, Niyazi Yıldırım, 103
geology, 5, 13, 16, 21, 97–98, 123–24, 130, 198n9; fault zones in Anatolia, 95–96; petroleum geologists, 61–62. *See also* earthquakes
Germany (pre-1945), 20, 29–30, 213n37. *See also* Germany, Federal Republic of; Germany, German Democratic Republic
Germany, Federal Republic of, 126, 167–69, 235n94, 235n95
Germany, German Democratic Republic, 167, 236n96
Gharraf Project. *See* Kut Barrage
Ghazi ibn Faysal, 66–67
Gilmartin, David, 33, 57, 208n58
Green Revolution, 19, 55, 61
Gulf War (1991), 91–92, 162
Göle, Nilüfer, 148
Güneydoğu Anadolu Projesi (GAP). *See* Southeast Anatolia Project
Gürpınar, Doğan, 129

al-Habbaniyya: airbase, 74, 216n81; depression and flood escape, 31, 34, 66, 74, 88, 213n42; salinization, 93
Haigh, Frank Fraser, 72–81; biography of, 72; Haigh Commission (*see* Irrigation Development Commission); vision for southern marshes, 74–76, 90–91. *See also* Development Board (Iraq); floods
Halfeti, 184
Haraway, Donna, 139
Harran, 191–92
Harris, Leila M., 183, 185
Hasan, Muhammad Salman, 83
Hasankeyf, 1, 139. *See also* Ilısu Dam
Haydar, Rustam, 65
Hindiyya Barrage, 30–31, 34, 59–60, 62 (fig.), 65, 205n20, 206n35, 207n41, 211n18. *See also* Ottoman Empire; Willcocks, William
historical materialism: contrasted with new materialism, 10–11, 193–94
Hodge, Joseph Morgan, 19

Howell, Evelyn, 36–37
Husain, Faisal, 9, 23, 200n28, 208n61
Hussein, Saddam, 5, 91

Ilısu Dam, 1, 136. *See also* dams; engineering; reservoirs
imagined communities, 8
India, 27, 32, 38, 112, 206n32; irrigation expertise, 29, 33–36, 57, 71, 73, 78. *See also* Britain; Haigh, Frank Fraser; Iraq; Willcocks, William
Indus River, 1, 33–34, 57, 72
infrastructure, 5, 7, 9, 139, 187, 190, 194, 209n79, 221n8, 223n36, 238n2; as civilizational dreams, 142, 145, 157, 164, 190, 193; embedded narratives and symbolic roles, 189; foreign and geopolitical influence, 30–31, 169; in the political ecology of salt, 55, 90–91, 93; role in state power, 110, 121, 151, 155, 170, 185; social and political transformation, 33, 56, 111, 153, 172, 178, 182. *See also* civilizational dreaming; regeneration; technical civilization ; techno-poetics
International Bank for Reconstruction and Development. *See* World Bank
Iraq: agriculture and resources, 55, 58, 61, 69, 78, 82–84, 94, 101, 212n23, 218n93 (*see also* oil; salinization; political ecology of salt); development and economy, 69, 85–86, 90, 217n86, 237n1 (*see also* Development Board; Iraq Petroleum Company; Dujayla Land Settlement Project; Irrigation Development Commission); international relations: with Syria, 166–67, 170; with Turkey, 167; with the United States, 71; with the Soviet Union, 167; politics and governance: 1920 revolt, 26, 32–33, 38; mandate government, 20–21, 28–29, 32, 38, 44–45, 48–51; post-mandate politics, 65–66, 79–81, 213n35, 213n37; 1958 coup, 56, 81, 89–90; supposed invention of, 26–28; Irrigation Directorate, 33–37, 41, 50, 209n79, 210n84; environmental activism, 194–95; wars and conflict: Iraq-Iran War, 91, 219n115; Gulf War and sanctions, 91–92. *See also* floods; Islamic State; marshlands; Mosul Dam
Iraq Petroleum Company (IPC), 61–65; 80, 219n113. *See also* Development Board (Iraq); Irrigation Development Commission (Iraq); oil; salinization
irrigation: ecological effects, 52–53, 55, 57, 78, 89, 94, 135 (*see also* salinization; drainage); historical contexts, 3, 35, 54–55, 191, 208n58; methods, 210n85: canals, 34, 59, 97, 186, 215n72; culverts, 41; pumps, 49–51, 58, 60–61, 66, 209n81, 226n75; tunnels, 186 (*see also* Irrigation Development Commission); policies, 49–50, 60, 77, 79–80, 158, 185, 209n79, 210n84, 212n23; projects, 29–32, 66, 70–71, 93, 118, 165, 205n25, 209n79, 214n62, 215n65, 235n90, 238n5. *See also* Development Board (Iraq); Dujayla Land Settlement Project; Kut Barrage; Southeast Anatolia Project
Irrigation Development Commission (Iraq), 72–80. *See also* Development Board (Iraq); drainage; Haigh, Frank Fraser; Iraq; Iraq Petroleum Company (IPC); irrigation; Main Outfall Drain; Mosul Dam; salinization
Islamic State, 5–8, 198n14, 199n16
Israel, 155, 165–69, 174, 235n92, 236n98
Italy, 127, 164, 234n81
İnan, Kamuran, 191
İnönü, İsmet, 146

Jabr, Salih, 66, 79, 85–86, 217n83
al-Jamali, Muhammad Fadhil, 79

al-Jazira. *See* Euphrates River; Mesopotamia; Tigris River
al-Jomard, Atheel, 83
al-Jundi, 'Abd al-Karim, 167

Karakaya Dam, 181, 227n90
Karasu River, 1, 96, 98, 220n3
karst, 17, 97, 124, 130, 139, 220n1. *See also* geology; Mosul Dam
Keban, 96, 100, 101, 223n41
Keban Dam, 21–22, 100, 226n75, 227n81; archaeological rescue project, 126–27, 227n82; as economic object, 111, 119–22; as prototype for the GAP, 182–83 (*see also* Southeast Anatolia Project); as symbol of social order, 110–11, 116, 122–23 (*see also* Demirel, Süleyman; technical civilization); cultural construction of, 101–10, 180, 192 (*see also* techno-poetics); displacement and labor disputes, 127–28, 158; financing and Cold War politics, 144, 155, 160, 163–64, 169–70; leaks, 130–32; physical description of, 97–98; politics of feasibility report, 110–14, 116; reservoir, 98, 124, 130–32, 134–36, 137 (fig.), 156; temporality and supposed completion, 124–25, 128–33, 196, 228n92. *See also* dams; engineering; reservoirs
Keban Dam Lake. *See* Keban Dam: reservoir
Kemal, Mustafa (Atatürk), 65, 129, 145–47, 154, 183, 188, 221n17. *See also* Turkey
Khudayyir, Muhammad, 195, 240n40
Komer, Robert, 162
Koç, Atilla, 185
Kurds: access to resources and displacement, 61–62, 93, 157–58, 175, 185; conflict, resistance and political movements, 5, 111, 115–16, 170, 190; local militias in eastern Anatolia, 114–15; oppression, colonization, and state relations, 66, 116, 118–19, 122–23, 126, 157–58, 184, 191, 224n49. *See also* Armenian Genocide; displacement; Southeast Anatolia Project; Turkey
Kut, 44, 56, 64, 67, 69, 210n85. *See also* Kut Barrage
Kut Barrage, 45, 50, 56, 65–69, 77, 90. *See also* Dujayla Land Settlement Project; political ecology of salt

labor, 11, 24, 33–34, 44, 50, 70, 79, 82–83, 86, 90, 122, 124–25, 128, 183, 209n81, 217n86; strikes and organized labor, 27, 80, 85, 127, 156
Lake ar-Razzaza. *See* Abu Dibs depression
Lake Assad, 134–36, 144, 174–79
land policy, 27–28, 48–51, 66–67, 77–79, 82–84, 115, 174, 209n79, 209n83
land tenure. *See* land policy
Latour, Bruno, 11–12, 109, 223n30. *See also* actor-network theory; entanglement
Lefebvre, Henri, 156
limestone. *See* karst
Lombardo, Joseph D., 127–28

Al-Maghmūrūn, 144, 174–78, 180
Mahdi, Kamil, 61, 217n86
Main Outfall Drain, 90–94. *See also* drainage; Iraq; salinization
Makhus, Ibrahim, 167
al-Mala'ika, Nazik, 72–73
Mandate System, 20, 26, 32–33, 61, 65, 141. *See also* Britain; Iraq
Marsh Arabs. *See* Ma'dān
marshlands, 2–4, 23, 36, 57, 75–76, 90–92, 97, 194–96, 219n115; draining of, 4, 75–76, 90–92
Marx, Karl: historical materialism, 10–11, 193–94, 201n45

Ma'dān, 2, 76, 91–92. *See also* marshlands; Iraq
Memişoğlu, Fikret, 102
Menderes, Adnan, 112, 151, 160, 162. *See also* Turkey: politics and governance
Mesopotamia, 2, 22, 25, 27, 30, 32, 38, 62 (fig.); ancient civilizations, 1–4, 9, 18, 23, 54, 141, 191, 201n46; regeneration of, 21, 34–37, 187–89, 191–92. *See also* civilization; civilizational dreaming; regeneration
al-Midfa'i, Jamil, 65
Mitchell, Timothy, 12, 201n44, 212n31, 229n8
modernization, 4–5, 14, 27, 45, 79–80, 86, 104, 112, 141, 152, 165, 178–79, 186, 200n26, 215n64. *See also* civilizational dreaming; Green Revolution
Morrison Knudsen, 149, 152
Mosul, 5, 25, 40, 45, 61, 63 (fig.), 64
Mosul Dam, 5–7, 16–17, 181, 237n1. *See also* dams; engineering; reservoirs
Murat River, 1, 96, 98, 220n3

Nasser, Gamal Abdel, 166, 174, 235n92
nationalism, 4, 8–9, 16, 26, 126, 188; Arab, 79, 81, 179, 187, 192; Iraqi, 26–27, 44, 86, 90; Kurdish, 111, 157, 184, 190; Syrian, 164, 169, 172, 180; Turkish, 100, 102–7, 111, 114–15, 127–29, 147, 169, 180, 187–88, 192, 221n17, 224n49. *See also* imagined communities; Iraq; Kurds; Syria; Turkey
new materialism, 10–14, 28–29, 56, 68, 100, 106, 109–10, 126, 132, 140–41, 155, 159, 189–90, 193–94, 200n35, 229n8. *See also* agency; Barad, Karen; Bennett, Jane; entanglement
Noah: story of, 89
North Atlantic Treaty Organization, 112, 155, 160–64, 169, 234n77, 234n78

oil, 7, 10, 21, 55–57, 69, 79–81, 217n84, 219n113, 219n115; contestation over rights to, 61–62, 63 (fig.); dependence on international markets for food, 69, 84–85, 92; exploration in Iraq, 61–64; financing of hydraulic works, 65–68, 77, 90, 93. *See also* geology; Iraq Petroleum Company; political ecology of salt
Orontes River, 178
Ottoman Empire, 9, 20–21, 25, 27–34, 36–38, 44, 49, 57, 61–62, 71, 104–5, 114–15, 146–47, 151, 162, 174, 188, 191, 205n20, 205n21, 206n28, 208n61, 209n83, 215n66. *See also* Armenian Genocide; Turkey
Özal, Korkut, 148
Özal, Turgut, 148, 191

al-Pachachi, Hamdi, 70–71, 76
Palestine, 64, 80–81, 168, 174, 235n92
Persian Gulf, 2, 24, 29, 32, 54, 64, 69, 91, 94, 195
petroleum. *See* oil
Phillips, Doris G., 83
plate tectonics. *See* earthquakes
poetry, 20, 44, 69, 72–73, 89, 102–5, 173, 221n12, 221n17. *See also* technopoetics
political ecology of salt, 68–69, 80–81, 83, 85, 89, 90–94, 185; definition of, 56. *See also* drainage; irrigation; Main Outfall Drain; oil; salinization
Pritchard, Sara, 12, 139–40
pumps. *See* irrigation: pumps
Pursley, Sara, 13–14, 27, 67–68; Dujayla Land Settlement Project, 67–68; invention of Iraq, 27; temporality of development, 13–14

Qazzaz, Sa'id, 88

regeneration, 18, 21–22, 36–37, 143, 181, 187–93, 215n69. *See also* civilizational dreaming; Mesopotamia

reservoirs, 2, 17, 21–22, 26, 90, 100, 105, 134–43, 157, 180–82, 184, 186, 188, 228n2, 229n13; as spatial and political control, 156–59; diplomatic, financial, and military, 141, 144, 153–71; flood mitigation and salinity, 31, 74, 91, 93–94 (*see also* dams; salinization); of oil and salt, 55–56, 61. *See also* displacement; Euphrates River; Keban Dam: reservoir; Lake Assad; Southeast Anatolia Project; Tigris River

Reynolds, Nancy: rockscape, 98, 100–101, 107, 128, 132

rivers, 108, 117, 128; disappearance of, 136–41; dreaming civilization through, 141–43, 154, 170–71, 180, 190, 193 (*see also* civilizational dreaming; technical civilization; techno-poetics); history and historiography of, 4–5, 8, 10, 13, 15, 19–20, 193, 196, 200n28, 200n36, 222n23. *See also* Euphrates River; Tigris River

rock, 6, 17, 95–98, 111, 117, 123–26, 134, 180, 193; in history, 98, 100–101; instability of, 98, 101, 125–26, 130–33 (*see also* Mosul Dam). *See also* earthquakes; geology

rockscape. *See* Reynolds, Nancy

Saddam Dam. *See* Mosul Dam
Safwa, Najda Fathi, 65
al-Said, Nuri, 66, 85, 213n41, 214n63, 217n84
salinization: adaptations to, 53–54, 59, 83; conceptual framings, 10, 13, 16–17, 21, 52–53, 89–90 (*see also* political ecology of salt); effects on society, economy, and environment, 50, 52–57, 59–61, 69–70, 82–85, 88–90, 93–94, 104, 217n86; historical contexts and awareness, 54–60, 77, 210n1, 211n10; in modern development plans and dam projects, 66–71, 73–78, 80–82, 85, 90–94, 97, 185–87, 212n23, 215n67, 218n94. *See also* Dujayla Land Settlement Project; Irrigation Development Commission (Iraq)
Salter, Arthur, 86
Sassoon, Joseph, 71
Satia, Priya, 27, 38
Scott, James C., 7, 109, 156
Senemoğlu, Bahattin, 105
Seyhan Dam, 116, 149, 160–62, 233n68
al-Shabibi, Muhammad Rida, 44
Shatt al-Arab, 2, 58, 61, 94
Shatt al-Gharraf, 64–65, 67. *See also* Kut Barrage
Shawkat, Naji, 44–45, 65–66
Shiva, Vandana, 55
silt, 23, 30, 87, 97, 132, 140, 196, 207n48
Sneddon, Christopher: concrete revolution, 19, 116, 150. *See also* dams
Sousa, Ahmed, 72, 219n110
Southeast Anatolia Project, 22, 150, 157, 182–93; and regeneration, 22, 187–93; social, economic, and environmental effects, 183–87. *See also* dams; Keban Dam; reservoirs; Turkey
sovereignty, 27, 65, 164, 169–70, 180, 192
Soviet Union, 147; development planning and relations with Turkey, 112, 160; disputes over Euphrates water, 167–68; involvement in Iraq, 5, 91, 213n37; relations with Syria, 158, 164, 166–70, 235n89, 235n90, 235n94; similarity to United States, 18, 155–56
State Hydraulic Works (Turkey), 20, 116, 128, 131, 149, 151. *See also* Demirel, Süleyman; Five-Year Development Plan (Turkey); Keban Dam; Southeast Anatolia Project

State Planning Organization (Turkey), 112, 152, 190. *See also* Five-Year Development Plan (Turkey); Southeast Anatolia Project

Suez Crisis, 166

Suleiman, Khalid, 194–95

Sumeria, 23, 54

al-Suwaydi, Tawfiq, 80

Syria: agriculture and resources, 135, 173–74, 187; development and econonmy, 64, 97–98, 188, 193, 237n1; dam financing, 160, 164–70; international relations: with Britain, 165–66; with East Germany, 164, 167; with Egypt, 166–70; with Iraq, 166–67; with Israel and Jordan, 168; with the Soviet Union, 155, 158, 166–70, 235n89, 235n90, 235n94; with Turkey, 103, 166–67; with the United States, 167; with West Germany, 167–69, 235n94; politics and governance: Ba'th Party, 6–7, 155, 157, 171–73, 178–80, 187, 192; United Arab Republic, 159, 167; military coup of 1966, 168, 171; dual imperialism, 158; displacement of Kurds, 157–58, 175, 237n116; wars and conflict: civil war, 6–8, 237n115; suppression of Muslim Brotherhood, 170; with Israel, 165–69. *See also* Ba'th Party (Syria); Euphrates River; al-Tabqa Dam; United Arab Republic

al-Tabqa Dam, 6–8, 22, 97–98, 100, 134–36, 138, 140, 143–44, 155–58, 160, 164–81, 187, 192, 231n52; as a symbol of revolution, 171–72, 174–80, 187, 192; as landscape of war, 160, 170; as rock and representation, 100, 140, 143; as technical civilization, 164; cause of international disputes, 170; displacement of Kurds, 157–58, 175; financing of, 144, 155, 164–69; in art, film, and literature, 172–80; in Cold War politics, 155–56, 169–70, 231n52; Islamic State takeover of, 6–8; physical description of, 97–98; switch from Yusuf Pasha site, 166. *See also* dams; engineering; reservoirs

Taurus Mountains, 1–2, 23, 25, 95–96, 119, 136

technical civilization, 150–60, 164, 169–71, 180–81, 188; to assert sovereignty, 169–70; as an imperial ontology, 159; definition of, 150–51. *See also* civilization; civilizational dreaming; entanglement; Mesopotamia

techno-poetics, 19, 101, 108–11, 118–19, 126–28; and environmental imaginaries, 101, 108, 119; definition of, 19; related theoretical frameworks, 109–10. *See also* entanglement; poetry

Thomas, Nicholas, 15

Tigris River: conceptual and cultural framings, 19, 22, 35–37, 136, 139–42, 154, 187–93 (*see also* Demirel, Süleyman; Mesopotamia); control, conflict, and contestation, 5–8, 32; ecology and environmental effects, 3, 21, 23–26, 34, 39–48, 58, 69, 72–73, 87–88, 94, 135, 195 (*see also* floods; salinization; dust storms; marshlands); geography and regime, 1–4, 23–24, 34–35, 64–65, 96–97; global and transnational connections, 18–19; historical transformation and continuities, 4, 29 (*see also* Euphrates River; Iraq; reservoirs; Turkey); historiographical position, 9–10, 15, 17; technical engineering, 2, 4–7, 16–17, 21, 31, 40 (fig.), 50, 56–57, 65–67, 74, 136, 139, 182, 186; tributaries: Diyala River, 31, 34, 41, 45, 58, 72, 87, 198n14, 237n1; Greater and Lesser Zab Rivers, 62, 237n1. *See also*

Tigris River *(cont.)*
   Ilısu Dam; Kut Barrage; Mosul Dam; Southeast Anatolia Project
Tripartite Aggression (1956). *See* Suez Crisis
Tsing, Anna, 14. *See also* entanglement
Turgut, Hulusi, 150–51
Turgut, Mehmet, 116
Turkey: agriculture and resources: hydroelectricity, 117, 182; cotton, 186–87; development and economy: 21, 127, 136, 145, 231n48 (*see also* Keban Dam); planning, 106, 112–15; dam financing, 22, 160–64, 190; economy as object, 119–22; dams as social order, 122–23, 180, 183; Southeast Anatolia Project, 182–93; international relations: with Britain, 160–61, 163; with Syria, 166–67; with NATO and "the West," 144, 150, 160–64, 169–70, 190; with the United States, 149–50, 160, 162, 164; with the Soviet Union, 112; politics and governance: dual imperialism, 158; parliamentary debates over planning, 113–14, 116–17; Democrat Party, 112, 151, 160; May 1960 coup, 159; Justice Party, 152; power of engineers, 147–48; 2016 coup, 127; Turkification, 115–16, 191–92; Kemalism, 145–47; Ottoman legacies, 162, 188; wars and conflict: policies toward Kurds, 115–16, 157, 170, 184–85, 190 (*see also* Kurds); involvement in Syrian Civil War, 6; domination of Anatolia, 128–29. *See also* Five-Year Development Plan (Turkey); nationalism: Turkish; Southeast Anatolia Project

Ulbricht, Walter, 167
unintention: in relation to agency, 15, 68, 100, 132, 139, 196; unintended consequences of development, 68–69, 124, 140. *See also* agency; entanglement

United Arab Republic, 167. *See also* Egypt; Syria
United Kingdom. *See* Britain
United Nations, 80, 91–92, 212n23
United States, 126–27; economic competition and development aid, 71, 80, 122, 144, 148, 155–58, 160–64; engineering companies and activities, 5, 107, 117–19, 126, 148–49; international relations: with Iraq, 5, 7, 71, 91–92; with Syria, 155, 164, 167–68, 231n52; with Turkey, 105–6, 149, 160–62, 170; military operations in the Middle East, 5–7, 162, 167, 170, 233n74; similarity to Soviet Union, 18, 155–59, 169–70; water management practices, 19, 57, 117–18, 148–50, 153–55, 216n78
al-'Ujayli, 'Abd al-Salam, 174, 179–80, 237n116; *Al-Maghmūrūn*, 144, 174–78, 180
Urfa, 186, 191

virtual water, 84, 186, 218n96. *See also* Allan, Tony; Iraq
Viswanath, B., 58, 77
vital materialism, 10–13. *See also* agency; Bennett, Jane; entanglement

Wadi al-Tharthar, 31, 45, 74, 93, 215n65
Walton, H., 35–37
Ward, T. R. J., 33, 206n35
water: and identity, 104, 178, 183, 195; borne diseases, 29, 67; power and politics, 15, 168, 235n92; security, 9, 84, 94, 135; sharing agreements, 167–68; table and waterlogging, 67–68, 76–77, 83, 215n72. *See also* drainage; irrigation; salinization
al-Wathba uprising, 79–80, 216n81
Webster, J. F., 58–60, 77
Wedeen, Lisa, 172
wetlands. *See* marshlands
White, Richard, 11–12, 139

Willcocks, William, 29–32, 34–37, 45, 57, 60, 66, 70–71, 73, 93, 205n21, 205n23, 214n62, 215n66; description of salinization, 57, 60; plans for flood escapes, 29, 66; view of delta and causes of flooding, 34–35; vision for Mesopotamian agriculture, 31–32. *See also* floods; Irrigation Development Commission (Iraq); Mesopotamia
Williams, Raymond, 193–94
Wilson, Arnold, 32, 36

Wittfogel, Karl A., 10, 13, 201n46
Wohlwend, Wolfgang, 184
World Bank, 161–63, 166, 212n23, 215n67
worldmaking, 14, 189, 193–94
Worster, Donald, 10, 150

Yeni Fırat, 102, 104–5
Yusuf Pasha: dam site in Syria, 165–66

Zeyzoun Dam, 178